Paediatric Patient and Family-Centred Care: Ethical and Legal Issues

INTERNATIONAL LIBRARY OF ETHICS, LAW, AND THE NEW MEDICINE

VOLUME 57

For further volumes:
http://www.springer.com/series/6224

Randi Zlotnik Shaul
Editor

Paediatric Patient and Family-Centred Care: Ethical and Legal Issues

Springer

Editor
Randi Zlotnik Shaul
Dept. of Bioethics
The Hospital for Sick Children
Toronto
Canada

ISSN 1567-8008
ISBN 978-1-4939-5358-5 ISBN 978-1-4939-0323-8 (eBook)
DOI 10.1007/978-1-4939-0323-8
Springer New York Heidelberg Dordrecht London

Printed on acid-free paper

Springer is part of Springer Science+Business Media (www.springer.com)

Foreword

This collection of essays marks an important step in the evolution of the concepts or models of paediatric patient-centred care and family-centred care. While there is still no consensus in the healthcare literature on the definition of these terms, there is sufficient recognition of relevant values, to allow for the consideration of definitional options and their application across Canadian geography and healthcare settings. In an increasingly regulated healthcare system, providers, administrators, patients, families and regulators must be attuned to the legal and ethical tensions and synergies between concepts and models publicly endorsed to serve as guidance for care.

In this text, traditional themes in health law and paediatric bioethics are engaged in an attempt to provide an in-depth appreciation of the opportunities and challenges posed by the application of patient-centred care and family-centred care in a variety of clinical paediatric contexts.

Together the authors represent an exceptional group of clinical ethicists, lawyers, researchers, parents, healthcare providers, counselors and trainees with both theoretical and real life hospital-based experience. While one chapter revisits my earlier collaboration exploring the relationship between patient-centred care and family-centred care in the context of consent to research, all other chapters were written specifically for this text.

Contributors were asked to consider key strengths and challenges of working with the models of paediatric patient-centred care and family-centred care. Are there tensions and synergies between these models? How are these concepts understood and used within specific contexts? Based on experiences to date, are there any issues that warrant further research? Through their essays, the contributors provide instructive reflections on these issues from their unique vantage points, further developing both the theory and application of the models.

This collection contributes to a constructive discussion about the future of paediatric patient-centred care and family-centred care. These engaging and provocative essays reflect these models of paediatric healthcare within a particular social and historical period in Canada, and within a variety of distinct treatment settings. Notwithstanding the fact that the chapters were written as independent entities, the common goal of providing accurate current descriptions of these models in practice as well as rigorous analysis of the ethical and legal implications of their application, is a connecting feature that threads through them all.

While this book focuses on the Canadian context, the case descriptions and analyses are pertinent internationally. To date, most of the patient-centred care and family-centred care literature has come from the United States and England. Because the values these concepts represent are relevant in international healthcare contexts, there is much benefit in considering their nature and application within a distinct legal system and socio-political healthcare setting. While many countries approach the legal and ethical issues addressed in ways that are similar to Canada, there are some distinct approaches grounded in the laws and policies unique to each jurisdiction. Readers engaged with paediatric healthcare outside of Canada will need to be cognizant of laws specific to their own setting and use the Canadian examples as inspiration and at times cautions. Examples of two relevant and distinguishing features of the Canadian context are the quasi-socialized funding of healthcare services in Canada as well as the legal frameworks of informed consent.

This book is intended to be both theoretically and practically useful. The theoretical aims are to help readers tease out the meaning and significance of these models, the extent to which it is defensible to interpret them as conceptually consistent, the extent to which real life circumstances reveal their inconsistencies and contexts where the separate models can combine as a hybrid model of care. The chapters include references to clinical and research-based cases that highlight the theoretical concepts being discussed. The cases discussed, allow readers to engage with the real-life circumstances in which the models are applied. The models of paediatric patient-centred care and family-centred care were never intended to be purely theoretical abstractions but rather to serve as frameworks to guide those working in healthcare organizations.

This book was prepared with the needs of practicing clinicians, allied health professionals, policy developers, hospital administrators and educators in mind. On a practical level, the book aims at maximizing the potential benefits of these models and addressing the ethical and legal challenges they can create. The goal is to support those working at applying these models in healthcare organizations and to serve as an educational resource for trainee education. It may also be helpful for researchers and students in non-clinical disciplines interested in philosophy, law, health policy, health administration and management. The book provides a unique window into how operationalizing values at the clinical interface, has implications for theory and organizational frameworks. While the focus of this book is the Canadian paediatric context, the relationships, duties and lessons discussed are germane across the age spectrum.

This book is made up of nineteen chapters divided into four parts. The chapters can be read sequentially linked thematically or used as a reference by those interested in specific topics.

The chapters in Part I provide an introduction to patient and family-centred care that reviews its evolution to both distinct and hybrid models of care. In Chap. 1 David B. Nicholas, Krista Keilty and Karima Karmali examine the evolution and literature on paediatric patient-centred care with a focus on child and family-centred care. While the authors acknowledge that evidence for child and family-centred care is limited, they demonstrate how the literature increasingly reveals the positive impact it has on outcomes and processes related to general health, mental health, well-being, satisfaction and family functioning. In Chap. 2 Linda Sheahan, Michael Da Silva, Christine Czoli and I discuss the values underpinning two major trends in paediatric bioethics. We propose a model for obtaining consent for paediatric research in a context that recognizes the family relationships of paediatric research participants, as well as the participants' emerging autonomy and rights. In Chap. 3 Jennifer Gibson explores the implications of paediatric patient and family-centred care on resource allocation processes and decisions at the micro (clinical) and meso (organizational) levels in paediatric healthcare settings. In this chapter she argues that the core principles of paediatric patient and family-centred care are not only consistent with many new approaches to resource allocation, they may also assist healthcare organizations in better mobilizing their limited resources toward providing high quality healthcare.

The chapters in Part II focus on legal, ethical and cultural considerations. In Chap. 4 Rani Srivastava examines paediatric family-centred care through the lens of religious and cultural diversity. She contends that reframing our conceptual understanding of family-centred care and its corresponding competencies through the lens of cultural competence offers significant promise for paediatric healthcare. In Chap. 5 Julija Kelečević uses narratives reflecting Aboriginal and non-Aboriginal perspectives as the basis for exploring the common ethical and legal challenges that Aboriginal people face in healthcare decision-making. She also compares the concepts of Western bioethics and Aboriginal values and discusses them within the developing models of patient and family-centred care in the paediatric setting. Kelečević presents paediatric patient and family-centred care as a mechanism for providing care that is consistent with patients' wishes. In Chap. 6 Lee A. Chapman examines the tension that can surface when the more collaborative nature of family-centred health care confronts the representation of individual legal rights, particularly the rights of children in legal disputes. She highlights for readers the significant distinction in law between an impartial and impersonal point of view. In Chap. 7 Joan M. Gilmour traces some of the ways in which the law both affirms and limits the model of paediatric patient and family-centred caregiving. She considers the legal issues that result when responsibility for care is shared or delegated to parents, family members or the patient and highlights important considerations to promote sustained delivery of safe, quality care.

The chapters in Part III provide 2 examples of internationally recognized hospital-based implementation of paediatric patient and family-centred care, a chapter highlighting the vantage point of the parent and another highlighting the vantage point of healthcare providers. In Chap. 8 Krista Keilty, David B. Nicholas, Karen Sappleton and Karima Karmali present an overview of how Child and Family-Centred Care is currently implemented at The Hospital for Sick Children (SickKids). It describes historical influences and impactful innovative practices led by staff and families at SickKids. In Chap. 9 Bonnie Nicholas describes the Thunder Bay Regional Health Sciences Centre's Patient and Family Centred Care model, using paediatrics to illustrate how Patient and Family-Centred Care and Patient and Family Advisors, in particular have significantly improved patient and family care. In Chap. 10 Janis Purdy provides a firsthand account of her family's experience with paediatric patient and family-centred paediatric care told from her vantage point of being a parent of a paediatric patient. In Chap. 11, Christy Simpson discusses the need for both changes to, and transitions in, how healthcare is provided for youth, so as to correspond to the development of their abilities to make their own health decisions. This chapter explores several issues related to this transition period both in the context of providing care for youth patients with chronic illness and from the perspective of the role of healthcare providers and their relationships with youth patients and parents.

The chapters in Part IV examine the current application of paediatric patient and family-centred care in a variety of complex health care contexts. In Chap. 12 Nancy Walton discusses general guidelines in place in Canada for the ethical conduct of research, and how the two approaches of patient-centred care and family-centred care may provide some valuable insights to deal with ethical concerns regarding recruitment, informed consent and protection of privacy in paediatric research. In Chap. 13 Cheryl Shuman and Riyana Babul-Hirji discuss genetic counseling in the context of patient-centred and family-centred care and whether patient-centred care is just a point on the family-centred care spectrum. They highlight the need for a healthcare model that in the genomic era of personalized medicine, can best shape clinical practice guidelines for optimal patient and family care. In Chap. 14 Jonathan Hellman explores how the recognition of the stress endured by parents of infants in the neonatal intensive care unit has led to the ready adoption of a family-centred care approach to neonatal care. He emphasizes the clarity arising out of mutually derived decisions in which all parties are empowered, the preferences established and where dialogue and negotiation have achieved resolution. In Chap. 15 Rebecca A. Greenberg, Michael Campbell and Linda Wright, critically examine ethical issues in the domain of paediatric organ transplantation through the lens of paediatric patient and family-centred care. They provide rich examples from the field of tranplant. They highlight tensions between the models of care and circumstances where they may work synergistically in pursuing the goal of detemining when it is medically and ethically appropriate to offer transplantation to children. In Chap. 16 Lori d'Agincourt-Canning, Lucinda Kunkel and Karen Dixon explore ethical considerations related to treatment refusal by adolescents and/or their parents, including respect for autonomy, parental decision-making authority and the duty to protect patients from harm. With a focus on anorexia nervosa, they provide practical approaches to promote continued engagement with the patient and family when there are discrepant views of what is in the best interest of the adolescent. In Chap. 17 Maria L.Z. McDonald and Lucie Wade explore the challenge of providing treatment within a patient and family-centred care model, when a capable adolescent with a disability disagrees with her mother about a proposed service. Their analysis of the legal and ethical obligations of health practitioners, adolescent patients and family includes attention to adolescent human rights and the social model of disability. In Chap. 18 Christine Newman, Adam Rapoport and Gurjit Sangha use cases created by merging actual clinical experiences, to explore the potential divergences created by a patient-centred and family-centred approach to the care of a dying child, while offering suggestions on how to resolve these conflicts.

I feel honoured to have worked with such an esteemed group of contributors in preparing this collection of essays. My hope is that this text will be useful in furthering the theoretical development of paediatric patient-centred care and family-centred care, practically useful for those in healthcare organizations working to operationalize these models and that it will act as a catalyst or springboard to further study in Canada and internationally.

Acknowledgements

I extend my sincere appreciation to all who made this project possible.

Inspiration for this book came from many sources: (1) dedicated colleagues across disciplines bringing their best efforts to care for and do right by patients, families, healthcare providers and society at large, (2) healthcare institutions working to develop and operationalize visions and missions that translate into optimal results for patients and families, (3) academic institutions where interdisciplinary work is being pursued and challenged through almost a methodological reflective equilibrium, and (4) my confidence in the relevance of ethics and law to the paediatric healthcare context.

I particularly thank the outstanding contributing authors. Each one generously shared their unique perspectives and brought enthusiasm, patience, cooperation and commitment to the project. I am appreciative of the unique sense of community that developed around this book. The Hospital for Sick Children, The University of Toronto Joint Centre for Bioethics and the Canadian bioethics community are replete with talented colleagues, many of whose insights will be shared with readers through the chapters that follow.

I thank The Hospital for Sick Children for its institutional support. I thank Parviz Manji for her enthusiastic engagement and highly valued administrative assistance. I thank David Heti for the care, resilience, insights and meticulous attention to detail he brought to his review of chapters.

I thank Springer Publishing and its evolving stellar team of Megan Curtis, Christopher Wilby, Floor Oosting and reviewers.

I thank those who have mentored and supported my work in bioethics including: Bernard Dickens, Christine Harrison, Barbara Secker, Joan Gilmour and Katherine Boydell.

I thank several additional academics and practitioners who provided inspiration, encouragement and/or opportunities that played pivotal roles in inspiring chosen paths and processes: Charles Taylor, David Weisstub, Gilbert Sharpe, Peter Singer, Bill Harvey, Wayne Sumner, Raisa Deber, Mary Rowell, Becky Greenberg, Jonathan Hellmann, Linda Wright, Jennifer Gibson, Denis Daneman, Mary Jo Haddad, Margaret Keatings, Pam Hubley and James Wright.

I thank the many patients, families, colleagues, students and trainees who provide me with inspiration and opportunities for learning on a daily basis. Motivation for taking on a project such as this, comes from their energizing experiences and insights.

An enormous thank you goes to close family and friends who have sustained years of interest and encouragement. My parents have been powerful models of putting all that you can into whatever you undertake. My husband Richard and children Dylan and Eden regularly take on the role of high-level sounding board, providing much appreciated critical reflection - always with support, good humour, excellent perspective and love.

Contents

Part III Implementation and Lived Experience

Part IV In Context

Contributors

Riyana Babul-Hirji Clinical and Metabolic Genetics, The Hospital for Sick Children, Molecular Genetics, University of Toronto, Toronto, Canada

Michael Campbell The Joint Centre for Bioethics, University of Toronto, Bioethics Programme, University Health Network Toronto, Ontario, Canada

Lee A. Chapman PBLO (Pro Bono Law Ontario), The Hospital for Sick Children, Toronto, Ontario, Canada

Christine Czoli The Hospital for Sick Children, Toronto, Ontario, Canada

Lori d'Agincourt-Canning Ethics Services, BC Children's Hospital & BC Women's Hospital, K4-161, Vancouver, BC, Canada

Michael Da Silva Rutgers University, New Brunswick, New Jersey, United States

Karen Dixon Provincial Specialized Eating Disorders Program for Children and Adolescents, BC Children's Hospital, Vancouver, British Columbia, Canada

Rebecca Greenberg Bioethics Department, The Hospital for Sick Children, University of Toronto, Toronto, Ontario, Canada

Jennifer L. Gibson Institute of Health Policy, Management & Evaluation, and Joint Centre for Bioethics, University of Toronto, Toronto, Canada

Joan M. Gilmour Osgoode Hall Law School, Toronto, Canada

Jonathan Hellmann Bioethics Department, The Hospital for Sick Children, Toronto, Ontario, Canada

Karima Karmali Centre for Innovation & Excellence in Child and Family-Centred Care, The Hospital for Sick Children, Toronto, Ontario, Canada

Lucinda Kunkel Provincial Specialized Eating Disorders Program for Children and Adolescents, BC Children's Hospital, Vancouver, British Columbia, Canada

Krista Keilty Centre for Innovation & Excellence in Child and Family-Centred Care, The Hospital for Sick Children, Toronto, Ontario, Canada

Julija Kelečević Alberta Health Services, Alberta, Canada

Christine Newman Paediatric Advanced Care Team (PACT), The Hospital for Sick Children, Toronto, Canada

Bonnie Nicholas Thunder Bay Regional Health Sciences Centre, Thunder Bay, Ontario, Canada

David B. Nicholas The University of Calgary, Calgary, Alberta, Canada

Maria L. Z. McDonald University of Toronto Joint Centre for Bioethics, Toronto, Ontario, Canada

Adam Rapoport Paediatric Advanced Care Team (PACT), The Hospital for Sick Children, Toronto, Canada

Karen Sappleton Centre for Innovation and Excellence in Child & Family-Centred Care, The Hospital for Sick Children, Toronto, Ontario, Canada

Gurjit Sangha Paediatric Advanced Care Team (PACT), The Hospital for Sick Children, Toronto, Ontario, Canada

Linda Sheahan Joint Centre for Bioethics University of Toronto, Toronto, Ontario, Canada

Cheryl Shuman Clinical and Metabolic Genetics, The Hospital for Sick Children, University of Toronto, Toronto, Ontario, Canada

Christy Simpson Department of Bioethics, Dalhousie University, Halifax, Nova Scotia, Canada

Rani H. Srivastava Centre for Addiction and Mental Health, Toronto, Ontario, Canada

Janis Purdy Office of the Provincial Advocate for Children and Youth, Toronto, Ontario, Canada

Lucie Wade Bioethics, Jewish General Hospital, Montreal Quebec, Ontario, Canada

Nancy Walton Research Ethics Board, Daphne Cockwell School of Nursing, Ryerson University, Toronto, Ontario, Canada

Linda Wright Bioethics, University Health Network, University of Toronto, Toronto, Ontario, Canada

Randi Zlotnik Shaul Bioethics Department, The Hospital for Sick Children, University of Toronto, Toronto, Ontario, Canada

About the authors

Riyana Babul-Hirji MS, CGC received a Master of Science in Genetic Counselling from McGill University in 1993 and works as a genetic counsellor in the Division of Clinical and Metabolic Genetics at The Hospital for Sick Children. She is an Assistant Professor in the Department of Molecular Genetics at the University of Toronto and also holds a Project Investigator appointment in the Research Institute at SickKids due to her involvement in clinical research (process of genetic counseling and adolescent genetic counselling studies).

Michael Campbell MHSc, BA (Hons) Senior Fellow in Organ Donation and Transplantation Ethics, University Health Network (UHN). Michael joined the University Health Network Bioethics programme as a Senior Fellow in Transplantation Ethics in July, 2010. Michael earned a Bachelor's degree in Philosophy from York University and a Master of Health Science degree in Bioethics from the University of Toronto. He completed a clinical fellowship in Bioethics at the Joint Centre for Bioethics, University of Toronto in 2010.

Lee Ann Chapman B.A. J.D. is the onsite lawyer at PBLO at SickKids. The Program offers free legal assistance to low income families to address non-medical issues that may impact a child's health or a family's capacity to care for a sick child. The medical legal partnership with Pro Bono Law Ontario and The Hospital for Sick Children and two law firms McMillan and Torkin Manes, is the first of its kind in Canada. Prior to this she was a legal advocate at the legal clinic Justice for Children and Youth for 10 years.

Christine D. Czoli BSc is a PhD student in the School of Public Health and Health Systems at the University of Waterloo. While pursuing her Bachelor's degree at the University of Toronto, she completed a minor in Philosophy and worked as a research assistant in the Bioethics Department at The Hospital for Sick Children. Her current research focuses on examining tobacco control policies in the context of chronic disease prevention.

Lori d'Agincourt-Canning MSc, MA, PhD is Clinical Assistant Professor, Pediatrics, UBC and Ethicist, Children's & Women's Health Centre of British Columbia. She received her PhD in Interdisciplinary Studies from the UBC in 2003, specializing in bioethics and feminist ethics. She has a key role in clinical ethics consultation, ethics education and capacity-building. Her main research areas include adolescent decision-making, social justice and access to health care, moral agency in interdisciplinary practice and research ethics.

Michael Da Silva JD is presently studying philosophy at Rutgers University. He worked as a research assistant in the Bioethics Department of The Hospital for Sick Children while pursuing his J.D., which he received from the University of Toronto Faculty of Law in 2012. He then served as a foreign law clerk at the Supreme Court of Israel before resuming his philosophical studies. His main academic interests include ethics, political philosophy, philosophy of law, (comparative) constitutional law and epistemology. His individual work has appeared in several journals, including the Canadian Journal of Law and Jurisprudence.

Karen Dixon MSW, RSW is a Social Worker at BC Children's Hospital Provincial Specialized Eating Disorders Program for Children and Adolescents. She is also a nurse by training. She completed her Social Work training at the University of Calgary and her nursing training at the University of Alberta. She is passionate about empowering parents of youth with eating disorders through family, group, and multiple family therapy.

Jennifer L. Gibson PhD is Associate Professor in the Institute of Health Policy, Management, & Evaluation and Associate Director of the Joint Centre for Bioethics at the University of Toronto. Jennifer has broad research, consulting, and policy experience in resource allocation ethics. She has worked with health organizations across Canada to address common priority setting challenges and has advised government and other health sector agencies on a range of resource-related issues. Her current research focuses on: ethics and health system integration, public engagement in cancer control, and values-based decision-making.

Joan M. Gilmour B.A., LL.B. (Toronto), J.S.M., J.S.D. (Stanford) is a Professor at Osgoode Hall Law School in Toronto, and a member of the Bar of Ontario. She is Director of Osgoode's Master's program specializing in Health Law, and has served as Acting and Associate Director of the York University Centre for Health Studies (now the York Institute for Health Research). She has consulted for Health Canada and the Ontario Law Reform Commission. She researches legal issues in paediatric care, health systems restructuring, professional regulation, health care human rights challenges, withdrawing life support, and the effects of the tort system on patient safety.

Rebecca Greenberg RN PhD is a Bioethicist at The Hospital for Sick Children. She is an assistant professor in the Department of Family and Community Medicine, and a member of the Joint Centre for Bioethics at the University of Toronto. Rebecca has a Bachelor of Arts (psychology) from the University of Manitoba, and a Bachelor of Science in Nursing and PhD (bioethics) from the University of Toronto. Prior to her work in ethics Rebecca practiced as a Registered Nurse in the area of general medicine. She has completed an ethics internship and a post-doctorate fellowship in Bioethics.

Jonathan Hellmann MBBCh, FCP (SA), FAAP, FRCPC, MHSc (bioethics) is a medical graduate of the University of Witwatersrand, Johannesburg , South Africa. Following his Pediatric residency he completed a Fellowship in Neonatology at the Hershey Medical Center, Pennsylvania State University, USA. He is a Professor of Pediatric s in the Faculty of Medicine at the University of Toronto, and a Staff Neonatologist in the Neonatal Intensive Care Unit (NICU) at The Hospital for Sick Children. He was the Clinical Director of the NICU for 17 years. He completed a Masters in Bioethics at the Joint Centre for Bioethics (JCB) in Toronto and has been on its teaching faculty for the past 7 years. His is currently the Medical Bioethics Associate in the Bioethics Department at The Hospital for Sick Children, Toronto.

Karima Karmali RN, BScN, MBA is the Director of the Centre for Innovation & Excellence in Child- and Family-Centred Care at The Hospital for Sick Children (SickKids) in Toronto. In this role, she provides strategic leadership for the advancement of child- and family-centred care at the clinical and health systems level through the implementation of innovative programming and strong patient and family engagement. Her focus includes improving health equity and accessibility of care for vulnerable populations. Karima has over 20 years of leadership experience in health care and in leading organization and system-wide change.

Krista Keilty NP, PhD Candidate, is a Nurse Practitioner located in the Centre for Innovation & Excellence in Child and Family-Centred Care and Project Investigator through the Research Institute at The Hospital for Sick Children. Krista's clinical practice has informed her research interest in the experience of family caregivers of children with acute and chronic illness. Krista is a PhD Candidate at the Bloomberg Faculty of Nursing, University of Toronto. Her doctoral dissertation, will examine sleep disturbance in family caregivers of children with medical complexity. This prospective matched-cohort investigation is expected to be of interest to child health clinicians, researchers and policy makers.

Julija Kelecevic MD, PhD(ABD) works as a clinical ethicist Alberta Health Services, supporting Calgary and South Alberta. Prior to joining AHS, Julija worked as a bioethicist in Thunder Bay and the Northwest Ontario. Julija holds a medical degree from the University of Belgrade, Serbia and is finishing her PhD at the University of Alberta. Her primary research interests include cross-cultural ethics, interprofessional ethics and evaluation of clinical ethics consultations. Julija continues to teach health care ethics and medical humanities.

Lucinda Kunkel MD FRCPC is a psychiatrist at BC Children Hospital's Provincial Specialized Eating Disorders Program for Children and Adolescents where she works in the inpatient, outpatient and Day Treatment units. She completed her psychiatry residency through the University of Toronto. Interests include bringing mindfulness into eating disorders treatment.

Maria McDonald JD MHSc works as an independent consultant in bioethics, privacy and health law. Previously, she worked as the Bioethicist at Holland Bloorview Kids Rehabilitation Hospital from 2007 to 2013, and as the Privacy Officer from 2010 to 2013. Since 2007, she has taught "Legal Approaches to Bioethics" in the MHSc Program in Bioethics at the Joint Centre for Bioethics, and is appointed Adjunct Professor to the Faculty of Law and the Institute of Medical Sciences at the University of Toronto. Maria earned a Juris Doctor from Osgoode Hall Law School and MHSc in Bioethics from the University of Toronto.

Christine Newman MD FRCPC is a paediatric palliative care physician with the Paediatric Advanced Care Team (PACT) at The Hospital for Sick Children in Toronto and an assistant professor of paediatrics at the University of Toronto. A graduate of the University of Toronto School of Medicine with postgraduate training in paediatrics and neonatology, Chris joined the Division of Neonatology at SickKids and served as Clinical Director of the NICU for 6 years. After further study in palliative care Chris joined PACT providing consultative care to children with life-limiting illnesses and their families. In 2010 Chris was awarded the Robert Salter Humanitarian Award following nominations from co-workers, paediatric trainees and patients' families.

Bonnie Nicholas RN, CNCC(C), CPTC As the Patient Family Centred Care Lead, Patient Advocate, Bonnie has led the organization on their Patient Family Centred Care journey and developed a robust Patient Family Advisory Program that integrates the voice of the patient and their families in "everything we do". This collaboration with the patients, families, staff, physicians, volunteers and community has led to improved patient and staff satisfaction and Thunder Bay Regional Health Sciences Centre to the first and only organization in Canada to receive a "Patient Family Centred Care Leading Practice" designation from Accreditation Canada in 2011.

David B. Nicholas RSW, PhD is an Associate Professor in the Faculty of Social Work at the University of Calgary in the Edmonton Division. He is cross-appointed as a Senior Associate Scientist in Child Health Evaluative Sciences at The Hospital for Sick Children's Research Institute, with a focus on child and family-centred care. His research addresses individual and family quality of life and social support in child health and disability. Dr. Nicholas has received multiple grants from the Canadian Institutes of Health Research, Social Sciences and Humanities Research Council of Canada, and other national and international funding bodies.

Janis Purdy MM is a Child and Youth Advocate with the Office of the Provincial Advocate for Children and Youth in Ontario. She is Co-Chair of the Special Needs Committee and holds a portfolio in Peadiatric Health Advocacy. As a volunteer she works with The Sick Kids NICU Family Care Committee and with Scouts Canada as a Beaver Scout Leader. Janis has a Masters in Management from McGill University in Montreal and a post graduate certificate in Infant Mental Health from York University in Toronto. In 2006 Janis gave birth to her second son, Jonathan, who was born with a Giant Omphalocele and Tetralogy of Fallot. She lives in Toronto with her wife, Diane Flacks, and her two sons, Eli and Jonathan Purdy-Flacks.

Adam Rapoport MD, FRCPC, MHSc completed his pediatric residency in Toronto from 2002-2006. During a 2-year academic fellowship, he obtained a Masters of Health Sciences in bioethics at the University of Toronto Joint Centre for Bioethics. Since 2010 Adam has held the position of Pediatric Undergraduate Ethics Coordinator at the University of Toronto Medical School. He is the pediatric consultant at the Temmy Latner Centre for Palliative Care at Mount Sinai Hospital and in 2011 he became the first Medical Director of the Paediatric Advanced Care Team at The Hospital for Sick Children.

Karen Sappleton MSED, MSW, RSW is the Manager for Child & Family Centred Care and Health Equity in the Centre for Innovation and Excellence in Child & Family Centred Care at The Hospital for Sick Children in Toronto, Ontario. She was a social worker and an educator at SickKids for 7 years and has extensive experience working with children, youth and families in community and hospital settings. She now promotes child and family centered care initiatives and practices at the hospital. She is also the manager of Interpreter Services Department and cultural competence education and training for staff and trainees.

Gurjit Sangha RN MN CHPCN(C) holds a Bachelor of Science (University of Waterloo), a Bachelor of Science in Nursing and a Masters of Nursing (University of Toronto). She was Project Manager for the Best Practice Spotlight Organization (BPSO) initiative leading the implementation of Best Practice Guidelines across the Hospital for Sick Children. Her clinical work has been in pediatric oncology as a staff nurse, Clinical Support Nurse and Interlink Community Cancer Nurse. Currently Gurjit is the Clinical Nurse Specialist with the Pediatric Advanced Care Team (PACT). Gurjit is passionate about providing excellence in nursing care to children and their families.

Linda Sheahan MD(Hons), M. Bioethics, FRACP is a Palliative Medicine Specialist and Bioethicist from Sydney Australia. She completed medicine with Honours through the University of Sydney, and went on to specialist training in internal medicine through the Australian College of Physicians, earning her from Monash University, and a Fellowship in Clinical Bioethics through the Joint Centre of Bioethics in Toronto 2008. She is an expert reviewer for the CMAJ, has published in both medicine and ethics, and is an Australian 2012 Churchill Fellow.

Cheryl Shuman MS CGC received a Master of Science degree in Genetic Counselling from Sarah Lawrence College in 1979, following which she worked in the Genetics-Birth Defects Center at the Medical University of South Carolina in Charleston. In 1981, she joined the Division of Clinical and Metabolic Genetics at The Hospital for Sick Children (SickKids). In 1998, she was appointed Program Director of the MSc Program in Genetic Counselling at the University of Toronto and in 1999, was appointed Director, Genetic Counselling at SickKids. As well, she holds a Project Investigator appointment in the Research Institute at SickKids based on her involvement in clinical research (genotype/phenotype correlations in overgrowth syndromes with cancer predisposition, most specifically Beckwith-Wiedemann syndrome) and in research related to the provision of genetic counselling.

Christy Simpson PhD is Head and Associate Professor in the Department of Bioethics at Dalhousie University. She is the coordinator for the ethics-based collaborations between the Department, Capital Health and the IWK Health Centre, as well as the Nova Scotia Health Ethics Network. Her primary responsibilities include ethics education and capacity-building, policy development and review, and support for clinical and organizational ethics consultations. Christy completed her doctorate in philosophy, specializing in bioethics, at the Department of Philosophy, Dalhousie University in 2001. Her research interests include rural health ethics, paediatric ethics, organizational ethics and the role of hope in health care.

Rani Srivastava RN PhD is the Chief of Nursing & Professional Practice at the Centre for Addiction & Mental Health (CAMH), Assistant Professor at the Lawrence Bloomberg Faculty of Nursing, University of Toronto and Adjunct Professor, York University, Toronto. She is a recognized expert in the areas of cultural competence and health care both at the clinical care level as well as the organization and systems level. She is the author and editor of The Healthcare Professional's Guide to Clinical Cultural Competence (2007) which outlines a framework for cultural competence in care. Dr. Srivastava holds a bachelors degree from Dalhousie University and her graduate and post-graduate studies were completed at the University of Toronto.

Nancy Walton PhD is an Associate Professor in the Daphne Cockwell School of Nursing at Ryerson University in Toronto and Chair of the Ryerson University Research Ethics Board since 2004. Nancy has a PhD in Bioethics from the University of Toronto Collaborative Program in Bioethics. She has served on the National Council on Ethics in Human Research, is a member of the Research Ethics Board at The Hospital for Sick Children and was a founding member of the Research Ethics Board at the Ontario College of Art and Design. She is a consulting ethicist for the Cardiac Care Network of Ontario.

Lucie Wade is Clinical Ethicist at the Jewish General Hospital, Montreal and Associate Investigator with NeuroDevNet, Canada. She has worked as a Senior Ethics Fellow and completed an Academic Fellowship in Clinical and Organizational Ethics, both at the University of Toronto Joint Centre for Bioethics, supporting the ethics programs of Toronto Central Community Care Access Centre, Humber River Regional Hospital, Holland Bloorview Kids Rehabilitation Hospital and Baycrest Geriatrics Centre. Lucie holds an MSc Experimental Medicine (Specialization

Bioethics) from McGill University and a combined honours degree in Marine Biology and Contemporary Studies from Dalhousie University and the University of King's College.

Linda Wright MHSc, MSW, RSW is Director of Bioethics at the University Health Network (UHN), Toronto, assistant professor in the Department of Surgery, Faculty of Medicine, and a member of the Joint Centre for Bioethics at the University of Toronto. Linda provides clinical and organizational ethics consultation and teaching, has research expertise in organ transplantation ethics and writes on current issues in transplantation ethics. Her research focuses mainly on ethical issues of living organ donors. In addition to working on national and international projects in donation and transplantation, Linda is ethics section editor of the journal Progress in Transplantation.

Randi Zlotnik Shaul JD, LLM, PhD is Director of the Bioethics Department and Associate Scientist in Child Health Evaluative Sciences at The Hospital for Sick Children. Randi provides ethics consultation services and policy support, teaches health care professionals and bioethics students, and conducts bioethics research. Randi is an Associate professor in the Department of Paediatrics at the University of Toronto and is a member of the University of Toronto's Joint Centre for Bioethics. Randi's research focuses on paediatric health care challenges at the intersection of ethics and law. Before graduate school Randi practiced law at the Ontario Ministry of Health.

Part I
Introduction to Patient and Family-Centred Care

Chapter 1
Paediatric Patient-Centred Care: Evidence and Evolution

David B. Nicholas, Krista Keilty and Karima Karmali

1.1 Introduction

Paediatric patient-centred care as a model of care delivery has been influenced by advances in clinical practice, research and evaluation, along with shifting societal values towards greater public input in healthcare design and delivery. In this chapter, a review related to paediatric patient-centred care is described, and is grounded by a broad scan of the peer-reviewed evidence-based literature, select gray literature and other targeted sources (e.g., relevant organizations and associations). Based on this review, this chapter presents applied definitions of both patient-centred care and child and family-centred care (C&FCC), and focuses on the key considerations in support of understanding the background to, and current state of, C&FCC as a preferred framework for application in paediatric patient-centred care. Descriptions of multiple influences on the development of C&FCC, conceptual and theoretical considerations, and a review of select studies examining the impacts of exemplary C&FCC practices, are also presented. On balance, the literature and related sources of evidence strongly endorse C&FCC, including evidence of enhanced care processes, and positive patient and family experiences. Operational and ethical tensions exist in the application of C&FCC, many of which are discussed later in this book.

D. B. Nicholas (✉)
Faculty of Social Work, The University of Calgary, Calgary, AB, Canada
e-mail: nicholas@ucalgary.ca

K. Keilty · K. Karmali
Centre for Innovation & Excellence in Child & Family-Centred Care, The Hospital for Sick Children, Toronto, ON, Canada
e-mail: krista.keilty@sickkids.ca

K. Karmali
e-mail: karima.karmali@sickkids.ca

R. Zlotnik Shaul (ed.), *Paediatric Patient and Family-Centred Care: Ethical and Legal Issues,* International Library of Ethics, Law, and the New Medicine 57,
DOI 10.1007/978-1-4939-0323-8_1,

1.2 What is Patient-Centred Care?

Patient-centred care has been located mainly in the adult delivery system. It is now considered an essential component to the organization and delivery of healthcare services. Originating in the late 1980's, the primary idea purported about patient-centred care was that patients should be included in all aspects of decision-making about their health and healthcare. This early attempt to define patient-centred care has evolved significantly, moving beyond decision-making to include the ethical principle of autonomy and attention to the individual patient's unique preferences, values, traditions and family situation (Sine and Sharpe 2011). In recent years, the integration of this approach in the design and delivery of healthcare services has become well accepted, but also raises some challenges in implementation and ethics.

Early definitions of patient-centred care referred to the primary importance of the relationship between the patient and physician (Henbest and Stewart 1989). More recently, patient-centred care models have evolved to be more inclusive of the full care team and thus operational definitions of patient-centred care have changed to include patients (and their families), clinicians (including clinical ethicists) and health systems (Epstein and Street 2011).

Health policy spanning the last decade has developed to situate patient-centred care as a guiding framework for advancing the organization and delivery of quality care across most western jurisdictions. The American-based Institute of Medicine's *Crossing the Quality Chasm: A New Health System for the 21st Century* report (2001, p. 6) directed that healthcare reform be built around the core need for healthcare to be, among other directives, patient-centred and thus, "providing care that is respectful of and responsive to individual patient preferences, needs, and values, and ensuring that patient values guide all clinical decisions". The report further emphasized the imperative that reforms to the American health sector must place the patient as the source of control, and care needs to be customized according to patient needs and values. This seminal work has continued to be a source of reference influencing policy and practice changes towards a more patient-centred healthcare delivery system.

In Canada, where universal access to healthcare is considered a right of all Canadians, the province of Ontario has positioned patient-centred care as one of the core principles of recent healthcare legislation. The Ontario Ministry of Health and Long-term Care's *Excellent Care for All Act* (2011) was passed at a time when Canadians sought to ensure that advances in the healthcare sector would not put technology, healthcare providers or hospitals ahead of patient care needs. The *Act* explicitly states that regulations and policies that follow will, "put patients first by improving the quality and value of the patient experience through the application of evidence-based health care" (Ontario Ministry of Health and Long-term Care 2011). Among the list of stated imperatives for Ontario are requirements to promote patient experience, systematically assess patient satisfaction with healthcare, and publish results in the public domain.

While these and other policies have been helpful in situating patient-centred care as a requirement for healthcare administrators and clinicians, a critique of the policy literature in patient-centred care remains. McClimans et al. (2011) suggest that policy related to patient-centred care provides inadequate theoretical justification for why and how patient-centred care should be valued. This lack of clarity can situate healthcare administrators and providers at odds with patients, and result in hospital personnel seeking guidance from clinical ethicists and others about how to most effectively apply these principles in challenging circumstances.

1.3 What is Child and Family-Centred Care (C&FCC)?

C&FCC is recognized as central to the delivery of paediatric patient-centred care for children and their families. Fundamentally, C&FCC is reflective of respect for, and partnership with, the child and family in the delivery of the child's healthcare. It aims to support the integrity of the family and individualize care to promote both individual health and normal family functioning when a child is unwell (Palmer 1993). This principle, applied in paediatric patient-centred care, acknowledges that child-rearing is a unique child and family process in which the family is central to the child's life (American Academy of Pediatrics 2003). Shields et al. (2006, p. 1318) complement this perspective by suggesting that, "family centred-care is a way of caring for children and their families within health services which ensures that care is planned around the whole family, not just the individual child/person, and in which all the family members are recognized as care recipients".

Paediatric care policy has influenced, and been influenced by, C&FCC reforms. An example is the *Joint Positions Statement on Family-Centred/Maternity Newborn Care in Hospitals* (American College of Obstetricians and Gynecologists 1978) and the U.S. congressional amendment to broaden the scope of services for children with special needs i.e., the U.S. Congress' *Education of All Handicapped Children Act* (1975). Originally published in 1987 and currently in its fourth edition, the Public Health Agency of Canada (2000) published national guidelines for maternal and newborn care in Canada (Family-Centred Maternity and Newborn Care: National Guidelines [http://www.phac-aspc.gc.ca/hp-ps/dca-dea/publications/fcmnc-smnnpf/index-eng.php]). The guidelines identify principles of family support as well as participation and choice.

The definition of C&FCC, while differing across sources, is commonly associated with elements of partnership and knowledge exchange. Despite this relative lack of unanimity of definition, consensus seems to indicate that C&FCC opens 'space' for individual encounters between stakeholders involved in care (e.g., the patient, family, healthcare providers) via enhanced communication strategies, sensitivity to family values and customs, and provision of timely information by healthcare team members. C&FCC is associated with elements of partnership related to the care plan for the child. Accordingly, families and professionals share in the development and implementation of the plan, recognizing that the family is important to the child's well-being. Essential to C&FCC is providing to children and families information tailored to their unique needs (e.g., diagnostic information conveyed in ways that are developmentally-appropriate) and communicated in a way that accounts for their preferences in how they will be involved in care decisions.

Despite C&FCC being well accepted, myths have emerged in the healthcare community and in some circumstances prevailed, with the risk that these myths might undermine the uptake and application of C&FCC. In a series of informational fact sheets prepared for broad circulation, the Canadian child disability clinical and research network entitled *CanChild* (2010, p. 2), has highlighted common myths about C&FCC in paediatric health services (http://www.canchild.ca/en/childrenfamilies/resources/FCSSheet2.pdf): (1) "Families always get what they want. Service providers may have to go against what they believe professionally to satisfy a family's request," and (2) "Families must be responsible for making decisions or doing things that they would rather have service providers do". Rebuttal to the ideas presented in these myths is located throughout the peer-reviewed C&FCC literature; nonetheless, tensions can arise in practice when the same understanding of C&FCC is not shared among clinicians and families.

In summary, it is widely accepted that despite ongoing debates about the definition of C&FCC and the strength of data that supports its effectiveness (which is further explored later in this chapter), C&FCC is strongly valued and an ongoing element of care. It has become wide-spread and is described by leading paediatric authors as, "ubiquitous as a way of delivering care" (Shields et al. 2012, p. 1).

In the pages that follow in this chapter, the influences, processes and outcomes of C&FCC will be explored in terms of their applicability to paediatric healthcare services. This review will position C&FCC as highly regarded, if not yet fully established, as a comprehensive model of healthcare in paediatrics. The review will highlight where accomplishments in meeting the principles of C&FCC have been achieved through selected care practices, and where some challenges in the delivery of C&FCC continue to present themselves.

1.4 What Informs C&FCC?

C&FCC has been informed by history, theory, research and practice. Key to the development of this understanding has been the foreground position that children and family members offer important commentary to healthcare providers and policymakers about what is most important to children and families. Many of these influences will be described in this section of the chapter, with a particular emphasis on contemporary sources of information.

1.5 Theoretical Foundations of C&FCC

Despite increased interest directed to C&FCC in paediatric practice and planning over recent years, there appears to be a lack of conceptual and theoretical clarity related to this construct. Without this grounding, practice and evaluation of C&FCC processes and outcomes may be impeded. Early authors in this area identified family-centred care (FCC) as a construct rather than a well-developed and tested theory (Dickoff et al. 1968). Since then, concept analysis has been used to explore meanings associated with C&FCC. Subsequent fieldwork (Hutchfield 1999) has informed a conceptual analysis of FCC that has identified both antecedents and attributes of FCC with relevance for advancing both theory and practice, yet a lack of clarity related to FCC has continued to be identified. Hutchfield (1999) presented alternate models of FCC that demonstrate strong associations with concepts of partnership with parents, parental participation, and care by parents; concepts that remain current today.

Toward developing a theoretical underpinning to enable model development and testing of C&FCC, the importance of the family as a system and set of subsystems (individual, individual in family, family in environment) emerges as central. Family systems theory, for instance, recognizes the individual as a part (subsystem) of a whole (e.g., family system), such that elements affecting the individual (e.g., illness and care) have a bearing on the whole (functioning and well-being of the family) and vice versa. Recognizing the person as part of a larger system becomes integral to understanding the importance that is ascribed to considering the child *and* family in C&FCC, and thereby emphasizing the processes and procedures of care that have a bearing on individual and family experience.

Below is a description of select studies that have used a theoretical model in applying concepts of C&FCC. These perspectives consistently push against medical model elements that may have traditionally focused primarily upon health deficits as the focal point for assessment and intervention.

In a critical examination of FCC in a critical care setting, Cannon (2011) utilizes a family systems approach in considering the interdependence of parts of the system. In examining experiences and resources, the family is described by this author as a unit, with points of participation, interaction and shared information. Harrison (2010) discusses the cyclical nature of relationships, in that children both affect, and are affected by, those around them. By adhering to a C&FCC philosophy, a framework is provided for practitioners to view patient care as supporting health and wellness across physical and psychological domains.

Hjorngaard (2011) applies the social model of disability in demonstrating how structures and discourses may influence attitudes and ultimately limit participation in society, and thereby disadvantage persons and families affected by disability and other health conditions. Respecting and responding to families' perspectives is argued as crucial, as illustrated in Hjorngaard's (2011, p. 244) comments:

> The rationale for therapy, or any intervention, is critical to understanding what a successful outcome may be for every unique family. Gathering information about the family's values, ethics and fundamental view of their child's disability (as a "problem/burden" or a "special gift", or any number of interpretations), is crucial to providing appropriate service centered on the family as a whole, as opposed to an expectation that the family conform to the practice.

Hjorngaard (2011) posits that a long-term commitment to the developing child in the context of the family is needed, whereby disability exists on a continuum of care between the medical and social models of practice.

Another theory-informed model of C&FCC is John Rolland's *Family-Systems-Illness Model,* which offers a biopsychosocial map of the multiple, imposing and intersecting demands on families affected by illness or disability, including life cycle demands, intergenerational experiences, and beliefs (Rolland and Walsh 2006). This approach calls attention to nuanced family support, including the recognition of various demands and losses associated with childhood health conditions. Moreover, this approach highlights and grounds an integral role for C&FCC in supporting and fostering family navigation through the complex healthcare experience.

1.6 Other Influences on C&FCC

The engagement of families (largely parents) in articulating the principles of C&FCC has been integral to the pace at which C&FCC has advanced. Historically, parents of children with complex and chronic illness have engaged actively with clinicians, researchers and policy makers to advance the field, by helping to document and design C&FCC. Paediatric healthcare providers have also been early adopters of C&FCC principles, and face the reality that they cannot effectively care for children needing health services without partnering with the family in their assessments (e.g. most often parents provide information about their child and her/his health condition) and treatments (e.g., health information to parents is required to optimize the child's health and welfare).

A number of authoritative statements about C&FCC have been authored by professional and patient (child and family) representative groups, including: the Maternal and Child Health Bureau (Arango 2011; Kuhlthau et al. 2011); a joint collaboration between the American Academy of Family Physicians, the American Academy of Pediatrics, the American College of Physicians, and the American Osteopathic Association (American Academy of Family Physicians et al. 2011); the Institute for Patient- and Family-Centered Care (Adler 2009; Johnson et al. 2009); the American Academy of Pediatrics (American Academy of Pediatrics 2012; Gooding et al. 2011); and the Association for the Care of Children's Health (Harrison 2010). These statements are grounded in perspectives that inherently value the patient and their family as central in children's healthcare journey, and these positions appear to have influenced healthcare policy and practice.

The Institute for Patient and Family-Centered Care (IPFCC) was established in 1992 to take over for, and build upon, the substantial contributions to C&FCC by Shelton et al. (1987) and the Association for the Care of Children's Health (ACCH) that was first established in 1965. Since its formation, the IPFCC has been tasked to advance the practice and understanding of FCC by serving as a central resource to child health policy makers, healthcare administrators and educators, paediatric providers, and family leaders. One of the major contributions of the IPFCC has been the affirmation of Shelton et al.'s (1987) initial work in their development of key elements of C&FCC, and the initiation of a framework for assessment, implementation and evaluation of FCC based on four core concepts: dignity and respect, information sharing, participation, and collaboration (Adler 2009; Johnson et al. 2009). These concepts underscore a philosophy of FCC articulated by the IPFCC, which upholds the family as constant in the child's life. It asserts that all members of the family are affected by the illness of a family member, which implies a family systems approach in which the family unit is integrally connected to individual patient experiences of illness and care. Collaboration between the healthcare team and the family, and respect for diversity, emerge as other key components (Frost et al. 2010). The American Academy of Pediatrics (AAP), widely recognized as an authoritative voice in paediatrics (Gooding et al. 2011; American Academy of Pediatrics 2012), offers guiding principles for C&FCC that include the following core components:

- Respect for each child and family—honouring of the diversity and its effect on the family's perception and experience of care
- Recognition of, and commitment to, the strengths of the child and family, particularly in difficult situations
- Support and facilitation of choice regarding decision-making and approaches to care for the patient and family
- Flexibility in organizational policies, procedures and provider practices that best suit the needs, beliefs and values of the family
- Information-sharing, including unbiased and honest information to families throughout the care process, and providing it in a way that is useful and reaffirming
- Collaboration with families at all levels of healthcare
- Provision of formal or informal support, as needed, that empowers the family and child to make choices about their health, while capitalizing on their strengths.

In summary, multiple consumer and professional bodies and organizations have advanced the evolution of C&FCC by bringing together key stakeholders and building consensus about the most significant components of a C&FCC model. Each of the frameworks share commonalities such as reference to collaboration, partnership, respect and decision-making, and serve to provide a basis from which to operationalize and evaluate C&FCC.

1.7 Evidence for C&FCC

The evidence in support of C&FCC is limited in terms of scope and quality. Historically, the C&FCC literature has been narrative in nature, with published studies mainly including uncontrolled descriptive research designs that offer limited information about the sociodemographic profile of the sample in the respective study. These gaps render challenges in applying study findings more broadly to diverse populations. Systematic reviews offer insights into the strengths, but especially the limitations, of the empirical evidence in support of C&FCC. Descriptive studies about specific C&FCC practices inform both a broader understanding of C&FCC and specific elements or practices therein (e.g., individual interventions related to peer support, patient and family advisory committees, etc.). In the following section, selected commonly cited sources that offer evidence in support of C&FCC are reviewed on the basis of their contributions and limitations to our evolving understanding of the C&FCC literature.

1.7.1 Systematic Reviews of C&FCC

Over the years, a number of authors have published reviews and critiques of C&FCC that together solidify and add depth to our understanding and application of this concept. In a recent update to an earlier systematic review of FCC based on established standards from the Cochrane collaboration, an internationally-recognized repository of health evidence review, the effectiveness of family-centred models of care compared to standard models of care was examined for hospitalized children from birth to 12 years of age (Shields et al. 2012). Studies were included and reviewed if they explicitly demonstrated that provider and child/family collaboration was a key component of the intervention. Single intervention studies were excluded, and only randomized-controlled trials (RCTs) that met stringent criteria were included. The following scoring criteria ranked the degree of family-centredness relative to each study's expressed orientation to principles of FCC: (1) family as a constant, (2) culturally responsive (sensitive to the cultural context of the family), and (3) supportive of family individuality and need for different types of family support. Results of this systematic review suggest that out of 271 potentially relevant studies examined, only one study (Bolton 2004) met criteria for inclusion (Shields et al. 2012). The review reported that FCC models showed no indication of harm and suggested the presence of minimal, moderate quality evidence (as noted above, only RCTs were reviewed), demonstrating impacts of C&FCC on children's clinical care, parental satisfaction and costs. The authors concluded that more studies are needed to examine the effectiveness of C&FCC on hospitalized children, and offered critique about the overall body of evidence.

In a recent review of qualitative studies in C&FCC (Shields et al. 2006), a total of 11 studies were identified that met the following FCC-related criteria: (1) an intervention is described that promoted the FCC model of care for the hospitalized child, (2) clear evidence of child/family involvement in care is provided, (3) timely collaboration between providers and the child/family is demonstrated, (4) FCC models are compared to standard (i.e. medical) models of care. The following key themes emerged across these diverse studies that reported on a range of between 5 and 76 participants: (1) negotiation is important between staff and families, and (2) the importance of roles for all. Results of these studies suggested that parents felt imposed upon if their participation in care was taken for granted.

The process of how nurses negotiate with parents in a C&FCC model was the focus of a review of the literature published from 1990 to 2005 (Corlett and Twycross 2006). This review of 11 diverse studies described the following three themes: (1) role negotiation and whether it occurred in practice, (2) parental expectations of participation in their child's care, and (3) issues relating to power and control. The authors reported that on balance, but not always, the literature suggests that parents and nurses manage to negotiate how each will take a role in the child's care. It also described that parents are not consistently involved in decision-making about their child's care. The study concluded by acknowledging a major limitation of this literature in that the perspective of children was only investigated in one of the 11 studies. Recommendations for future research included the inclusion of children's perspectives and as well as the development of curricula for healthcare provider education.

Synthesis-based studies of children with special healthcare needs (including those with chronic conditions that result in extraordinary care needs) related to C&FCC, were examined. In a review of the published literature that examined evidence for FCC with a focus specifically on family-provider partnership, 24 studies met the authors' criteria for review (Kuhlthau et al. 2011). The studies were diverse in terms of their methods (mostly cross-sectional, with 7 RCTs), sample characteristics (ranging from children with asthma to those with complex multi-system conditions including those with disabilities), and measures (most used validated instruments, but not all). Overall, the authors concluded that the available evidence links FCC with improved outcomes in several domains including health, mental health and well-being, satisfaction, healthcare utilization or access, communication, systems improvements, family functioning and costs to family. Limitations of this systematic review included the lack of study representation beyond only American studies, and publication bias favouring studies that have demonstrated a positive association with FCC (Kuhlthau et al. 2011).

Overall, C&FCC has been the focus of numerous systematic reviews. Each of the reviews described in this chapter conclude that there do not appear to be any harms in C&FCC interventions, but generally cite that it is challenging to define which studies should be included for analysis based on limited descriptions of study intervention details and a lack of agreement about what constitutes C&FCC. Positive outcomes, however, are described across all of the reviews, signaling the merits of C&FCC, even in the absence of strong evidence for its overall effectiveness.

In the next section, studies of interventions targeting individual elements or practices of C&FCC will be reviewed. Cumulatively, these studies help to inform a broad understanding of how C&FCC is operationalized and implemented 'on the ground'.

1.7.2 Evidence for Specific Child & Family-Centred Care Practices

While there are many specific activities and processes that illustrate elements of C&FCC, we have chosen some common examples along with selected studies that demonstrate their application. These examples more narrowly identify specific processes of care that ultimately may contribute to programs that are patient and family-centred.

1.7.2.1 Family-Centred Rounding

Family-centred rounds are based on the premise that patients and families bring important information for optimal care planning, and have the intrinsic right to information that is generated in patient rounds. Sisterhen et al. (2007, p. 320) state that,

> [the] key difference in family-centered rounds from traditional bedside teaching is the active participation of the patient and family in the discussion. Family-centered care encompasses the understanding that patients and their families need open, honest, and unbiased communication with all health care providers.

However, as cited by Muething et al. (2007), healthcare providers may have concerns about the inclusion of families in rounds, such as a loss of time and disruption of workflow due to the extra effort needed or potential communication problems.

As noted in a case report from an American hospital (Muething et al. 2007), family-centred rounds were instituted with the following core principles: families chose how and to what level they participated; families decided if they wished to be awakened for rounds; and introductions to key team members (or all team members if the group was small enough) were made at rounds. In these rounds, efficiency was of interest to all, with team members having pre-assigned roles. The intern or student explained the purpose of rounds, invited the family to play an active role, and summarized the medical status and treatment options. The family was invited to participate in daily plans and goals, and all participants were ideally made aware of, and comfortable with, the treatment plan for the day. The authors suggested that in C&FCC rounds, teachers (e.g., attending physicians) can model appropriate behaviour for students, and parents can be asked about their preferences for additional teaching to be done in the room or elsewhere (Muething et al. 2007).

Reported study outcomes related to trainee teaching, staff time and patient confidentiality (Muething et al. 2007). While initially believing that teaching would be impeded, residents reported improved teaching and learning. Family-centred rounds reportedly took approximately 20 % longer, but staff reported that this time was used efficiently and saved time later in the day. Families were less likely to question the care plan and had a greater understanding of discharge plans. Issues of confidentiality (e.g., concern that rounds may be overheard by others in the room) were explained to families, and families indicated their preferences about the degree to which they chose to participate. Overall, while aware of the risks, approximately 85 % of families sought involvement in rounds when given the choice.

1.7.2.2 Peer Support

Peer support has been identified as a resource for coping with life challenges including illness. It involves the opportunity to engage with another who shares elements of a similar experience. Gooding et al. (2011) identified the potential for support groups, parent-to-parent support, and online and technology-based support in the context of a Neonatal Intensive Care Unit. Parents who made contact with peers (e.g., other parents) who had successfully navigated the hospital experience reported that the contact was a source of information, hope, advice and support for parents. Support groups provided a forum to share feelings and receive advice and information, and positively affected parent-staff communication, parental coping, and interactions between parents and children. In a prospective pre-post intervention study (Nicholas and Keilty 2007), parents of children with chronic lung disease were matched in parental dyads for the purpose of exchanging and evaluating support. Over a 4-month period, parents who were paired according to similarities in terms of children's age, diagnosis and care requirements (e.g., home oxygen use), engaged in peer-support by sharing their adjustment experiences, and mutually offering and receiving support. This peer support reportedly reduced social isolation, increased knowledge, and led to parents feeling more understood by others. Studies in other clinical areas have demonstrated benefits from support in multiple delivery modalities including online applications (e.g., Gooding et al. 2011; Nicholas et al. 2009). However, Gooding et al. (2011) have raised caution about the importance of avoiding potential detriments to peer support from persons who have not been adequately screened, oriented, trained or they themselves supported.

1.7.2.3 Navigation of Care

In an adult-based study, Ferrante et al. (2010, p. 736) defined patient navigation as, "the process of helping patients to effectively and efficiently use the health care system". Assisting families in the navigation of their care was described to commonly be supported by a "patient navigator" (Ferrante et al. 2010, p. 736) or "family care coordinator" (Howitt 2011, p. 107). Cancer care institutions have utilized individuals from a variety of backgrounds as patient navigators in order to provide education and support for patients as they interact within the healthcare system. Reported beneficial outcomes include improved efficiencies, such as decreased duplication and gaps in care, increased screening rates, greater communication and collaboration, improved adherence to aftercare instructions, higher patient satisfaction, and decreased patient anxiety (Campbell et al. 2010; Howitt 2011; Lasser et al. 2009; Palmieri et al. 2009; Wells et al. 2008). In a commentary about this role within a Canadian paediatric hospital, Howitt (2011) suggested that navigators serve as communicators among staff and families, and manage partnerships between various stakeholders. Reported benefits include: heightened continuity and accessibility of care, improved quality of care and patient safety, development of trusting and respectful relationships, increased team efficiencies, and optimized patient and family outcomes (Howitt 2011). Yet challenges were also identified, such as time limitations, recruitment/retention difficulty, professional boundary determination, and potential burnout deleteriously impacting the navigator.

1.7.2.4 Patient and Family-Centred Facility Design

The literature increasingly identifies the potential role of physical facilities and interior design in constructing healthcare environments that foster patient and family well-being (e.g., Adams et al. 2010). As an example, Coyne et al. (2011, p. 2567) present qualitative data in which nurses advocated for, "improved internal and external hospital facilities for extra play and school facilities for both child and siblings", in ultimately fostering patient and family well-being. Recommended improvements to amenities also included parking access, green space and purpose-built environments (Coyne et al. 2011).

In an innovative analysis of a large eight storey-high entrance and central atrium area within The Hospital for Sick Children in Toronto, Canada, Adams et al. (2010) drew on C&FCC-compatible metrics of a sense of comfort, socialization, wayfinding, interface with nature, and respite from medicalized spaces. Data were collected from key stakeholders, including paediatric patients via interviews, observations, and image and document analysis.

Findings from this study linked architectural and aesthetic design features with patient and family experience. Accordingly, this paediatric space was assessed in the context of its perceived function as a healing environment, and a reflection of the surrounding urban context and the prevailing sociocultural ethos of consumerism, with the conclusion that interior design has a bearing on health-generating environments (Adams et al. 2010). Findings further illuminated children as important commentators, among planning teams, for designing innovative spaces for paediatric care.

To date, the ways that healthcare environments affect ill children and their families remain largely unknown (Adams et al. 2010). This leads to the recommendation that facility planners and space allocation committees in healthcare settings further consider well-being as a determinant of facility design (Adams et al. 2010).

1.8 Evidence of Impact of C&FCC on Health Status

Several studies indicate the positive impact of C&FCC on the patient. As an example, literature has increasingly emerged that links FCC practices to enhancements for patients in a number of areas: adherence (Stewart et al. 1999), health status (Stewart et al. 2000, 2007), satisfaction (Krupat et al. 2000), and physiological well-being (Greenfield et al. 1988; Kaplan et al. 1989; Stewart et al. 1999; Griffin et al. 2004). Early family interventions can positively impact the outcomes of a child's development. For example, Bugental et al. (2010) found that a cognitively-based home visitation program lowered children's cortisol levels while bolstering their short term memory.

The quality of family members' sleep during hospitalization is another area that has recently been linked with, and identified as an outcome of, some C&FCC practices (Stremler et al. 2008; Zupanec et al. 2010). In a survey of Canadian and American hospitals regarding parents' quality of overnight sleeping, most responding hospitals identified some C&FCC policies in support of family sleep and respite but not all related practices were reported to be family-centred (Stremler et al. 2008). For example, it was reported that hospitals commonly placed restrictions on the number of family members who could stay overnight with the patient (typically one family member was permitted to stay). These authors concluded that, "future exploration of potential benefits of parental overnight stays, such as decreased stress and anxiety for both hospitalized children and their parents, is needed" (Stremler et al. 2008, p. 296). Sleep in children with cancer and their parents, was examined by Zupanec et al.(2010) who reported on initiatives that may potentially help children with cancer to fall asleep, which could be implemented within the context of C&FCC. In a qualitative study of sleep in hospitalized children and their parents, Stremler et al. (2011) reported that excessive light, noise and interruptions (regardless of sleeping area in hospital) affected parents' ability to achieve quality sleep. Also in this study (Stremler et al. 2011), it was reported that parents described several areas that negatively impacted their quality of sleep that may warrant future investigation and inform C&FCC policies and practices.

Overall, the evidence related to C&FCC generally links health benefits to a C&FCC approach. In some cases, direct health outcomes have been interwoven within evaluation of intermediate or process-based outcomes.

1.9 Barriers, Gaps and Limitations in the Evidence for C&FCC

While generally positive, the literature also points to barriers to C&FCC, such as a lack of time, negative attitudes or perceptions (including healthcare providers' concern about losing their role), insufficient system support, lack of communication skills, and difficulties with role and relationship negotiation (Harrison 2010). Concerns among healthcare providers include an assumption that families might impede the efficiency of care, resulting in resistance to parental involvement in C&FCC processes such as bedside rounding (Mortez 2010). The sustainability of focus on C&FCC was raised as a potential challenge, inviting continued healthcare provider and family education (Frost et al. 2010; Harrison 2010), as well as managerial and organizational support for the establishment and ongoing advancement of C&FCC (Coyne et al. 2011).

Despite increasing evidence pointing to the value of C&FCC, the advancement of a solid theoretical and empirical base of C&FCC is needed in solidifying the efficacy of C&FCC as well as having a greater understanding of its salient component processes. Concerns have been raised about inconsistencies in the actual practice of C&FCC in institutional and community care settings. These considerations invite investment not only in effective partnerships between the child, family and healthcare provider, but also enhancement and innovation in guideline and program development, quality management, and evaluation. It seems well understood that effective partnerships and processes are integral to this advancement of paediatric care.

1.10 Summary

In this chapter, C&FCC has been conveyed to be an emerging ethical imperative in paediatric care. There is increased consensus about the importance of C&FCC and emerging agreement about its core concepts. Although, as noted in this chapter, evidence of outcomes is yet limited, the literature generally points to promising outcomes. Beyond evidence, practice experience and core values in societal and healthcare practice broadly endorse an ethic of C&FCC. Accordingly, C&FCC seems important for healthcare not just on the basis of evidence-based links to outcomes, processes and efficiencies of care, but also on the basis of socially- and culturally-imbued values such as consumer and stakeholder inclusion, and transparent and accountable healthcare.

A tighter link to outcomes and costs associated with C&FCC continues to be needed, given pressing healthcare needs, ethical considerations and limited resources in healthcare. Evaluation methodologies are invited that utilize inclusive designs that are consistent with child and family-centred philosophies, address the complexities of C&FCC (e.g., accommodate the breadth of potentially relevant variables, etc.), and are rigorous.

In conclusion, C&FCC is an approach to strengthen the relationship between the patient, family and healthcare provider in the aim of improving the experience and quality of paediatric care. This growing body of literature highlights that C&FCC is worthy of ongoing pursuit and evaluation, in ultimately optimizing healthcare for children and their families.

References

Adams, A., D. Theodore, and E. Goldenberg, et al. 2010. Kids in the atrium: Comparing architectural intentions and children's experiences in a pediatric hospital lobby. *Social Science & Medicine* 70:658–667. doi:10.1016/j.socscimed.2009.10.049.

Adler, J. 2009. Establishing patient and family-centered care in a non-academic pediatric intensive care unit. *Physician Executive* 35 (5): 50–52.

American Academy of Family Physicians, American Academy of Pediatrics, American College of Physicians, et al. 2011. Guidelines for patient-centered medical home (PCMH) recognition and accreditation programs. http://www.acponline.org/running_practice/pcmh/understanding/guidelines_pcmh.pdf. Accessed 1 Sept 2012.

American Academy of Pediatrics. 2003. Family pediatrics: Report of the task force on the family. *Pediatrics* 111:1541–1571.

American Academy of Pediatrics. 2012. Patient and family-centered care and the pediatrician's role. *Pediatrics* 129 (2): 394–404. doi:10.1542/peds.2011-3084.

American College of Obstetricians and Gynecologists. 1978. Joint position statement: The development of family-centered maternity/newborn care in hospitals—Interprofessional task force on health care of women and children *Journal of Obstetric, Gynecologic, & Neonatal Nursing* 7 (5): 55–59.

Arango, P. 2011. Family-centered care. *Academic Pediatrics* 11 (2): 97–99.

Bolton, C. M. 2004. An evaluation of the quality of different forms of early postoperative care in children following tonsillectomy. Diss., University of Melbourne.

Bugental, D. B., A. Schwartz, and C. Lynch. 2010. Effects of an early family intervention on children's memory: The mediating effects of cortisol levels. *Mind, Brain, and Education* 4:159–170.

Campbell, C., J. Craig, and J. Eggert, et al. 2010. Implementing and measuring the impact of patient navigation at a comprehensive community cancer center. *Oncology Nursing Forum* 37 (1): 61–68.

CanChild (Centre for Childhood Disability Research). 2010. Myths about family-centred service. http://www.canchild.ca/en/childrenfamilies/resources/FCSSheet2.pdf. Accessed 20 Aug 2012.

Cannon, S. 2011. Family-centered care in the critical care setting: Is it best practice? *Dimensions of Critical Care Nursing* 30 (5): 241–245. doi:10.1097/DC.0b013e3182276f9a.

Chauke, H. I., and R. C. Pattinson. 2006. Ward rounds: Bedside or conference? *South African Medical Journal* 96:398–400.

Corlett, J., and A. Twycross. 2006. Negotiation of parental roles within family-centred care: A review of the research. *Journal of Clinical Nursing* 15:1308–1316.

Coyne, I., C. O'Neill, and M. Murphy, et al. 2011. What does family-centered care mean to nurses and how do they think it could be enhanced in practice. *Journal of Advanced Nursing* 67 (12): 2561–2573. doi:10.1111/j.1365-2648.2011.05768.x.

Dickoff, J., P. James, and E. Wiedenbach. 1968. Theory in a practice discipline: Part 1. Practice oriented theory. *Nursing Research* 17 (5): 415–435.

Epstein, R. M., and R. L. Street. 2011. The values and value of patient-centered care. *Annals of Family Medicine* 9 (2): 100–103. doi:10.1370/afm.1239.

Ferrante, J. M., D. J. Cohen, and J. C. Crosson. 2010. Translating the patient navigator approach to meeting the needs of primary care. *Journal of the American Board of Family Medicine* 23 (6): 736–744. doi:10.3122/jabfm.2010.06.100085.

Frost, M., A. Green, B. Gance-Cleveland, et al. 2010. Improving family-centered care through research. *Journal of Pediatric Nursing* 25 (2): 144–147. doi:10.1016/j.pedn.2009.09.001.

Gooding, J. S., L. G. Cooper, and A. I. Blaine, et al. 2011. Family support and family-centered care in neonatal intensive care unit: Origins, advances, impact. *Seminars in Perinatology* 35 (1): 20–28. doi:10.1053/j.semperi.2010.10.004.

Greenfield, S., S. H. Kaplan, and J. E. Ware, et al. 1988. Patients' participation in medical care: Effects on blood sugar control and quality of life in diabetes. *Journal of General Internal Medicine* 3 (5): 448–457.

Griffin, S. J., A. L. Kinmonth, and M. W. Veltman, et al. 2004. Effect on health related outcomes of interventions to alter the interaction between patients and practitioners: A systematic review of trials. *Annals of Family Medicine* 2 (6): 595–608.

Harrison, T. M. 2010. Family-centered pediatric nursing care: State of the science. *Journal of Pediatric Nursing* 25 (5): 335–343. doi:10.1016/j.pedn.2009.01.006.

Henbest, R. J., and M. A. Stewart. 1989. Patient-centredness in the consultation: A method for measurement. *Family Practice* 6 (4): 249–253.

Hjorngaard, T. 2011. Family-centred care: A critical perspective. *Physical & Occupational Therapy in Pediatrics* 31 (3): 243–244. doi:10.3109/01942638.2011.589728.

Howitt, M. J. 2011. The family care coordinator: Paving the way to seamless care. *Journal of Pediatric Oncology Nursing* 28 (2): 107–113. doi:10.1177/1043454210377331.

Hutchfield, K. 1999. Family-centred care: A concept analysis. *Journal of Advanced Nursing* 29 (5): 1178–1187.

Institute of Medicine, Committee on Quality of Health Care in America. 2001. *Crossing the quality chasm: A new health system for the 21st century*. Washington: National Academy.

Johnson, B. H., M. R. Abraham, and T. L. Shelton. 2009. Patient and family-centered care: Partnerships for quality and safety. *North Carolina Medical Journal* 70 (2): 125–130.

Kaplan, S. H., S. Greenfield, and J. E. Ware Jr. 1989. Assessing the effects of physician-patient interactions on the outcomes of chronic disease. *Medical Care* 27 (3 Suppl): S110–S127.

Krupat, E., S. L. Rosenkranz, and C. M. Yeager, et al. 2000. The practice orientations of physicians and patients: The effect of doctor-patient congruence on satisfaction. *Patient Education and Counseling* 39 (1): 49–59.

Kuhlthau, K. A., S. Bloom, and J. Van Cleave, et al. 2011. Evidence for family-centered care for children with special health care needs: A systematic review. *Academic Pediatrics* 11 (2): 136–143. doi:10.1016/j.acap.2010.12.014.

Lasser, K. E., J. Murillo, E. Medlin, et al. 2009. A multilevel intervention to promote colorectal cancer screening among community health center patients: Results of a pilot study. *BMC Family Practice* 10 (37). doi:10.1186/1471-2296-10-37.

McClimans, L. M., M. Dunn, and A. Slowther. 2011. Health policy, patient-centered care and clinical ethics. *Journal of Evaluation in Clinical Practice* 17 (5): 913–919. doi:10.1111/j.1365-2753.2011.01726.x.
Mortez, J. G. 2010. Strengthening patient and family-centered care: Learning through webinars. *Pediatric Nursing* 36 (3): 168–170.
Muething, S. E., U. R. Kotagal, and P. J. Schoettker, et al. 2007. Family-centered bedside rounds: A new approach to patient care and teaching. *Pediatrics* 119 (4): 829–832.
Nicholas, D. B., and K. Keilty. 2007. An evaluation of dyadic peer support for care giving parents of children with chronic lung disease requiring technology assistance. *Social Work in Health Care* 44 (3): 245–259.
Nicholas, D. B., G. Picone, and A. Vigneux, et al. 2009. Evaluation of an online peer network for adolescents with chronic kidney disease. *Journal of Technology in Human Services* 27 (1): 23–33.
Ontario Ministry of Health and Long-term Care. 2011. The excellent care for all act. www.health.gov.on.ca/en/legislation/excellent_care. Accessed 10 Sept 2012.
Palmer, S. 1993. Care of sick children by parents: A meaningful role. *Journal of Advanced Nursing* 18 (2): 185–191.
Palmieri, F. M., E. R. DePeri, and B. A. Mincey, et al. 2009. Comprehensive diagnostic program for medically underserved women with abnormal breast screening evaluations in an urban population. *Mayo Clinic Proceedings* 84 (4): 317–322. doi:10.1016/S0025-6196(11)60539-9.
Public Health Agency of Canada. 2000. Family-centred maternity and newborn care: National guidelines. http://www.phac-aspc.gc.ca/hp-ps/dca-dea/publications/fcmnc-smnnpf/index-eng.php. Accessed 10 Aug 2012.
Rolland, J. S., and F. W. Walsh. 2006. Facilitating family resilience with childhood illness and disability. *Current Opinion Pediatrics* 18 (5): 527–538.
Shelton, T., E. Jeppson, and B. Johnson. 1987. *Family-centered care for children with special health care needs*. 2nd ed. Washington: The Association for the Care of Children's Health.
Shields, L., J. Pratt, and J. Hunter. 2006. Family centred care: A review of qualitative studies. *Journal of Clinical Nursing* 15 (10): 1317–1323.
Shields, L., H. Zhou, J. Pratt, et al. 2012. Family-centered care for hospitalised children aged 0–12 years. *Cochrane Database of Systematic Reviews* 10:CD004811. doi:10.1002/14651858.CD004811.pub3.
Sine, D. M., and V. A. Sharpe. 2011. Ethics, risks, and patient-centered care: How collaboration between clinical ethicists and risk management leads to respectful patient care. *Journal of Healthcare Risk Management* 31 (1): 32–37.
Sisterhen, L. L., R. T. Blaszak, M. B. Woods, et al. 2007. Defining family-centered rounds. *Teaching and Learning in Medicine* 19 (3): 319–322.
Stewart, M., J. B. Brown, and H. Boon, et al. 1999. Evidence on patient-doctor communication. *Cancer Prevention and Control* 3 (1): 25–30.
Stewart, M., J. B. Brown, and A. Donner, et al. 2000. The impact of patient-centered care on outcomes. *Journal of Family Practice* 49 (9): 796–804.
Stewart, M., J. B. Brown, and J. Hammerton, et al. 2007. Improving communication between doctors and breast cancer patients. *Annals of Family Medicine* 5 (5): 387–394.
Stremler, R., L. Wong, and C. Parshuram. 2008. Practices and provisions for parents sleeping overnight with a hospitalized child. *Journal of Pediatric Psychology* 33 (3): 292–297. doi:10.1093/jpepsy/jsm096.
Stremler, R., Z. Dhukai, and L. Wong, et al. 2011. Factors influencing sleep for parents of critically ill hospitalised children: A qualitative analysis. *Intensive and Critical Care Nursing* 27:37–45.
U.S. Congress. 1975. Public Law 94-142, Education for all handicapped children act, 29 Nov 1975.
Wells, K. J., T. A. Battaglia, and D. J. Dudley, et al. 2008. Patient navigation: State of the art or is it science? *Cancer* 113:1999–2010.
Zupanec, S., H. Jones, and R. Stremler. 2010. Sleep habits and fatigue of children receiving maintenance chemotherapy for all and their parents. *Journal of Pediatric Oncology Nursing* 27:217–228.

Chapter 2
A Canadian Perspective on a Child's Consent to Research Within a Context of Family-Centred Care: From Incompatibility to Synergy

Linda Sheahan, Michael Da Silva, Christine Czoli and Randi Zlotnik Shaul

The content of this chapter was previously published April 24, 2012: Sheahan L, Da Silva M, Czoli C, Zlotnik Shaul R (2012) A Canadian Perspective on a Child's Consent to Research within a Context of Family-Centered Care: From Incompatibility to Synergy. J Clinic Res Bioeth 3:132. doi:10.4172/2155-9627.1000132.

2.1 Introduction

The recent development of paediatric bioethics has seen two dominant trends evolve simultaneously:

1. the framework of family-centred care and
2. recognition of the emerging autonomy and rights of children.

L. Sheahan (✉)
Joint Centre for Bioethics University of Toronto, Toronto, Ontario, Canada
e-mail: lindas@med.usyd.edu.au

M. Da Silva
Rutgers University, New Brunswick, New Jersey, United States

University of Toronto, Ontario, Canada
e-mail: Michael.e.dasilva@gmail.com

C. Czoli
The Hospital for Sick Children, Toronto, Ontario, Canada
e-mail: christine.czoli@gmail.com

R. Zlotnik Shaul
Bioethics Department, The Hospital for Sick Children,
University of Toronto, Toronto, Ontario, Canada

R. Zlotnik Shaul (ed.), *Paediatric Patient and Family-Centred Care: Ethical and Legal Issues,* International Library of Ethics, Law, and the New Medicine 57,
DOI 10.1007/978-1-4939-0323-8_2,

The former led to the development of a "family-centred" care healthcare delivery model while the latter is consistent with a "patient-/child-centred" care model. Both healthcare delivery models are fundamental to caring for children in Canadian hospitals and each is accordingly increasingly pursued as a vehicle for guiding the delivery of healthcare in the best interests of children. While these recent trends have resulted in a kind of hybrid model of care on a functional level, they hold an inherent tension in establishing the ultimate decision-maker in health-related issues. This tension is particularly relevant in obtaining consent for participation in paediatric research.

In the last 25 years, family-centred care has become a familiar component of paediatric clinical practice in North America. Children in early paediatric hospitals were often kept apart from their parents, who had little to no say in the care given to their children; eventually simple moves, like allowing mothers to stay with their breastfed infants, expanded into a wider recognition of the importance of parents in paediatric healthcare. The family-centred care movement, as we know it today, emerged post-World War II and crystallized after the 1970s (Shields 2011). Today, many healthcare executives view patients and families as important decision-making partners and many healthcare institutions include family participants at the executive meeting level (Conway 2008). The movement is located in multiple, distinct disciplines. In the academic world, it has been ingrained in children's nursing and therapy curricula in particular (Carter 2008). Its bases however, can also be found in social work, and it "fits well with a social work perspective that understands individuals in the context of their family system and greater environment" (Kovacs et al. 2006, p. 13).

The traditional triad has the parent(s) in the role of substitute decision maker, reflecting that the parent(s) is in the best position to appreciate what is in the child's best interests. The language of family-centred care, while intended as a vehicle for bringing benefit to the child, considers the interests of those beyond the triad-focused patient.

Along with the rise of family-centred care, paediatric bioethics has also seen the rise of a framework that emphasizes the development of children's autonomy and rights. This model of healthcare, focused on the unique needs and wishes of the patient, is often referred to as "patient-centred." In a paediatric context, this approach is often referred to as "child-centred." Paediatric clinical research now requires recognition of children's developing capacity and eventual respect for their assent or dissent. Recognition of children's developing capabilities has grown such that Canadian policy recently began to explicitly recognize that children's assent and dissent should be determinative in the research consent process as well as in the context of care. Canadian ethics and law have thereby entrenched a patient-centred model of care that emphasizes respect for the autonomy of individual patients and the developing autonomy and rights of children in the context of treatment decision making.

Prima facie, these two perspectives may conflict in their designation of the primary ethical locus of decision-making. According to some interpretations, family-centred care can emphasize the full family unit as the primary focus of ethical consideration, while the children's emerging autonomy/rights perspective clearly puts ultimate authority in the hands of the child. The existence of conflict may increase risk to patients, physicians, and the institutions in which they work. Should physicians fall short of certain duties when giving priority to other duties and/or undermine the integrity of clinical care, this may result in compromised trust in the physician-patient or physician-family relationship. While one could argue that the different foci of patient/child-centred care and family-centred care do not have normative significance that would justify having a physician deviate from his or her primary fiduciary duty to the patient, the lack of clarity on how these distinct models are to be integrated in practice, finds the practical application of the models threatened by inconsistency or destined to only offer purely theoretical value.

This chapter explores the compatibility of these perspectives, and offers a guide as to how these views may be integrated in the context of consent to research (Sheahan et al. 2012).

2.2 "Family-Centred Care"

Family-centred care acknowledges "the importance of family participation in healthcare" (Williams 2006, p. 203). Although there are many broad definitions and conceptualizations of family-centred care, one used as the basis of a systematic review of the effectiveness of family-centred care describes it as "a way of caring for children and their families within health services which ensures that care is planned around the whole family, not just the individual child/person, and in which all the family members are recognized as care recipients" (Shields et al. 2006, p. 1318).

A formal role for parents in their children's health really only began to emerge in the 1950s and 1960s (Palmer 1993; McGonigel 1998). The concept of family-centred care in North America emerged through a strong advocacy movement in the late 1960s led by parents of children with special healthcare needs (MacKean et al. 2005). The movement argued against the dominant expert model, with parents advocating for more involvement in their child's healthcare and for the healthcare system to recognize the influence of the family on a child's health and wellbeing. Throughout the next 40 years, family-centred care began to influence health policy in the North American paediatric setting. For instance, paediatric hospitals changed their visiting policies from a very restrictive set of visiting hours each day to an open visiting policy allowing parents to stay with their children 24 h a day (MacKean 2005).

Recent bioethics literature suggests family-centred care only received a "nominal definition" that does not touch the "real nature" of what is being described to date (Shields 2011, pp. 144–145), but some preliminary features still prove useful. Fostering family-friendly environments, along with acknowledging parental expertise in providing care and encouraging collaboration between parents and the healthcare team emerged as key elements of family-centred care (MacKean 2005). Indeed, more recent definitions give "all the family members" of a paediatric patient the status of care recipients (Shields et al. 2006, p. 1318). This arguably gives all family members a sense of ethical concern that could be translated to the research context with the possible aforementioned implication of multiple family members having a sense of authority in their own right rather than as individual proxies for individual patients.

Contrasted with the medical professional-centred care model, a merging of competing interests into a hybrid "patient- and family-centred care" model, which recognizes the perspectives of both children and their families, has developed (IPFCC 2011). This hybrid "is an approach to the planning, delivery and evaluation of healthcare that is governed by collaborative partnerships among healthcare providers, patients and families" (Sodomka 2006, p. 7). The paediatric hybrid approach of "child- and family-centred care" reflects a simultaneous commitment to the focus on patients and recognition of the fact that a paediatric patient is generally embedded in a family unit. It is described as an "equal partnership" between providers and recipients of healthcare (Sodomka 2006, p. 7). In the research context, the principles of patient- and family-centred care require designing, conducting and evaluating research in collaboration with parents as well as respecting the diversity and privacy of families (Johnson et al. 1992; McGonigel 1998). Such collaborative research programs have been widely successful, with programs at schools like the University of Kansas, Stanford University and others proving productive in the domain of evidence-based research (Johnson 2008).

Unfortunately, broad concepts and a lack of empirical evidence make further development and evaluation of family-centred care difficult. Although family-centred care is widely accepted in children's healthcare, little work has been done to evaluate its effect on child and family outcomes (Shields et al. 2006; Franck and Callery 2004). This leads some to suggest more research is needed to ensure family-centred care is "being properly implemented" (Shields et al. 2006, p. 1317). Others argue that it is not obvious that family-centred care is "intrinsically good and therefore worthwhile pursuing and getting right" (Carter 2008, p. 2092). Some authors suggest that improvements to the concept are required, and that the voice of the child is notably missing in the debate (Lambert 2009).

One area where effectiveness may be questionable is in relation to the tensions between the interests of children and of the family. As Franck and Callery note, "there may be important differences between the perspectives and objectives of children and of their families…[P]arents may not be best placed to assess symptoms and quality of life from their children's point of view" (Franck and Callery 2004, pp. 268–269). Semantically, family-centred care and patient-centred care clearly identify different parties as holding the ultimate position of privilege. Just as there can only be one true centre of a circle, there can only be one ultimate position of privilege. Family-centred care developed due to an understanding that "a child's illness has the ability to impact all aspects of family life" (Locsin 2003, p. 203). It does not in any way minimize the fact of the child's illness, nor does it explicitly denigrate their decision-making capability. Placing the family at the semantic 'centre', however, could result in considerations relevant to family members taking priority. So, in relation to paediatric research, family-centred care potentially opens the possibility of research being done on a patient based primarily on the interests of his or her family, while not being held to a strict standard upholding the individual's interests. This risk is of particular concern in a research setting where the benefits to the patient are often less obvious to define.

A strict model of patient-centred care, however, would view the patient as the primary focus of ethical consideration. Inclusion of the child at each stage of the research process and giving the child ultimate authority to at least decide not to participate in a study would be necessary. A strict reading of patient-centred care could be taken further such that a capable child could participate in a research program without parental permission.

One may wonder if an increased role for children's decision making is desirable; some parents may wish for their consent to be determinative regardless of a child's viewpoint presuming that as adults they will likely have a greater understanding of a given research program. In addition, parents are held legally responsible for their children. They must provide them with both education and the necessaries of life (Criminal Code, R.S.C. 1985). It may seem inconsistent to allow children to act in a manner that would undermine parents' ability to provide that which they are statutorily obligated to provide.

Currently, clinicians and researchers accept a kind of ill-defined hybrid model for making decisions in paediatric medicine. A hybrid position does not provide us with adequate answers. The Institute for Patient- and Family-Centered Care's core concepts (dignity and respect, information sharing, participation and collaboration) include both patients and family members at each step of the process, but do not clearly articulate who the ultimate decision-maker should be (IPFCCND 2011). Amongst equal partners, it is difficult to find an ultimate decision-maker. A strict family-centred approach would make the third party permission determinative, but such an approach appears incongruous with our increased recognition of the importance of child's rights and developing capacity, including an explicit recognition that children's assent or dissent could be determinative of their participation in a research study.

Franck and Callery suggest that "the difference between 'child-centered' and 'family-centered' care is one of emphasis: neither term can exclude the other, because child-centered care must take account of the social environment in which children live and FCC [family-centered care] must be primarily concerned with the health of children" (Franck and Callery 2004, p. 269). Under this rubric, there is no merger or hybridization, only an important relationship between related positions. This may not solve the problem entirely. Where one places his or her *primary* emphasis can nevertheless still lead to substantial differences. While the hybrid model does encourage one to consider the values underpinning each model, it offers little guidance on how to balance the interests or perspectives of the patient/child with the family and how to balance the range of clinical duties that flow from each model, especially in the context of dispute and/or differences between the perspectives of patients (children), legal guardians (families) and/or physicians about what should be done.

2.3 Children's Rights to Consent Under the Tri-council Policy Statement

The increased importance of the child's rights and child's developing capacity framework for the Canadian paediatric research consent context can be best observed by examining Canada's primary research standards document, the Tri-Council Policy Statement (TCPS) (Canadian Institute of Health Research et al. 2010).

In the absence of statutes devoted to research ethics or common law standards, guidelines, regulations and policy statements are often the best resources for determining the standard of conduct in a given area. Outside of Quebec, where at least research consent standards are proscribed within the *Civil Code,* the TCPS is the leading research ethics resource in Canada. Canada's three main research fund-granting agencies, the Canadian Institutes of Health Research, the Natural Sciences and Engineering Research Council of Canada, and the Social Sciences and Humanities Research Council of Canada, originally adopted the TCPS in 1998. The TCPS serves a soft law function in Canadian jurisprudence and helps to establish standard of care for Canadian research.

Under the original TCPS, children were recognized as vulnerable persons (Canadian Institute of Health Research et al. 1998). Their participation was contingent on their assent or dissent (Simpson 2003). While children still needed to reach a certain lower capability threshold in order for their assent or dissent to be relevant (i.e., newborns could not give their assent to research), *prima facie,* respect for children's assent or dissent was an important component of an ethically permissible research study. Whether a valid parental consent would lead to research participation in the presence of a patient's dissent was questionable. Even where assent and dissent were seen as determinative, the lack of definition of these terms and the lack of clarity on when one is capable of giving determinative assent or consent were problematic. Accordingly, Christy Simpson recommended expanding the role of the child in decision-making about research, clarifying the language surrounding this role and outlining the respective roles and responsibilities of parents, researchers, and Research Ethics Boards in the decision-making process (Simpson 2003).

Children's roles in the TCPS-mandated consent process changed with the December 2010 adoption of its 2nd draft, also known as the "TCPS 2" (Canadian Institute of Health Research et al. 2010). Recognition of the fact that consent is an ongoing process now explicitly entails recognition of the developing capacity of children. Where third party proxy consent is initially given, researchers must gain valid consent from someone who either:

a. reaches the age of consent (in the case of children) or
b. acquires or regains capacity (in the case of all incapable individuals, including incapable children) (Canadian Institute of Health Research et al. 2010, p. 30, art. 3.9(e)).

The TCPS 2 notes that "the determination of capacity in research is not a static determination" (Canadian Institute of Health Research et al. 2010, p. 41).

Under TCPS 2, a child is either capable of consent or of assent/dissent only. However, the TCPS 2 does suggest that individuals with "diminished capacity [may] still be able to decide whether to participate in certain types of research" (Canadian Institute of Health Research et al. 2010, p. 41). One wonders if a developing child may be able to consent to some types of research, even if he or she is only capable of giving assent or dissent elsewhere.

Researchers now have a responsibility to seek a child's assent or dissent. The TCPS remains in an assent-dissent framework for those who are incapable of giving fully valid consent on their own, but the necessity of assent or dissent from a child capable of giving it as a component of a valid consent is now explicit:

> Where children have not yet attained the capacity to consent for themselves to participate in research, researchers shall seek consent from an authorized third party while ascertaining the child's assent or dissent.… While their assent would not be sufficient to permit them to participate in the absence of consent by an authorized third party, their expression of dissent or signs suggesting they do not wish to participate must be respected. (Canadian Institute of Health Research et al. 2010, p. 42, 50)

The threshold for when a child's assent or dissent is determinative is difficult to parse. Article 3.10 requires "some ability to understand the significance of the research" (Canadian Institute of Health Research et al. 2010, art. 3.10). Emphasizing the word 'some' suggests a low standard. If, however, one focuses on 'significance', what one needs to understand may be rather robust; it may approximate an appreciation standard. Clinical assent, by contrast, only requires understanding what is being proposed, not appreciating it (Diekema et al. 2011).

In either case, the explicit recognition of the importance of a capable child's assent or dissent brings the TCPS closer in line, in spirit if not in consequence (given the potentially different standards), with the Ontario *Health Care Consent Act's* capacity-based analysis. This consistency is desirable given that it can be difficult to draw a sharp divide between clinical care and clinical research in certain circumstances While the trend in paediatric research practice may reflect the notion that child assent/dissent is required for participation in paediatric research, most regulations make it possible for a child to be in a study entirely against their expressed wishes.

> both U[nited] S[tates] and ICH [International Conference on Harmonization of Technical Requirements for Registration of Pharmaceuticals for Human Use] regulations specifically allow for a child's participation in a study, against their [meaning the child's] wishes, if it provides them some benefit (which is related to their health). The E[uropean] U[nion], on the other hand, requires only that the dissent be 'considered'. Thus, all three regulations make it possible for a child to be in a study against their wishes It presents sits there as what is sometimes colloquially called an "orphan". (Blake et al. 2011, p. 73)

2.4 Seeming Incompatibility?

The trends in paediatric bioethics of family-centred care and recognition of emerging autonomy and child rights focus on two different values and can lead to different results in a given decision-making process. It has been argued that the ethical underpinnings of family-centred care are insufficiently grounded; consequently, the model has failed and "it is ethically untenable to continue to apply it when caring for children and their families" (Shields 2011, p. 152). The basic underlying (pragmatic) argument for its value nonetheless appears to be rooted in the simple suggestion that a child is, generally speaking, embedded in a family that will have the best sense of what is in the child's best interests. Of course, critics argue against this point. For instance, some suggest that the movement is based on a neoliberal, idealized understanding of the family (Breen 2009). It is possible that family-centred care developed on the basis of, and in conjunction with, recognition of the importance of the value of beneficence. Increased recognition of children's rights and the wider patient-/child-centred care model, by contrast, partially developed as increased respect for children's autonomy.

While a hybrid approach to paediatric research most often results in acceptable outcomes in terms of patient and family interests, it nonetheless fails to give patients and families a clear and coherent understanding of their roles and responsibilities. Researchers thus require a clear model for obtaining consent in paediatric research. The model needs to be broadly applicable, and based around clearly delineated values and principles which clarify the relative weights of potentially conflicting values. An example would be where parental autonomy comes into conflict with the emerging autonomy of an individual child.

According to Richard Miller, the distinctive feature of the paediatric model of medical ethics is that the norm of beneficence has general priority to the norm of respect for autonomy (Miller 2003). Prioritizing beneficence essentially means acting in a child's 'best interests'. The beneficence standard has a number of unique components in paediatrics. Most importantly, a child's 'best interests' are intimately tied to his or her social unit, and we must therefore recognize the importance of the family unit. This value or right is not synonymous with the value or right of patient autonomy, and is therefore accorded a different status in the decision-making process.

The importance of the family unit, as is enshrined in family-centred care, thus elevates the idea of parental autonomy, where parents may be considered the ultimate locus of decision-making. The obligation to recognize the significance of parental autonomy is contingent on parents fulfilling their obligations toward children. Where parents fail to act in the best interests of their children, outsiders may justifiably intervene in a family's domestic affairs to ensure that healthcare decisions are made in accordance with the child's best interests and where children may be in need of protection. The focus of ultimate ethical consideration may be the child in a beneficence model, but a substitute decision-maker, often a family member, is the ultimate ethical decision-maker. While consideration of the interests of the child is to be the focus of the family's decision-making rights, care decisions are centred on what the parents decide.

The first question for the researcher in paediatrics, therefore, becomes: 'Can this child give autonomous consent to participate in this project?' For the purposes of consent to research in paediatrics, autonomy can be seen as having two essential components: decision-making autonomy and executional autonomy. Decision-making autonomy refers to the ability and freedom to make decisions without external coercion, and executional autonomy refers to the ability one has to implement decisions made (Collopy 1988). In both research and treatment models for adults, decision-making autonomy tends to 'trump' all other relevant values.

The TCPS 2 marks an important moment in which children's autonomy is increasingly recognized. Children develop capacity for self-determination in decision-making as they mature. This creates an obligation to respect both the developing autonomy of young people, and their full capacity and independence when it is reached. Paediatric patients, therefore, should gradually accrue the rights accorded to adults in light of their full capacity for self-determination. In the research context they develop a right to dissent even prior to the full right to consent accorded to capable persons in the clinical context. Unlike in the previous TCPS, this dissent is now clearly determinative.

Different foci of ethical consideration (family interests vs. children's rights) can lead to different conclusions in a given decision-making process. It is not enough to say that which is good for a patient is good for his family and vice versa. Parents and children do not always agree about what is best for a child.

2.5 Privacy: A Third Concern?

A further logistical problem arises when initially gaining consent: Who does one approach first and what can one share with the other party? Respect for autonomy entails respect for children's decisions as to what medical information is shared with others. Under Canadian law, capable youths have a right to have their information kept private (Personal Health Information Protection Act 2004). For the competent child who is thinking about research, approaching parents about the research may result in a lack of respect for this position. Approaching parents with information about a study may violate privacy if inclusion/exclusion criteria are included in the discussion. Study information provision would thereby give away medical information children may want to keep private. Approaching parents about the theoretical idea of a study only, however, is less than adequate information provision for making a third party proxy decision.

On the other hand, asking children for permission may require explaining the research purpose to the children prior to telling their family about it. Family-centred care demands the inclusion of parents in the decision-making process. It is understandable that recruiters may feel uneasy about discussing research with children without parental presence. In order to ascertain permission to share information, however, such a discussion will need to take place. Discussing consent to disclosure of information without knowing what the disclosure is all about runs into the same theoretical problems as the theoretical discussion of a study mentioned above. Operationalizing both autonomy and family-centred care will accordingly require consideration of how to work around related privacy concerns.

2.6 An Analogous Case?

TCPS 2 does provide another example of synergy between group interests and individual interests within a TCPS 2 research context. The TCPS 2 devotes a chapter on how to promote such synergy in Aboriginal communities. Chapter 8 of the TCPS 2 is identified by the Tri-Council funding agencies as a "*significantly changed chapter*" (Interagency Advisory Panel on Research Ethics 2010, p. 3). Drawing on public consultations, Chapter 8 "serves as a framework for the ethical conduct of research involving Aboriginal peoples, premised on respectful relationships and encouraging collaboration between researchers and research participants, and community engagement" (Interagency Advisory Panel on Research Ethics 2010, p. 3). Community leaders must be consulted before research is conducted in Aboriginal communities, though the leaders do not necessarily recruit individuals for the research program.

The example of Aboriginal communities has many parallels with the concept of consent in the paediatric context, and can provide instruction on how to integrate family-centred care with the patient-centred care in a paediatric research setting. Just as community for this group has a heightened significance around consent to research, family should be recognized as serving a similar role for children. The family member's permission is important, but so too is the individual patient's decision not to participate in something of which the parent approves. Additionally, just as article 9.6 states that there are diverse interests in a community, such that a certain leader's decision is not dispositive (Canadian Institute of Health Research et al. 2010), so too is there a wide breadth of interests in a family, such that a parent's view is important, and whereby other family interests remain relevant. This is not an instance of individual consent waiver, but of an acknowledgment that cultural sensitivity is a prerequisite for even seeking it. Relatedly, one may wish to acknowledge the importance of a family context when seeking and examining the legitimacy of consent.

Community engagement is one aspect of the Aboriginal research context that demonstrates that model's ability to meet the demands of both individuals and the groups of which they are members. It is "a process that establishes interaction between a researcher or research team, and the Aboriginal community relevant to the research project" (Canadian Institute of Health Research et al. 2010, p. 108). Article 9.12 states that it is a collaborative process where even the nature of engagement is determined through collaboration and Article 2.9 states that:

> The nature and extent of community engagement in a project shall be determined jointly by the researcher and the relevant community, and shall be appropriate to community characteristics and the nature of the research. (Canadian Institute of Health Research et al. 2010, p. 111)

Community engagement is also highly contextual:

> [T]he engagement may take many forms including review and approval from formal leadership to conduct research in the community, joint planning with a responsible agency, commitment to a partnership formalized in a research agreement, or dialogue with an advisory group expert in the customs governing the knowledge being sought. (Canadian Institute of Health Research et al. 2010, p. 108)

All research, however, must be relevant and respond to the community's needs and priorities:

> The research should benefit the participating community… as well as extend the boundaries of knowledge. (Canadian Institute of Health Research et al. 2010, p. 124)

Combining Articles 9.11 and 9.16, one sees that operationalizing a research agreement between a community and researcher, as required by the TCPS 2, without violating privacy may be difficult, but is necessary:

> Researchers shall not disclose personal information to community partners without the participant's consent. (Canadian Institute of Health Research et al. 2010, p. 127)

While the demands in the paediatric research context differ, one may likewise wish to examine ways to meet both individual (patient) and community (family) demands in that context.

Synergy between community and individual interests, and the supremacy of the privacy of the latter, is now recommended by the TCPS 2. We propose a similar synergy below. It does not incorporate the Aboriginal model *in toto,* but does adopt the Aboriginal model's reflection of the fact that integrating group and individual research in the research context is a largely collaborative and contextual undertaking. One must respect both the group and the individual, but how to do so may change in different circumstances.

2.7 A Proposed Model for Operationalizing These Values

Given the seeming incompatibility of these trends, one is left wondering which should take precedence. Given the value of autonomy, which permeates consent discussion, the TCPS 2's recognition of children's rights to assent/dissent is to be applauded. The paediatric patient must be the centre of any decision-making discussion concerning health services provision for that child. Operationalizing this primary right, however, requires recognition of the fact that many children are embedded in a family relationship, which is deeply tied to their ability to provide consent and their interests. Beyond the logistical support that families must provide to children in research studies (i.e., arranging transportation to the hospital for children living at home), families also provide the context in which many children's identities are formed and decision-making capabilities are realized. Many children must be able to speak with their parents in order to gain full understanding of their decision. Many also respect their parents' opinions, even if they ultimately do not agree with them. Increased recognition of children's rights must be accompanied by recognition of the family-centred nature of their decision-making. However, presuming what is good for the family is good for the patient is insufficient. Ultimately, a patient's rights must triumph. Those rights, however, are deeply influenced by the family context.

That family-centred care is closely related to patient-centred care was clear even before the development of the hybrid position. Prior to the recognition that patients too must be partners in the clinical process, hybrid position and family-centred pioneers Johnson et al. were aware of the importance of children's rights in the research consent context. In 1992, they argued for guidelines for research on children in a psychosocial context, including:

1. seeking consent from parents and assent from children and
2. no participation requirement without assent, should apply in other contexts as well (Johnson et al. 1992, pp. 22–23).

Both parents and children were to be told that their participation was voluntary (Johnson et al. 1992). How this position was to be reconciled with the family, as the primary focus of ethical consideration, however, was left undeveloped.

In practice, the hybrid model of family- and patient-centred care often functions reasonably well despite its potential conflicts. Nevertheless, the parameters of this model have not yet been fully developed. The American Association of Paediatrics Committee on Bioethics identified many of the pertinent issues in their review in 1995 (Committee on Bioethics 1995), but stopped short of constructing an applicable model for consent in paediatric research. A delineation of the underlying values that are relevant in considering consent in paediatric research is required, along with a clearly defined model for applying those principles in research practice.

In order to fully operationalize these trends in concordance, we recommend the following principles:

1. *The priority of research ethics remains an obligation to protect the potential research subject from undue harm.* This responsibility lies with the researchers themselves, with the research ethics board, and with parents or guardians who are involved in the decision-making process.
2. *With incapable children, beneficence has presumed priority over other values, including autonomy per se.* Respect for a child's individual autonomy can be seen as one part of understanding and acting in accordance with beneficence or best interests.
3. *The relative weight of respect for individual autonomy in determining best interests or beneficence increases as the child develops decision-making capacity.* Along with accompanying executional autonomy or independence, this eventually leads to endorsement of a fully adult model, where it is presumed that the capable individual's choices in self-determination will be the best way to pursue that individual's best interests. Decision-making autonomy alone does not accord a paediatric patient complete independence or endorsement of the capable adult model, as executional autonomy is also required. Both components of autonomy are context specific, depending on the type of research proposed. The consent process for each study should be reviewed with this in mind.
4. *Family as a social unit in paediatric medicine holds a special place.* This is because of the necessary dependence of a child on their family unit, and because of a family's increased relative value to the child's best interests. These relational aspects create special obligations on behalf of parents toward their children. They also necessitate a greater obligation to respect the role of family in the context of paediatric research. Specifics for consideration in paediatric research include:
   - Family discordance can be understood as a harm that holds relatively more weight in paediatric research than in adult research.
   - The social value of parental values has comparably more weight in the paediatric environment than in the adult environment.

– Participation of a dependent child in research may result in specific responsibilities for a family, and as such will result in obligations on behalf of the family to their child. Participation may also result in potential harms to the guardian or family. The creation of these obligations and harms holds moral weight, and they are therefore important factors to consider in ethical deliberations regarding participation in research. In the case of a child who possesses decision-making autonomy, but not independence (as defined below), this should allow for a permission step to be included in the model, where the guardian accepts these burdens or potential harms. This should ideally be done in the presence of the capable child, providing an opportunity for them to exert independence or executional autonomy.

5. *When in doubt, err on the side of non-participation.* There are variable potential 'benefits' to participating in research for the child subject, which alter the harm-benefit ratio in accordance with both the nature of the research being conducted and the patient's specific context (Miller and Kenny 2002).
6. *Where paediatric subjects are unable to consent for themselves, respect for their developing autonomy requires us to provide information relevant to their developmental level.* Where a child is able to understand, his or her assent should be necessary to enroll in the research project. Unlike in treatment decisions, the benefits of research are not weighty enough to override dissent from a child who understands what is being proposed.

Variables for consideration in the harm benefit analysis for paediatric subjects therefore should include:

a. Nature of the research and its likely harms and benefits
b. Level of risk
c. Potential subjects' capacity/competence to make their own decision in regards to this specific project
d. Social or cultural components in determining benefits or 'best interests', including needs, values, priorities, and social expectations
e. Parent's current role and position in relation to the child, including action on behalf of their child's interests, and potential conflict of interests. This is relevant in light of the social value of parental decision-making, and in light of the extra obligations incurred by guardian and family by participating in research

Using these principles, the model represented by Fig. 2.1 is proposed. The following key is a guide to understanding terms included in the schema.

KEY:

'*Independent*' is:

1. As determined by law (e.g., by the age of majority in that province, where pregnant or have their own children/family, or where they are completely outside of parental control)
2. Where the potential subject has full executional autonomy in the specific area proposed for investigation (i.e., able to act on decisions and family not affected)

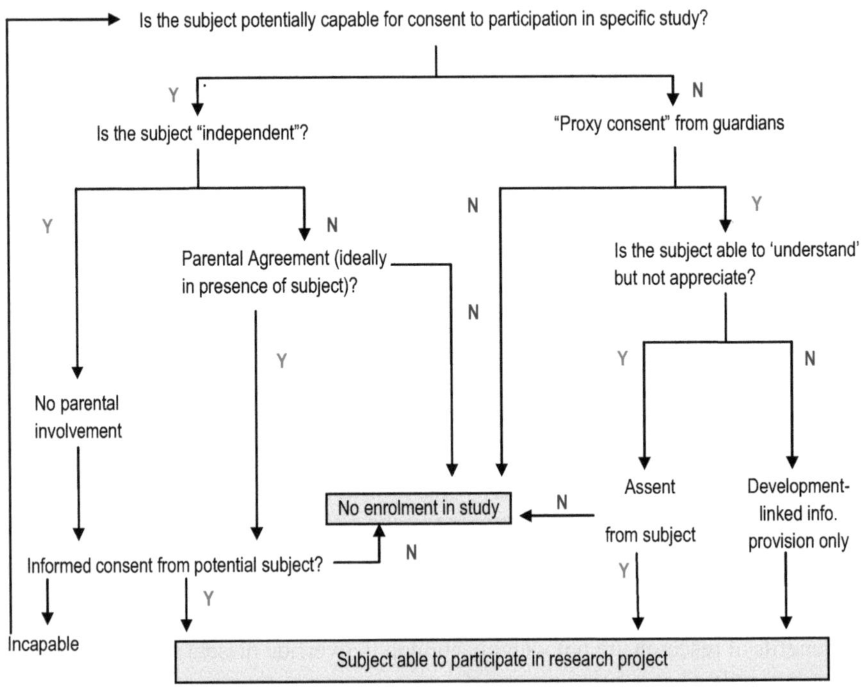

Fig. 2.1 Paediatric consent schema

3. Where the claims a parent/guardian has to decision-making authority are justifiably forfeited by their failure to execute their duties toward the child (i.e., basic interest protection)

'*Proxy Consent*' refers to authorization by a capable and legally responsible third party for a subject to participate in research. It should remain distinct from informed consent obtained from an individual subject, but still considered a valid authorization for participation in research.

'*Assent*' involves agreement to participate by an incapable subject, where the subject is able to understand but not fully appreciate the decision. It recognizes the developing capacity for decision-making, and thus allows children to have some authority over their participation in research. Valid assent requires the provision of information in terms understandable by the potential subject-participant.

2.8 Conclusion

Simultaneous recognition of individual and group interests is an important component of paediatric research. It helps to promote synergy between two dominant trends in contemporary paediatric research and research ethics: family-centred care and respect for children's rights and emerging autonomy. It thereby helps integrate seemingly incongruous family- and patient/child-centred care models. While the semantics surrounding the former prove problematic insofar as they move the moral locus away from the individual patient (who we argue is the ultimate primary source of ethical consideration), the role of the family in paediatric ethics is important. It is often the backdrop for autonomous decision-making.

The simultaneous recognition of individual and group interests would not be new to Canadian research ethics guidelines. The TCPS 2 already attempts synergy in the case of Aboriginals. We propose an analogous recognition of the competing interests in the paediatric context. As in the Aboriginal case, paediatric research ethics must be collaborative and contextual, as different circumstances will require different results. Our paediatric consent to research schema clearly articulates the important values at stake, and provides a reproducible and transferrable guideline for how to deal with our growing appreciation of issues surrounding consent in paediatric research.

References

Blake, V., S. Joffe, and E. Kodish. 2011. Harmonization of ethics policies in pediatric research. *The Journal of Law, Medicine & Ethics: A Journal of the American Society of Law, Medicine & Ethics* 39:70–78.

Breen, L. 2009. Early childhood service delivery for families living with childhood disability: Disabling families through problematic implicit ideology. *Australasian Journal of Early Childhood* 34:14–21.

Canadian Institutes of Health Research, Natural Sciences and Engineering Research Council of Canada, and Social Sciences and Humanities Research Council of Canada, *Tri-Council Policy Statement: Ethical Conduct for Research Involving Humans*, 1998 (with 2000, 2002, 2005 amendments).

Canadian Institutes of Health Research, Natural Sciences and Engineering Research Council of Canada, and Social Sciences and Humanities Research Council of Canada, Tri-Council Policy Statement: Ethical Conduct for Research Involving Humans (TCPS 2), December 2010.

Carter, B. 2008. Commentary on Shields L, Pratt J & Hunter J (2006) Family centered care: A review of qualitative studies. *Journal of Clinical Nursing* 17:2091–2093

Collopy, B. J. 1988. Autonomy and long term care: some crucial distinctions. *The Gerontologist* 28:10–17.

Committee on Bioethics. 1995. Informed consent, parental permission, and assent in paediatric practice. Committee on bioethics, American Academy of Pediatrics. *Paediatrics* 95:314–317.

Conway, J. 2008. Patients and families. *Healthcare Executive* 23:60–62.

Criminal Code, R.S.C. 1985. c. C-46, Section 215 rev 1991 (cited 27 March 2008). http://www.canlii.org/ca/sta/c-46/sec215.html. Accessed 30 Dec 2012.

Diekema, D. S., M. R. Mercurio, and M. B. Adam. 2011. *Clinical ethics in pediatrics: A case-based textbook*, 2–3. New York: Cambridge University Press.

Franck, L.S., and P. Callery. 2004. Re-thinking family-centered care across the continuum of children's healthcare. *Child: Care, Health and Development* 30:265–277.

Institute for Patient- and Family-Centered Care. ND Core concepts. Institute for Family-Centered Care, Bethesda, MD. http://www.ipfcc.org/pdf/CoreConcepts.pdf. Accessed 1 Feb 2013.

Institute for Patient and Family-Centered Care. 2011 Advancing the practice of patient- and family-centered care: How to get started. Institute for Patient- and Family-Centered Care, Bethesda, MD. www.ipfcc.org/pdf/getting_started.pdf. Accessed 1 Feb 2013.

Interagency Advisory Panel on Research Ethics. 2010. Interagency advisory panel on research ethics, table of concordance. 1st ed. 1998 to 2nd ed. 2010.

Johnson, B. 2008. *Partnering with patients and families to design a patient- and family-centered health care system: Recommendations and promising practices*. Institute for Family-Centered Care, Bethesda.

Johnson, B., E. Seale Jeppson, and L. Redburn. 1992. *Caring for children and families: Guidelines for hospitals*. Bethesda: Association for the Care of Children's Health.

Kovacs, P., M. Hayden Bellin, and D. Fauri. 2006. Family-centered care: A resource for social work in end-of-life and palliative care. *Journal of Social Work in End-of- Life & Palliative Care* 2:13–27.

Lambert, V. 2009. Response to Carter, B (2008) Commentary on Shields L, Pratt J, Hunter J (2006) Family centred care: A review of qualitative studies. Journal of Clinical Nursing 15, 1317–1323 in Journal of Clinical Nursing 17, 2091–2093. *Journal of Clinical Nursing* 18:623–624.

Locsin, R. 2003. Culture perspectives. *Holistic Nursing Practice* 17:ix–xii.

MacKean, G., W. Thurston, and C. M. Scott. 2005. Bridging the divide between families and health professionals' perspectives on family-centered care. *Health Expectations: An International Journal of Public Participation in Health Care and Health Policy* 8:74–85.

McGonigel, M. 1998. *Guidelines for family-centered research*. Washington, DC: ACCH.

Miller, R. 2003. *Children, ethics & modern medicine*. Indiana: Indiana University Press.

Miller, P. B., and N. P. Kenny 2002. Walking the moral tightrope: Respecting and protecting children in health related research. *Cambridge Quarterly of Healthcare Ethics: CQ: The International Journal of Healthcare Ethics Committees* 11:217–229.

Palmer, S. 1993. Care of sick children by parents: A meaningful role. *Journal of Advanced Nursing* 18:185–191.

Personal Health Information Protection Act. 2004. SO 2004, c 3, Sch A s. 23 of PHIPA.

Sheahan, L., M. Da Silva, C. Czoli, and R. Zlotnik Shaul. 2012. A Canadian perspective on a child's consent to research within a context of family-centered care: From incompatibility to synergy. *Journal of Clinical Research Bioethics* 3:132. doi:10.4172/2155-9627.1000132

Shields, L. 2011. The ethics of family-centered care for hospitalised children. In *Ethical and Philosophical Aspects of Nursing Children and Young People,* eds. G. Brykczyńska, and J. Simons, 144–154. Ames: Wiley-Blackwell.

Shields, L., J. Pratt, and J. Hunter. 2006. Family centered care: A review of qualitative studies. *Journal of Clinical Nursing* 15:1317–1323.

Simpson, C. 2003. Children and research participation: Who makes what decisions. *Health Law Review* 11:20–29.

Sodomka, P. 2006. Engaging patients & families: A High leverage tool for health care leaders. Hospitals and Health Network: 28–29

Williams, W. G. 2006. Advanced Practice Nurses in a Medical Home. *Journal for Specialists in Pediatric Nursing: JSPN* 11:203–206.

Chapter 3
Resource Allocation in Paediatric Patient and Family-Centred Care

Jennifer L. Gibson

3.1 Introduction

Paediatric patient and family-centred care (PPFCC) is a philosophy of care and organizational management committed to "the planning, delivery, and evaluation of health care that is grounded in mutually beneficial partnerships among health care providers, patients, and families" (IPFCC 2010; AAP 2012, p. 394). Since 2001, when the Institute of Medicine released its *Crossing the Quality Chasm: A New Health Care System for the 21st Century* report and identified 'patient-centredness' as one of six critical aims of a high functioning health system, health ministries, departments, and organizations across North America and elsewhere have begun explicitly to embrace a patient-centred care philosophy and to apply this philosophy in strategic and operational planning. To illustrate, in Ontario, Canada, where I reside, the provincial government has instituted a piece of legislation called the *Excellent Care for All Act*, which specifies patient-centred care as a constitutive principle and stipulates various requirements to engage patient and public values and perspectives in the design, delivery, and evaluation of patient care (Government of Ontario 2010). Patient-centredness is also one of nine attributes according to which Health Quality Ontario (an arms-length governmental agency) evaluates and reports annually on the performance of Ontario's health system overall and of its 14 local health integration networks (Health Quality Ontario 2012). There is an extensive and growing literature theorizing, documenting, and evaluating the evolution of patient-centred (and PPFCC) practices in health systems and organizations internationally. For many clinicians who work in paediatrics or other clinical disciplines serving uniquely vulnerable patient populations, which involves caring for patients within their family systems and tailoring care to optimize the health potential of each patient within his or her limitations, this new strategic and regulatory emphasis on patient-centred care is encouraging, but hardly a novel development.

J. L. Gibson (✉)
Institute of Health Policy, Management & Evaluation, and Joint Centre for Bioethics, University of Toronto, Toronto, Canada
e-mail: jennifer.gibson@utoronto.ca

R. Zlotnik Shaul (ed.), *Paediatric Patient and Family-Centred Care: Ethical and Legal Issues,* International Library of Ethics, Law, and the New Medicine 57,
DOI 10.1007/978-1-4939-0323-8_3,

Meanwhile, resource allocation is one of the most commonly identified ethical challenges encountered among health providers and managers in contemporary health systems and organizations (Gibson et al. 2008; Lemieux-Charles et al. 1993; Foglia et al. 2008; Cooper et al. 2002, 2004). Clinicians report moral distress when confronted with situations where they cannot meet standards of care given available resources (Bell and Breslin 2008). Although front-line staff perceive health leaders to be disconnected from the ethical impact of organizational resource decisions (Foglia et al. 2008), recent research has documented the moral distress of senior managers faced with "selling" decisions they questioned and worrying whether they made the "right decision" (Mitton et al. 2011). There is a vast literature on theoretical, methodological, and practical issues related to resource allocation, and in the last 15 years, there has been a notable shift toward interdisciplinary analyses and pragmatic approaches to resource allocation incorporating clinical, economic, ethical, legal, and policy lenses to address the complex dimensions of resource allocation in practice, particularly at the level of health organizations and systems (e.g., Coulter and Ham 2000; Gibson et al. 2006; Peacock et al. 2010; Norheim 2002; Biller-Andorno et al. 2004). Notably, over the last decade, considerable attention has been paid to developing, evaluating, and implementing methods for grappling explicitly and systematically with the multiple decision factors that may be contextually and ethically relevant to allocation decisions about limited health resources (e.g., multi-criteria decision analysis) and for eliciting and engaging patient and public perspectives on how health resources ought to be allocated.

Given the intensity of scholarship and practice in PPFCC and in resource allocation, and given the prevailing moral (and political) imperative in many health systems to provide high quality patient-centred and efficient care at reasonable cost, one might have expected a vigorous scholarly engagement in the literature. Indeed, when the editors of this collection invited me to write a chapter on ethics and resource allocation related to PPFCC, I thought it would be a straight-forward assignment. As an ethicist whose predominant area of scholarship and practice is healthcare resource allocation, I anticipated finding a robust literature on resource allocation in paediatrics and a comparably robust literature on PPFCC addressing the ethical and practical issues of resource allocation involving paediatric populations. It was surprising therefore to discover, on the one hand, a sparse literature on ethics and resource allocation in paediatric care *per se,* and on the other hand, a resounding silence about resource allocation issues in the patient and family-centred literature (including PPFCC). Indeed, in a number of recent collections on paediatrics ethics, resource issues are addressed implicitly (e.g., within discussions of the futility of intensive care) (Miller 2010), or when they are tackled directly, it is at the extreme of clinical care (e.g., in relation to triage decisions in public health emergencies) (Diekema et al. 2011). While I understand now why the editors would think it useful to include a chapter on resource allocation ethics and PPFCC, I must confess to a certain amount of trepidation about how to scope and advance an ethical discourse on resource allocation in paediatrics and, notably, in relation to PPFCC. Hence, in this chapter, I propose to open a discussion about resource allocation in PPFCC. My intent is not to survey the literature on resource allocation related to paediatric care, but rather to explore the implications of a PPFCC philosophy of care for decisions regarding the use of healthcare resources. Specifically, I will use two hypothetical cases to explore the ethical and practical implications of PPFCC in resource-limited (but not necessarily scarce) environments. Ultimately, I contend that the guiding principles of PPFCC are not only concordant with many current advances in resource allocation practice and scholarship, but also that they may enable health organizations to better mobilize their limited resources toward achieving higher-quality and more efficient healthcare. I conclude that further conceptual and empirical scholarship is needed to explore this promising alignment between best practices in both PPFCC and healthcare resource allocation.

3.2 Resource Allocation in PPFCC: The Elephant in the Room

In the paediatric resource allocation literature, the predominant ethical discourse is clustered around a narrowly-defined set of issues including: emergency critical care triage (e.g., Antommaria et al. 2011; Burns and Mitchell 2011; Christian et al. 2011; Mason et al. 2011; Patel and Maconochie 2008); allocation of neonatal intensive care resources, particularly in relation to considerations of medical futility at the end of life (e.g., AAP 1996; Campbell and Fleischman 2001; De Luca 2007; Fortune 2006; Gillam and Sullivan 2011; Racine and Shevell 2009); access to expensive life-saving therapies (e.g., Ham 1999; Young et al. 2004; Zlotnik-Shaul and Vitale 2009); allocation of paediatric organs for transplantation (e.g., Freeman 2006; Ladin and Hanto 2011; Toker and Salzer2012); and access to paediatric and neonatal intensive care resources in low- and middle- income countries (e.g., Basnet et al. 2011; Miljeteig et al. 2009; Vimpany 1991). Much of this discussion is concerned with defining ethically defensible criteria for setting limits on access to paediatric patient care. The ethical justification for and against using age as a rationing criterion and the relative priority of children compared to adults and older adults for access to scarce medical resources has also been explored in the bioethics and health policy literatures (Kerstein and Bognar 2010; Ladin and Hanto 2011; Lantos et al. 1997; Nord 2010; Persad et al. 2009; Williams and Evans 1997). A second cluster of literature is oriented to generating evidence-based data and decision models to guide resource decisions regarding paediatric services and interventions. A number of studies have described the patient characteristics, resource utilization patterns, and associated costs of healthcare of different paediatric patient populations, particularly in paediatric intensive care units and at the end of life (Cohen et al. 2012; González-Cortes et al. 2011; Murphy and Morris 2008; Sachdeva et al. 1996). For example, Cohen et al. (2012) found that although medically complex children constitute only 1 % of the total child population in Ontario, they account for one-third of all paediatric healthcare expenditures. With a view to improving the economic evaluation of different healthcare interventions as an input to resource allocation decision-making, paediatric utility values have been proposed based on children's and parents' perspectives of paediatric health outcomes (Carroll and Downs 2009; Stevens 2010; Ungar 2011). Nevertheless, with few exceptions, these literatures make scant reference to PPFCC.

When the PPFCC literature itself addresses considerations of healthcare resources, it is primarily to establish the business case for PPFCC. PPFCC is described as leading to a "wiser allocation of resources," and although its implementation may require an ongoing resource investment (e.g., staff time and training), it is argued that this investment generates dividends in the form of "better health outcomes," "greater patient and family satisfaction" and reduced overall financial and other costs (AAP 2012, p. 395; IPFCC 2010). Patient and family-centred care more generally has been associated in the literature with shortened length of stay, reduced rehospitalizations, decreased referrals to specialists, reduced use of diagnostic tests, lower cost per case, decreased adverse events, increased market competitiveness, improved staff recruitment and retention, fewer lawsuits, and reduced legal costs, all of which contribute to reduced resource consumption or more efficient use of available resources (AAP 2012; Anonymous 2003; Bertakis and Azari 2011; Charmel and Frampton 2008; Kuhlthau et al. 2011; Kuo et al. 2012; Shield et al. 2012; Stewart et al. 2011). In other words, patient and family-centred care is presented not only as good for patients and families, but also as 'good for business.'

Even if PPFCC is implemented effectively—and about this there is some question even among those who are proponents, particularly given both the lack of common definitions and the variable quality of research evidence (Carter 2008; Kitson et al. 2013; Lambert 2009; Mikkelsen and Frederiksen 2011; Shields 2010; Shields and Hunter 2008; Shields et al. 2006, 2012)—it is not entirely clear that its benefits will be fully realized without consideration of the context of care, including availability of health resources. On the one hand, there are conventional resource challenges related to the availability and affordability of needed therapies, i.e., it may not be possible to fund all therapies that we would wish to fund. On the other hand, there are resource constraints inherent in the PPFCC care delivery model itself. For example, in exploring the factors which influence case managers' resource allocation decisions in paediatric home care, Fraser et al. (2009, p. 342) reported the significant role of seven "family related factors" including: the number of children in the family, family beliefs, family support, marital status, coping, risk to the family as a unit, and socioeconomic status. The fact of variable family need and capacity means not only that PPFCC must be tailored, but also that resource requirements may differ from family to family. Although the language of partnership is commonly used to describe the relationship of family members and healthcare providers in the care of children, family members themselves may also be seen as a "valuable deployable resource" (Mason et al. 2011, S 158), on the one hand, or as a potential drain on health provider resources, on the other, especially if family need exceeds the healthcare provider's ability to meet it (Mikkelsen and Frederiksen 2011). Moreover, while many parents may be willing participants and partners in the care of their children, others may "resent being made to do what they [perceive] to be nurses' work" (Shields 2010, p. 2633) or be limited in their ability to partner effectively given competing demands within the family unit (e.g., the needs of other children at home; ability to take time off work) or other situational stressors associated with the family's ability to cope with the psychological, social, and financial burden of having a child in care (Callery 1997; DiFazio and Vessey 2011; DeRigne 2012.) It is often suggested that negotiation and good communication are essential elements of effective partnership among parents and health providers (Conway et al. 2006; Johnson et al. 2008), both of which are time-intensive and contingent on family-related factors. Hence, even having committed to PPFCC at the clinical level, resource considerations are an inevitable feature of its implementation in the day-to-day practice of clinical care.

At an institutional level, questions remain about how PPFCC implementation priorities ought be set, how much of available healthcare resources should be allocated toward these priorities, how much weight should be given to direct investments in PPFCC compared to other possible investments in the organization, and what decisions ought to be made if demand for PPFCC exceeds available resources. In its *Guide to patient and family engagement*, the Agency for Healthcare Research and Quality (2012) outlined three key strategies for involving patients and family members at an organizational level: establishing patient and family councils, including patients and families on corporate committees and as "experts" on other corporate initiatives (including those related to the institution's research and education mandate), and eliciting patient and family feedback to inform institutional policy and procedures. Interestingly, of the list of policy topics, corporate initiatives, and institutional functions for which patient and family participation is recommended, participation in institutional priority setting or resource allocation is not among them (AHRQ 2012; AAP 2012; Johnson et al. 2008; Piper 2010). It is not clear whether this exclusion is intentional or whether it is simply assumed that resource considerations are already embedded within these other activities. One possible explanation is that the PPFCC has been primarily defined in relation to its contribution to improving quality of care and its recognition of patients and families as partners in this effort. However, it does seem like a lost opportunity for patients and families, as well as for organizational decision-makers, not to draw on patient and family values, perspectives, and expertise in setting healthcare priorities or in allocating resources as is recommended by PPFCC proponents in relation to improving healthcare quality and the patient and family experience through their participation in program development, care delivery, institutional policy, facility design, and so on. It is in this respect that current research on patient and community engagement in healthcare priority setting may be particularly relevant. Since the late 1990s, there has been an increasing emphasis on developing methods for incorporating patient and community values into decision-making about health priorities and the allocation of health resources (e.g., Barham 2011; Born and Laupacis 2012; Coulter and Ham 2000; Daniels and Sabin 2002; Gibson et al. 2005a, 2005b; Gibson et al. 2011; Haigh 2008; Martin et al. 2002; Mitton and Donaldson 2003, 2004; Peacock et al. 2010; Sibbald et al. 2009; Staniszewska et al. 2012). Not only are the core principles of PPFCC concordant with many of these advances in resource allocation practice and scholarship, they may also enable health organizations to better mobilize their limited resources toward achieving high-quality healthcare.

3.3 Clinical-Level Resource Allocation

Case Vignette: PPFCC and Access to Expensive Therapies in Clinical Care

Dr. Jones has a young patient, Infant A, who was born with a rare lysosomal storage disorder arising from an inherited deficiency of a metabolic enzyme. Although the adult form of the disorder is more common, the infantile form is more severe with few infants surviving beyond the age of 2 years without treatment. A newly approved enzyme replacement therapy (ERT) is considered first-line treatment for the disorder, but it is not a cure. Given the rarity of the disorder, the ERT is very expensive, with a projected annual cost of $ 250K for the rest of Infant A's life. Other treatment alternatives support symptom-level management, but the clinical outcomes, particularly regarding cognitive development, are not promising. Currently, total treatment costs, including ERT, would be covered by the hospital. Following discharge, however, it will be up to Infant A's parents—through third-party insurance, out-of-pocket payments, or other sources—to pay for the treatment. Dr. Jones is concerned about how to broach the issue of treatment options, outcomes, and costs with the parents, who do not have the personal financial resources to pay for the ERT. Initial efforts to advocate with the manufacturer for compassionate access to the therapy have been unsuccessful. At the same time, the hospital is not able to cover these costs indefinitely without an adverse budgetary and clinical impact on the hospital's capacity to meet other patients' needs. Dr. Jones is committed to the philosophy of paediatric patient and family-centred care, but is unsure how to proceed in this case, where what may be in the clinical best interests of the child (i.e., to continue ERT indefinitely) may not be in her global best interests when taking into consideration the financial burden this will place on her family.

International codes of ethics for physicians stipulate that a physician's first and primary moral obligation is to his or her patient (AAP 2012; AMA 1994, 2001, 2003; CMA 2004; GMC 2006, 2012). However, this case underscores how this primary duty may come into tension, if not conflict, with other important moral duties. First, the best interests of the child are often tied inextricably to the interests of his or her family. Nevertheless, if there is a conflict between the child's and the family's interests, it is generally agreed that the child's interests should trump. Second, the child's best interests are not, however, without moral consequences for families. Clinicians are very often aware of the considerable psychological, social, and/or financial burden carried by families caring for a sick child. While it may be the case that the child's best interests should be primary, it may only be possible to achieve this by addressing the family's interests and needs. Families themselves may also face competing duties as, for example, if there are other children in the family or if the cost of treatment puts the longer-term financial sustainability of the family at risk. Finally, the case underlines the unique ethical difficulties of treatment decision-making "at the extremes," where a decision to allocate resources to treat one patient may have significant implications for the availability of resources to treat other patients. In economic terms, the opportunity cost of providing care to one patient means care cannot be given to other patients who are equally deserving of care. In ethical terms, this is an issue of distributive justice, i.e., the distribution of benefits and burdens across equally entitled parties.

As previously noted, recent studies have attempted to document the medical and non-medical costs incurred by families during the hospitalization of a child or while raising a child with special health needs at home (Callery 1997; DiFazio and Vessey 2011; DeRigne 2012). Non-medical out-of-pocket costs have received comparatively less attention in the literature. However, it is well-known that families may face loss of income due to employment changes, food and transportation costs associated with hospital visits, as well as child care costs where there are other children in the family. When the financial burden includes significant out-of-pocket medical costs, e.g., in the case of high cost therapies, some families may face a tragic choice between meeting the needs of a sick child and maintaining the integrity and sustainability of their family unit as a whole. In situations where the financial burden of care is borne disproportionately by the family—as in the case of high-cost therapies for rare disorders—clinicians may find themselves torn between their duty to act in the child's best interests and a profound awareness of the moral consequences of prescribing therapies that are out of a family's reach. In a recent survey (Jefford et al. 2005), a sizeable number of oncologists (as much as 41 %, depending on the scenario) reported that they would not disclose information to patients about a drug if the drug were not subsidized. The most common reasons given for non-disclosure were: not wanting to distress the patient and family or feeling badly about discussing a drug the patient could not afford. While the principle of respect for autonomy requires physicians to disclose this information in order for a patient or her family to make an informed decision about treatment options, their moral discomfort is understandable and no less so in a PPFCC environment that espouses partnership with patients and families.

Resource allocation decisions affecting individual patients are ethically challenging, particularly in relation to expensive therapies such as drugs for rare diseases. Largent and Pearson (2012, p. 29) observed that "orphan drug coverage decisions highlight the tension that can arise in health care between doing the most good possible with scarce health care resources and the desire to assist identifiable individuals regardless of cost"—in other words, there exists a tension between utilitarianism and the rule of rescue. Whereas some argue that the rule of rescue confers special status on rare disorders where the number of people affected are few, the condition is severe and there are no alternative treatments available (Hughes et al. 2005), others contend that "no principled argument could be identified that distinguished patients with rare disease from those with common conditions" (Burls et al. 2005, 1020) and that "arguments about the measurement and valuation of health outcomes apply equally to orphan drugs and drugs for more common conditions"

(McCabe et al. 2005, p. 1018). Interestingly, both arguments prescribe limits on utility considerations by invoking egalitarian reasoning—one based on non-abandonment of identifiable individuals in need (rule of rescue) (Hughes et al. 2005; Largent and Pearson 2012) and the other based on equal regard for capacity to benefit (Burls et al. 2005)—but reach different conclusions about the claim on health resources of rare disorders.

In a resource constrained environment, questions may arise about whether physicians should play a more active role in rationing care at the bedside by, for example, recommending a less costly treatment option even if a more costly one might be more effective. Ubel (2001) has argued that physicians have a legitimate role to play in helping to contain healthcare costs by offering patients less than the "best" treatment option in order to conserve societal resources. Bedside rationing is often rejected as unethical on the grounds that it violates physicians' primary duty to advocate for the best interests of their patients above other considerations, leads to arbitrary or discriminatory treatment decisions, and undermines patient and family trust (Schafer 2001; Sulmasy 2007). Nevertheless, Ubel's controversial position underscores the reality that treatment decisions are not without moral consequences to other stakeholders' interests, or to put it in economic terms, there are opportunity costs associated with treatment decisions, the magnitude of which may put physicians in a situation where they may experience a profound conflict of duty.

The PPFCC philosophy provides complementary concepts for understanding clinicians' duties in relation to resource allocation decisions. The concept of *information-sharing* accords with a duty to disclose information to families about treatment options, including risks, consequences, and financial and other implications, in order that they might exercise their autonomy and "effectively participate in care and decision-making" (Johnson et al. 2008, p. vi). The concept of *dignity and respect* empowers families to make decisions consistent with their values, beliefs, and circumstances, which may involve making informed choices about available treatment options with which clinicians might not necessarily agree but are considered "best" for the child and her family. From a clinician perspective, it may also mean helping to prevent clinician conflicts of duty by giving responsibility for resource limit-setting decisions to others in the organization and by putting explicit institutional policies in place to guide decision-making in resource-constrained clinical situations. The concept of *participation* encourages and supports patients and families to "participate in care and decision-making at the level they choose" (Johnson et al. 2008, vi). Faced with difficult treatment choices, effective participation may require institutional supports to both families and clinicians. Examples of such supports might include: interpreter services, opportunities to speak with other similarly situated families, training or other professional support of clinicians disclosing 'bad financial news,' linking to alternate funding or community resources, advocacy activities on behalf and in collaboration with the family, and so on. Finally, the concept of *collaboration* suggests the possibility of patient and family involvement in, among other things, shaping resource allocation policies (e.g., high-cost drug policy), contributing to professional education on resource issues from patient and family perspectives, or aiding in the development of useful programs, processes, or materials to support families faced with making treatment decisions or dealing with the consequences of limited access to needed care.

What implications does this discussion have for Dr. Jones? Dr. Jones and the hospital are co-fiduciaries in the care of Infant A (Chervenak and McCullough 2003) and both have important roles to play in fulfilling a commitment to PPFCC. Dr. Jones should be forthcoming with Infant A's parents about the treatment options, associated costs and the avenues available to access and fund these treatment options. Although her primary duty is to Infant A, her commitment to PPFCC would be reflected in her advocacy with and on behalf of Infant A's parents to access financial resources to pay for this therapy. Dr. Jones can play an important advocacy role on behalf of all patients and families by contributing to the development and implementation of institutional policies regarding resource allocation and access to care and of institutional supports in aid of families faced with difficult decisions about accessing expensive treatments for their children. As appropriate, Dr. Jones can also work with her professional association or college to advocate for broader systemic changes to funding or access policies. In the case where all avenues have been exhausted, Dr. Jones should continue to provide the best alternate level of care for as long as Infant A is in her care. The role of hospital leaders has a number of elements as well. First, hospital leaders can support the family in its advocacy efforts, e.g., by assisting in the identification of alternate funding sources, providing media training, etc. However, as this type of support ought to be accessible to all families similarly situated, the hospital ought to define clear policy and procedures related to its advocacy and these should be clearly communicated to staff and families. Related to this policy, the hospital should have a formal policy and procedure for the review of expensive therapies based on relevant criteria and a transparent decision-making process. Ideally, this policy would be developed with the involvement of clinicians and families. At a health system level, hospital leaders also have an important role to play in advocating for policies and programs that make needed treatments more accessible and affordable to patients and their families and in supporting efforts to educate and engage patients, families, and the public about the limits of healthcare resources.

3.4 Institution-Level Resource Allocation

Case Vignette: PPFCC in Hospital Budgeting

Lake Ontario Children's Hospital (LOCH) is a large specialty hospital and primary referral site for tertiary- and quaternary-level paediatric care in the region. A number of years ago, LOCH made a strategic commitment to PPFCC as a guiding philosophy of care. LOCH has implemented a number of key initiatives to institutionalize this commitment, including the creation of a Family Council, the involvement of Family Advisors on various corporate committees, and the introduction of PPFCC rounding in its clinical programs.

These efforts have led to significant improvements in clinical outcomes, staff retention, and resource utilization, including reductions in length of stay and re-hospitalizations. LOCH has become well-known for its success in implementing PPFCC and has hosted several delegations from other health systems and organizations seeking to learn from LOCH's experience. In the past, LOCH has been able to sustain a balanced budget position by a number of efficiency and cost-saving measures; when budget cuts were necessary, these were focused primarily on administrative service areas rather than clinical programs. However, at today's senior management team meeting, LOCH's chief financial officer reported that, given the current fiscal climate and in order to achieve a balanced budget next year, it may be necessary to reduce some clinical program budgets, to postpone funding increases to other clinical programs, or to re-allocate funding across clinical programs to service high-demand areas. LOCH's senior management team is concerned not only about how this may adversely affect patient care, but also how LOCH will be able to sustain a commitment to PPFCC in a resource-constrained environment.

Healthcare resource allocation is commonly framed as an ethical issue of justice: when demand for healthcare exceeds available resources, resources should be allocated to achieve a fair distribution of benefits and burdens among relevant stakeholders (Beauchamp and Childress 2009; Martin et al. 2008; Williams et al. 2012). What counts as fair depends on whether distributive justice is defined as achieving the greatest good for the greatest number with available resources (utility), ensuring equality of opportunity or access based on medical need (equity), preserving individual choice (liberty), and so on. However one might define principles of fair allocation, they share in common a fundamental commitment to formal justice, i.e., like cases should be treated alike.

Although the PPFCC literature does not offer specific guidance on how the PPFCC philosophy ought to inform resource allocation decisions, we might reasonably infer that a PPFCC approach would support allocation principles such as: (i) the needs of all children and their families should be considered in deciding how resources are allocated (i.e., formal principle of equality), (ii) resources should be used to generate a net positive impact on the care of children and their families (i.e., beneficence or utility), (iii) resources should not be wasted on interventions that do not generate a significant benefit (i.e., efficiency), (iv) where possible, patient and family choice ought to inform allocation decisions, and (v) resources should be allocated to take into consideration future needs as well as current demands for high-quality care (i.e., sustainability). By themselves, these higher-level principles are not sufficient to address the concrete challenges of resource allocation in practice. For example, these principles do not specify *how much priority* should be given to funding essential administrative services or *what priority* should be given to servicing the budget deficit rather than investing in new PPFCC interventions. Daniels (1994, pp. 27–28) spelled out four unsolved rationing problems, which are germane to the discussion of resource allocation in paediatric care:

- "How much should we favour producing the best outcome with our limited resources?" (Best outcomes/fair chances problem)
- "How much priority should we give to treating to the sickest or most disabled patients?" (Priority problem)
- "When should we allow an aggregation of modest benefits to larger numbers of people to outweigh more significant benefits to fewer?" (Aggregation problem)
- "When must we rely on a fair democratic process as the only way to determine what constitutes a fair allocation of resources?" (Democracy problem).

As Daniels points out, although we may "intuitively reject extreme positions [in our response to these questions, we] have no satisfactory theoretical characterization of an intermediary position" (Daniels 1994, p. 28). These are questions for which there are no 'simple solutions' (Holm 1998) and which must be worked out in a fair deliberative process.

Healthcare organizations must contend with the further difficulties of a complex mandate comprised of multiple goals and the need to balance their ethical obligations to *all* patients and families served by the organization. In the field of organizational ethics, an organization's mission and values are sometimes described as its "moral compass" (Pearson et al. 2003, p. 26). The idea is that an organization's decisions ought to be consistent with its mission-based commitments and true to its values. Organizational ethics, therefore, calls on health organizations "to define their core values and mission, identify areas in which important values come into conflict, seek the best possible resolution of these conflicts, and manage their own performance to ensure that it acts in accord with espoused values" (Pearson et al. 2003, p. 32). The devil is of course in the details. Resource allocation is the quintessential organizational ethics challenge (Gibson et al. 2008), which is all the more difficult when the stakes are high, both in reality and as perceived by those most affected by these decisions. It has been described as a "wicked problem"—that is, a "pernicious social [vs. technical] problem where the solution is unclear and requires a leadership [vs. managerial] response" (Williams et al. 2012, p. 110). Factors that make priority-setting a wicked problem include: lack of patient and public engagement, unclear criteria and processes, multiple objectives and values, and high stakeholder expectations. Given finite resources, resource allocation questions cannot be resolved simply with reference to the organization's core mission of patient care, or the aspirational ideals or core concepts of PPFCC.

Current research and practice in healthcare resource allocation internationally suggests two key ways to address the moral uncertainty about justice and the complex practical and ethical challenge of making resource allocation decisions in the context of multiple goals and objectives, competing stakeholder needs and interests, and often conflicting values. First, *multi-criteria decision analysis* (MCDA) is an increasingly common aid to decision-making, which takes explicit account of the range of decision factors, including values, evidence, and other considerations, that may be relevant to a decision. These decision factors are specified as discrete criteria, which can then be used to evaluate and weigh different budgetary options or alternative allocations of resources. This approach has the advantage of synthesizing diverse streams of relevant information in a way that makes decision-makers' value positions explicit, but does not replace deliberation or judgment (Williams et al. 2012). As Williams et al. (2012, p. 67) observe, MCDA approaches "sacrifice some of the *rigour* associated with the discipline of health economics [e.g., cost-effectiveness analysis] in order to maximize *relevance* to the decision-making context". MCDA is used in well-recognized priority-setting approaches such as program budgeting and marginal analysis (Mitton and Donaldson 2003; Peacock et al. 2009), and operational versions of the accountability for reasonableness framework (Gibson et al. 2005b, 2011). Second, the *accountability for reasonableness* (A4R) framework itself has had a tremendous impact on how priority-setting is understood from an ethical perspective in health institutions and systems. Given reasonable disagreement among fair-minded stakeholders, a fair process is needed to reach publicly defensible and socially acceptable decisions. Daniels and Sabin (2002) specified four conditions of a fair priority-setting process (Table 3.1), which, they argued, provides an answer to two important questions: (i) under what conditions would a stakeholder (e.g., patient, clinician, member of the public) have sufficient reason to accept a particular priority setting decision as fair? (fairness) and (ii) under what conditions should we accept the moral authority of those making rationing decisions? (legitimacy). Based on empirical research in a hospital setting, Gibson et al. (2005a) proposed a 5th condition of A4R—the empowerment condition—to address the reality of power differences among stakeholders. Although primarily a model of procedural justice, it is the deliberative activity entailed in the Relevance condition where the hard work of decision-making is accomplished. A third promising approach, as noted previously, is the emerging emphasis on *patient and public engagement* in institutional and health system decision-making about priorities.

What implications does this discussion have for Lake Ontario Shores Hospital with its looming budget deficit? These approaches shed light on a number of ways in which an institutional commitment to PPFCC might inform organizational resource allocation. Importantly, they enable health institutions to shift from PPFCC as an aspirational goal to PPFCC as an operating principle in meeting the ethical and practical challenge of advancing patient care needs within the material constraints of finite resources. Table 3.1 outlines some possible implementation strategies organized according to the conditions of A4R. Notably, the Relevance condition incorporates MCDA and elements of patient and family engagement can be found in each condition. Recently, the IWK Health Centre in Halifax, Nova Scotia, which provides tertiary-level care for children, youth, and families in the Atlantic region of Canada, implemented a PBMA-based annual budgeting process. 'Patient and family-centred care' was specified as an explicit criterion in assessing the health impact of different budget options and involved a member of the Patient and Family Advisory Council participating on the decision-making body (Campbell 2012). A formal evaluation of the budgeting process is underway, so it is not yet known how effective these strategies were. However, IWK Health Centre serves as a good example of how PPFCC principles can be applied to resource allocation in practice.

Table 3.1 Accountability for reasonableness (A4R)

Condition of A4R	Definition	Possible implementation strategies (adapted from Gibson et al. 2005b and 2011)
Relevance	Decisions should be based on reasons (e.g., principles, evidence, values) that affected stakeholders agree are relevant under the circumstances	Define clear and explicit decision criteria that address key aspects of PPFCC Consider incorporating PPFCC as an explicit criterion in a multi-criteria decision analysis Develop the decision criteria with input from patients, families, and staff involved in the delivery of PPFCC Collect relevant patient and family-based data and information related to the decision criteria (not limited to the PPFCC criterion)
Publicity	The rationale for decisions (i.e., the "reasons") should be publicly accessible	Develop a communication plan to reach affected stakeholders, including patients, families, and hospital staff Communicate the decisions and rationales in clear and accessible language linked explicitly to the decision criteria Conduct an open forum with the Family Council to engage questions and concerns
Revision	There should be opportunities to review and revise decisions and a mechanism to resolve dispute	Develop a formal mechanism by which patients and families can address their questions about allocation processes, decisions, and rationales Monitor impact of allocation decisions on patients and families and revise as necessary
Empowerment	Efforts should be made to minimize power differences in the decision-making context (Gibson et al., 2005b)	Engage program-level family advisors in identifying budgetary options/solutions at the program level Include a critical mass of family advisors (e.g., Family Council representatives) on the corporate-level decision-making body Provide training for family advisors to optimize effective participation in the budgeting process Support the budgeting process with change management strategies to ease the transition from decision to implementation
Enforcement	There should be a credible leadership commitment to ensure the other four conditions are met	Specify PPFCC and fairness as guiding principles and assumptions of the budget-planning process Conduct a formal evaluation of the budgeting process, including family members, to identify good practices and opportunities for improvement Disseminate key evaluation findings to expand knowledge within the field

Decision-making about healthcare resources can be emotionally and politically charged, which may make it difficult for some patients or families to participate effectively without training, confidence, and a critical mass of collegial support (Martin et al. 2002). It is important to underscore that neither PPFCC nor fair priority-setting processes are all-or-nothing phenomena. In health institutions with a pre-existing and effective PPFCC infrastructure, e.g., a well-functioning Family Council and a track record of effective partnership among health leaders and family advisors, it may be feasible to embrace a more participatory model of PPFCC in organizational resource allocation. In health institutions with less developed or very little PPFCC infrastructure at an organizational level, however, a more consultative or communicative model of PPFCC may be more appropriate and yet no less committed to either PPFCC or fair resource allocation.

3.5 Conclusion

In this chapter, I have begun to scope the overlapping terrain of ethical resource allocation and PPFCC and to explore the implications of PPFCC as a philosophy of care and organizational management on decisions regarding the use of health resources at both clinical and institutional levels. My analysis suggests that the guiding principles of PPFCC are not only concordant with many current advances in resource allocation practice and scholarship, but also that they may enable health organizations to better mobilize their limited resources toward achieving higher-quality and more efficient healthcare. My contention is that health organizations committed to PPFCC can and must grapple with the "wicked problem" of priority-setting and resource allocation (Dickinson et al. 2011). Failure to do so is to turn a blind eye to one of the most pressing ethical issues faced by patients, families, clinicians and managers today. In as much as providing patient and family-centred care is understood as an ethical imperative of clinicians and healthcare managers alike (Piper 2010), so too should ensuring fairness in the allocation of health resources. These ethical imperatives are compatible and consistent with the core principles of PPFCC, but have not been previously explored in the literature.

Health organizations espousing PPFCC should not only accommodate fairness as a guiding principle for resource allocation; they should embrace it as a concrete expression of their commitment to treating patients and families with respect and dignity and as a constitutive feature of what it means for patients and families to collaborate effectively with clinicians and organizational leaders. Hence, in addition to their participation in "policy and program development, implementation, and evaluation; in healthcare facility design; in professional education, [or] in the delivery of care" (Johnson et al. 2008, vi), patients and families can contribute to decision-making about healthcare priorities and the use of health resources. Moreover, by engaging patients and families actively in institutional priority-setting and resource allocation processes, quality of care might actually be enhanced. The more transparent decision-makers are about the decision factors that inform their allocation decisions and the better-informed they are about (i) the concrete realities of care delivery, (ii) the needs and expectations of patients and families receiving this care, and (iii) what decision criteria are relevant to them in assessing the fairness of a priority-setting decision, the more likely their decisions will be and be perceived to be driven by a commitment to quality, even in the midst of resource constraints (Daniels and Sabin 2002; Gibson et al. 2005a). And when difficult choices must be made, which may have a negative impact on patient care at the bedside, patients, families and clinicians will be more likely to be able to accept such decisions. To my knowledge, this is the first published discussion of PPFCC (and indeed, PFCC, more generally) in relation to resource allocation as a domain of scholarship and practice in healthcare. Future research is needed to further elucidate the normative and conceptual compatibility of PPFCC and resource allocation ethics and to investigate effective methods for patient and family participation in resource allocation in paediatric care environments committed to PPFCC.

References

Agency for Healthcare Research and Quality. 2012. Guide to patient and family engagement: Environmental scan report. http://www.ahrq.gov/qual/ptfamilyscan/. Accessed 12 Jan 2013.

American Academic of Pediatrics (AAP). 1996. Ethics and the care of critically ill infants and children. *Pediatrics* 98 (1): 149–152.

American Academic of Pediatrics. 2012. Patient- and family-centred care and the pediatrician's role: Policy statement. *Pediatrics* 129:394–404.

American Medical Association (AMA). 1994. Opinion 2.03—Allocation of limited medical resources. http://www.ama-assn.org/ama/pub/physician-resources/medical-ethics/code-medical-ethics/opinion203.page. Accessed 10 Feb 2013.

American Medical Association. 2001. Principles of medical ethics. In: Code of medical ethics. http://www.ama-assn.org/ama/pub/physician-resources/medical-ethics/code-medical-ethics/principles-medical-ethics.page. Accessed 10 Feb 2013.

American Medical Association. 2003. Opinion 2.09—Costs. http://www.ama-assn.org/ama/pub/physician-resources/medical-ethics/code-medical-ethics/opinion209.page. Accessed 10 Feb 2013.

Anonymous. 2003. Measuring the cost benefits of family-centred care. *Healthcare Financial Management: Journal of the Healthcare Financial Management Association* 57 (9): 118.

Antommaria, A. H., T. Powell, J. E. Miller, et al. 2011. Ethical issues in pediatric emergency mass critical care. *Pediatric Critical Care Medicine* 12 (6 Suppl): 163–168.

Barham, L. 2011. Public and patient involvement at the UK National Institute for Health and Clinical Excellence. *Patient* 4 (1): 1.

Basnet, S., N. Adhikari, and J. Koirala. 2011. Challenges in setting up pediatric and neonatal intensive care units in a resource-limited country. *Peds* 128 (4): e986–992.

Beauchamp, T. L., and J. F. Childress. 2009. *Principles of bioethics*. 6th ed. Oxford: Oxford University Press.

Bell, J. A. H., and J. Breslin. 2008. Healthcare provider moral distress as a leadership challenge. *JONAS HealthCare Law, Ethics and Regulation* 10 (4): 94–97.

Bertakis, K. D., and R. Azari. 2011. Patient-centered care is associated with decreased health utilization. *Journal of the American Board of Family Medicine: JABFM* 24 (3): 229–239.

Biller-Andorno, N., C. Lenk, and J. Leititis. 2004. Ethics, EBM, and hospital management. *Journal of Medical Ethics* 30:136–140.

Born, K., and A. Laupacis. 2012. Public engagement in Ontario's hospitals—opportunities and challenges. *Healthcare(Toronto, Ont)* 15(Special Issue):16–20.

Burls, A. D., Austin, D. Moore. 2005. Commissioning for rare diseases: view from the frontline. *BMJ: British Medical Journal/British Medical Association* 331 (7523): 1019–1021.

Burns, J. P., and C. Mitchell. 2011. Resource allocation in disasters and pandemics. In: *Clinical ethics in pediatrics: A case-based textbook*, eds. D. S. Diekema, M. R. Mercurio, and M. B. Adam, 199–204. Cambridge: Cambridge University Press.

Callery, P. 1997. Paying to participate, financial, social and personal costs to parents of involvement in their children's care in hospital. *Journal of Advanced Nursing* 25:746–752.

Campbell, M. 2012. Implementation of a rigorous & transparent priority setting framework. Paper presented at Priorities 2012–9th International Conference on Priorities in Health Care, Vancouver, British Columbia, 17–19 Sep 2012. http://www.priorities2012.com/documents/5c-2-campbell.pdf. Accessed 15 Feb 2012.

Campbell, D. E., and A. R. Fleischman. 2001. Limits of viability: dilemmas, decisions, and decision makers. *American Journal of Perinatology* 18 (3): 117–128.

Canadian Medical Association. 2004. CMA code of ethics. http://policybase.cma.ca/dbtw-wpd/PolicyPDF/PD04-06.pdf. Accessed 11 Jan 2013.

Carroll, A. E, and S. M. Downs. 2009. Improving decision analyses: parent preferences (utility values) for pediatric health outcomes. *The Journal of Pediatrics* 155 (1): 21–25.

Carter, B. 2008. Commentary on "Shields L, Pratt J, Hunter J (2006) Family centred care: A review of qualitative studies. J Clin Nurs 5:1317–23". *Journal of Clinical Nursing* 17 (15): 2091–2093.

Capen, K. 1997. Resource allocation and physician liability. *Canadian Medical Association Journal* 156 (3): 393–395.

Charmel, P. A, and S. B. Frampton. 2008. Building the business case for patient-centered care. *Healthcare financial management: Journal of the Healthcare Financial Management Association* 62 (3): 80–85.

Chervenak, F. A., and McCullough, L. B. 2003. Physicians and hospital managers as cofiduciaries of patients: rhetoric or reality? *Journal of Healthcare Management* 48 (3): 172–180.

Christian, M. D., P. Toltzis, R. K. Kanter, et al. 2011. Treatment and triage recommendations for pediatric emergency mass critical care triage. *Pediatr Crit Car Med* 12 (6 Suppl): 109–119.

Cohen, Eyal, Berry J. G, X. Camacho, et al. 2012. Patterns and costs of health care use of children with medical complexity. *Pediatrics* 130: e1463–1470.

Conway, J., B. Johnson, S. Edgman-Levitan, et al. 2006. Partnering with patients and families to design a patient- and family-centered health care system: A roadmap for the future. Institute of Family-Centered Care, Bethesda, MD. www.ihi.org/knowledge/Pages/Publications/PartneringwithPatientsandFamilies.aspx. Accessed 16 Feb 2013.

Cooper, R. W., G. L. Frank, C. A. Gouty, et al. 2002. Key ethical issues encountered in healthcare organizations: perceptions of nurse executives. *The Journal of Nursing Administration* 32 (6): 331–337.

Cooper, R. W., G. L. Frank, M. M. Hansen, et al. 2004. Key ethical issues encountered in healthcare organizations: the perceptions of staff nurses and nurse leaders. *The Journal of Nursing Administration* 34 (3): 149–56.

Coulter A., and C. Ham. eds. 2000. *The global challenge of health care rationing*. Philadelphia: Open University Press.

Daniels, N. 1994. Four unsolved rationing problems. A challenge. *The Hastings Center Report* 24 (4): 27–29.

Daniels, N., and J Sabin. 2002. *Setting limits fairly: Can we learn to share medical resources?* Oxford: Oxford University Press.

Dare, T, M. Findlay, P. Browett, et al. 2010. Paternalism in practice: informing patients about expensive unsubsidised drugs. *Journal of Medical Ethics* 36:260–264.

De Chalain, T. M. B. 1997. Ethical resource allocation and the quest for normalcy: is pediatric reconstructive surgery justified? *Plastic and Reconstructive Surgery* 99 (4): 1184–1191.

De Luca, D. 2007. Resources allocation for neonatal critical care: The ethical debate in neonatology. *Acta Paediatrica* 96 (5): 648–649.

DeRigne, L .2012. The employment and financial effects on families raising children with special health care needs: An examination of the evidence. *Journal of Pediatric Health Care* 26:283–290.
Dickinson, H, T. Freeman, S. Robinson, et al. 2011. Resource scarcity and priority setting: From management to leadership in the rationing of health care. *Public Money and Management* 31 (5): 363–370.
Diekema, D., M. R. Mercurio, and M. B. Adam. eds. 2011. *Clinical ethics in pediatrics: A case-based textbook*. Cambridge: Cambridge University Press.
DiFazio, R., and J. Vessey. 2011. Nonmedical out-of-pocket expenses: A hidden cost of hospitalization. *Journal of Pediatric Nursing* 26:78–84.
Foglia, M. B., R. A. Pearlman, M. M. Bottrell, et al. 2008. Priority setting and the ethics of resource allocation in VA healthcare facilities: Results of a survey. *Organizational Ethics* 4 (2): 83–96.
Fortune, P-M. 2006. Limiting and rationing treatment in paediatric and neonatal intensive care. *Best Practice & Research. Clinical Anaesthesiology* 20 (4): 577–588.
Fraser, K. D., C. Estabrooks, M. Allen, et al. 2009. Factors that influence case managers' resource allocation decisions in pediatric home care: An ethnographic study. *International Journal of Nursing Studies* 46 (3): 337–349.
Freeman, R. B. 2006. Mortality risk, behavior, and pediatric liver allocation. *Liver Transplantation* 12 (1): 12–15.
General Medical Council. 2006. Good medical practice: Guidance for doctors. http://www.gmc-uk.org/static/documents/content/GMP_0910.pdf. Accessed 10 Feb 2013.
General Medical Council. 2012. Leadership and management for doctors. http://www.gmc-uk.org/guidance/ethical_guidance/11788.asp. Accessed 10 Feb 2013.
Gibson, J. L., C. Mitton, and G. Dubois-Wing. 2011. Priority setting in the LHINs: Ethics and economics in action. *Healthcare Quarterly (Toronto, Ont.)* 14 (4): 35–46.
Gibson, J. L., D. K. Martin, and P. A. Singer. 2005a. Priority setting in hospitals: Fairness, inclusiveness, and the problem of institutional power differences. *Social Science & Medicine* 61:2355–2362.
Gibson, J. L., D. K. Martin, and P. A. Singer. 2005b. Evidence, economics, and ethics: Resource allocation in health services organizations. *Healthcare Quarterly (Toronto, Ont.)* 8 (2): 50–59.
Gibson, J. L., C. Mitton, D. Martin, et al. 2006. Ethics and economics: Does program budgeting and marginal analysis contribute to fair priority setting? *Journal of Health Services Research & Policy* 11 (1): 32–37.
Gibson, J. L., R. Sibbald, E. Connolly, et al. 2008. Organizational ethics. In: *The Cambridge textbook of bioethics,* eds. P. A. Singer and A. Viens, 243–250. Cambridge: Cambridge University Press.
Gillam, L., and J. Sullivan. 2011. Ethics at the end of life: Who should make decisions about treatment limitation for young children with life-threatening or life-limiting conditions? *Journal of Paediatrics and Child Health* 47 (9): 594–598.
González-Cortés, R., J. López-Herce-Cid, A. García-Figueruelo, et al. 2011. Ingreso prolongado en la unidad de cuidados intensivos pediátricos: mortalidad y consumo de recursos asistenciales [Prolonged stay in pediatric intensive care units: mortality and health care resource consumption]. *Medicina* Intensiva 35 (7): 417–423.
Government of Ontario. 2010. Excellent care for all act. http://www.e-laws.gov.on.ca/html/statutes/english/elaws_statutes_10e14_e.htm. Accessed 15 Jul 2012.
Haigh, C. A. 2008. Exploring the evidence base of patient involvement in the management of health care services. *Journal of Nursing Management* 16 (4): 452–462.
Ham, C. 1999. The role of doctors, patients and managers in priority setting decisions: Lessons from the 'Child B' case. Health Expectations: An International *Journal of Public Participation in Health Care and Health Policy* 21 (1): 61–68.
Health Quality Ontario. 2012. Quality Monitor 2012. http://www.hqontario.ca/portals/0/Documents/pr/qmonitor-full-report-2012-en.pdf. Accessed 15 Jan 2013.
Holm, S. 1998. Goodbye to the simple solutions: The second phase of priority setting in health care. *BMJ: British Medical Journal/British Medical Association* 317:1000–1002.
Hughes, D. A., B. Tunnage, and S. T. Yeo. 2005. Drugs for exceptionally rare diseases: do they deserve special status for funding? *QJM: An International Journal of Medicine* 98:829–836.

Institute of Medicine (IOM). 2001. *Crossing the quality chasm: A new health care system for the 21st Century. Institute of Medicine,* Washington, DC. www.iom.edu/Reports/2001/Crossing-the-Quality-Chasm-A-New-Health-System-for-the-21st-Century.aspx. Accessed 13 Dec 2012.
Institute of Patient- and Family-Centered Care (IPFCC). 2010. What is patient- and family-centered health care? Institute for Patient- and Family-Centered Care, Bethesda, MD. http://www.ipfcc.org/faq.html. Accessed 12 Jan 2013.
Jefford, M., J. Savulescu, J. Thomson, et al. 2005. Medical paternalism and expensive unsubsidized drugs. *British Medical Journal/British Medical Association* 331:1075–1077.
Johnson, B. H., M. Abraham, and J. Conway. 2008. Partnering with patients and families to design a patient and family centered health care system: Recommendations and promising practices. Institute for Family Centered Care, Bethesda, MD. http://www.ipfcc.org/pdf/PartneringwithPatientsandFamilies.pdf. Accessed 15 Feb 2013.
Johnson B. H., M. R. Abraham, and T. L. Shelton. 2009. Patient- and family-centered care: Partnerships for quality and safety. *North Carolina Medical Journal* 70 (2): 125–130.
Kenny, N., J. Downie, and C. Harrison. 2008. Respectful involvement of children in medical decision making. In: *The Cambridge textbook of bioethics,* eds. P. A. Singer and A. Viens., 121–126. Cambridge: Cambridge University Press.
Kerstein, S. J., and G. Bognar. 2010. Complete lives in the balance. *The American Journal of Bioethics* 10 (4): 37–45.
Kitson, A., A. Marshall, K. Bassett, et al. 2013. What are the core elements of patient-centred care? A narrative review and synthesis of the literature from health policy, medicine and nursing. *Journal of Advanced Nursing* 69 (1): 4–15.
Kuhlthau, K. A., S. Bloom, J. Van Cleave, et al. 2011. Evidence for family-centered care for children with special health care needs: A systematic review. *Academic Pediatrics* 11:136–143.
Kuo, D. Z., A. J. Houtrow, P. Arango, et al. 2012. Family-centered care: Current applications and future directions in pediatric health care. *Maternal and Child Health Journal* 16:297–305.
Ladin, K., and D. W. Hanto. 2011. Rational rationing or discrimination: Balancing equity and efficiency considerations in kidney allocation. *American Journal of Transplantation* 11:2317–2321.
Lambert, V. 2009. Response to "Carter B (2008) Commentary on 'Shields L, Pratt J, Hunter J (2006) Family centred care: A review of qualitative studies J Clin Nurs 15:1317–1323' in J Clin Nurs 17:2091-2093." *Journal of Clinical Nursing* 18 (4): 623–624.
Lantos, J. D., M. Mokala, and W. Meadow. 1997. Resource allocation in neonatal and medical ICUs. Epidemiology and rationing at the extremes of life. *American Journal of Respiratory and Critical Care Medicine* 156 (1): 185–189.
Largent, E. A., and S. D. Pearson. 2012. Which orphans will find a home? The rule of rescue in resource allocation for rare diseases. *The Hastings Center report* 41 (1): 27–34.
Lemieux-Charles, L., E. M. Meslin, C. Aird, et al. 1993. Ethical issues faced by clinician/managers in resource allocation decisions. *Hospital & Health Services Administration* 38 (2): 267–285.
Martin, D. K., J. Abelson, and P. A. Singer. 2002. Participation in health care priority-setting through the eyes of the participants. *Journal of Health Services Research & Policy* 7 (4): 222–229.
Martin, D. K., J. L Gibson, and P. A. Singer. 2008. Priority setting. In: *The Cambridge textbook of bioethics*, eds. P. A. Singer and A. Viens, 251–256. Cambridge: Cambridge University Press.
Mason, K. E., H. Urbansky, L. Crocker, et al. 2011. Pediatric emergency mass critical care: Focus on family-centered care. *Pediatric Critical Care Medicine* 12 (6 Suppl): 157–162.
McCabe, C., K. Claxton, and A. Tsuchiya. 2005. Orphan drugs and the NHS: should we value rarity? *BMJ: British Medical Journal/British Medical Association* 331 (7523): 1016–1019.
McFarlane, J., J. Riggins, and T. J. Smith. 2008. SPIKE$: A six-step protocol for delivering bad news about the cost of medical care. *Journal of Clinical Oncology* 26 (25): 4200–4204.
Mikkelsen, G., and K. Frederiksen. 2011. Family-centred care of children in hospital: A concept analysis. *Journal of Advanced Nursing* 67 (5): 1152–1162.
Miljeteig, I., S. A. Sayeed, Jesani, et al. 2009. Impact of ethics and economics on end-of-life decisions in an Indian neonatal unit. *Pediatrics* 124 (2): e322–328.

Miller, G. ed. 2010. *Pediatric bioethics*. Cambridge: Cambridge University Press.
Mitton, C., and C. Donaldson. 2004. *The priority setting toolkit: A guide to the use of economics in health care decision making*. London: BMJ Books.
Mitton, C., and C. Donaldson. 2003. Setting priorities and allocating resources in health regions: Lessons from a project evaluating program budgeting and marginal analysis. *Health Policy (Amsterdam, Netherlands)* 64:335–348.
Mitton, C., S. Peacock, J. Storch, et al. 2011. Moral distress among health system managers: Exploratory research in two British Columbia health authorities. *Health Care Analysis: HCA: Journal of Health Philosophy And Policy* 19 (2): 107–121.
Murphy, J., and K. Morris. 2008. Accounting for care: Healthcare resource groups for paediatric critical care. *Journal of Pediatric Nursing* 20 (1): 37–39.
Nord, E. 2010. Balancing relevant criteria in allocating scarce life-saving interventions. *The American Journal of Bioethics: AJOB* 10 (4): 56–57.
Norheim, O. F. 2002. The role of evidence in health policy making: A normative perspective. Health Care Analysis: HCA: *Journal of Health Philosophy and Policy* 10 (3): 309–317.
Patel, M., and I. Maconochie. 2008. Triage in children. *Trauma* 10:239–245.
Peacock, S. J., C. Mitton, A. Bate, et al. 2009. Overcoming barriers to priority setting using interdisciplinary methods. *Health Policy (Amsterdam, Netherlands)* 92 (2–3): 124–132.
Peacock, S. J., C. Mitton, D. Ruta, et al. 2010. Priority setting in health care: Toward guidelines for the program budgeting and marginal analysis framework. *Expert Review of Pharmacoeconomics & Outcomes Research* 10 (5): 539–552.
Pearson, S. D., J. E. Sabin, and E. Emanuel. 2003. *No margin, no mission: Health care organizations and the quest for ethical excellence*. Oxford: Oxford University Press.
Persad, G., A. Wertheimer, and E. J. Emanuel. 2009. Principles for allocation of scarce medical interventions. *Lancet* 373 (9661): 423–431.
Piper, L. E. 2010. The ethical leadership challenge: Creating a culture of patient- and family-centered care in the hospital setting. *The Health Care Manager (Frederick)* 30 (2): 125–132.
Racine, E., and M. I. Shevell. 2009. Ethics in neonatal neurology: When is enough, enough? *Pediatric Neurology* 40 (3): 147–155.
Sachdeva, R. C., L. S. Jefferson, and J. Coss-Bu, et al. 1996. Resource consumption and the extent of futile care among patients in a pediatric intensive care unit setting. *Journal of Pediatrics* 128 (6): 742–747.
Schafer, A. 2001. Bedside rationing by physicians: The case against. *Healthcare Papers* 2 (2): 45–52.
Shields, L. 2010. Questioning family centered care. *Journal of Clinical Nursing* 19 (17–18): 2629–2638.
Shields, L., and J. Hunter. 2008. Reply to "Carter B (2008) Commentary on 'Shields L, Pratt J, Hunter J (2006) Family centred care: a review of qualitative studies. J Clin Nurs 5:1317–1323.'" *Journal of Clinical Nursing* 17 (15): 2093.
Shields, L., J. Pratt, and J. Hunter. 2006. Family centred care: A review of qualitative studies. *Journal of Clinical Nursing* 5:1317–1323.
Shields, L., Z. Huaqiong, J. Pratt, et al. 2012. Family-centred care for hospitalized children aged 0–12 years. *The Cochrane Database of Systematic Reviews (Online)* 10 (3): CD004811.
Sibbald, S. L., P. A. Singer, R. Upshur, et al. 2009. Priority setting: what constitutes success? A conceptual framework for successful priority setting. *BMC Health Services Research* 9:43.
Staniszewska, S., K. L. Haywood, J. Brett, et al. 2012. Patient and public involvement in patient-reported outcome measures: Evolution not revolution. *Patient* 5 (2): 79–87.
Stevens, K. J. 2010. Working with children to develop dimensions for a preference-based, generic, pediatric, health-related quality-of-life measure. *Qualitative Health Research* 20 (3): 340–351.
Stewart, M., B. L. Ryan, and C Bodea. 2011. Is patient-centered care associated with lower diagnostic costs? *Healthc Policy* 16 (4): 27–31.
Sulmasy, D. 2007. Cancer care, money, and the value of life: Whose justice? which rationality? *Journal of Clinical Oncology* 25 (2): 217–222.

Toker, A., and L. Salzer. 2012. Pediatric liver transplantation—ethical dilemmas in a disabled patient. *Pediatric Transplantation* 16 (6): E257–260.
Ubel, P. 2001. Physicians, thou shalt ration: The necessary role of bedside rationing in controlling health care costs. *Healthcare Papers* 2 (2): 10–21.
Ungar, W. J. 2011. Challenges in health state valuation in paediatric economic evaluation: Are QALYs contraindicated? *Pharmacoeconomics* 29 (8): 641–652.
Vimpany, G. V. 1991. Resource allocation in contemporary paediatrics: The case against high technology. *Journal of Paediatrics and Child Health* 27 (6): 354–359.
Warren, N. 2012. Involving patient and family advisors in the patient and family-centered care model. *Medsurg Nursing: Official Journal of the Academy of Medical-Surgical Nurses* 21 (4): 233–239.
Williams, A., and J. Grimley Evans. 1997. The rationing debate. Rationing health care by age. *BMJ: British Medical Journal/British Medical Association* 314 (7083): 820–825.
Williams, I., S. Robinson, and H. Dickinson. 2012. *Rationing in health care: The theory and practice of priority setting.* Bristol/Chicago: The Policy Press.
World Health Organization (Western Pacific Region). 2007. Person-centred health care: A policy framework. http://www.wpro.who.int/health_services/people_at_the_centre_of_care/documents/ENG-PCIPolicyFramework.pdf. Accessed 15 Feb 2013.
Young, J., G. Flores, and S. Berman. 2004. Providing life-saving health care to undocumented children: Controversies and ethical issues. *Pediatrics* 114 (5): 1316–1320.
Zlotnik-Shaul, R., and D. Vitale. 2009. Can we afford it?: Ethical considerations of expensive drug treatment for neonates and infants. *Clinical Pharmacology & Therapeutics* 86 (6): 587–589.

Part II
Legal, Ethical and Cultural Considerations

Chapter 4
Culture, Religion, and Family-Centred Care

Rani H. Srivastava

4.1 Introduction

In contemporary healthcare discourse patient-and family-centred care is viewed as a requisite to quality care. However, making this vision a reality continues to be challenging, especially when religious and cultural diversity is added to the mix. The purpose of this chapter is to explore paediatric family-centred care through the lens of religious and cultural diversity. Cultural competence is generally recognized as an approach to care that integrates the patient's and family's needs. In theory, frameworks for paediatric family-centred care and cultural competence are complementary to the point where it is difficult to have one without the other. Both are philosophical approaches to care that are grounded in principles of respect, collaboration, and the context of the patient and family. This synergy, however, is largely theoretical and does not necessarily translate into quality care. In fact, religious and cultural diversity is often seen to get in the way of family-centred care (Pergert et al. 2007; MacKay 2009; Saskatchewan Ministry of Health 2009). The challenge is to transform practice such that, rather than being a barrier, culture is understood as foundational aspect of paediatric family- centred care. The aim of this chapter is to explore family-centred care through the lens of cultural diversity and illuminate the synergies, tensions and gaps that impact on care. This requires an understanding of the essential components of paediatric family-centred care as well as cultural competence. The chapter will conclude with a discussion of strategies to integrate cultural competence into paediatric family-centred care.

R. H. Srivastava (✉)
Centre for Addiction and Mental Health, Toronto, ON, Canada
e-mail: Ranisrivastava25@gmail.com

R. Zlotnik Shaul (ed.), *Paediatric Patient and Family-Centred Care: Ethical and Legal Issues,* International Library of Ethics, Law, and the New Medicine 57,
DOI 10.1007/978-1-4939-0323-8_4,

4.2 Patient and Family-Centred Care

At its core, paediatric family-centred care is an approach to care delivery that recognizes patients and families as partners in care. This approach shifts the paradigm of healthcare from being provider driven—where decisions are made by care experts, often unilaterally or with minimal patient and/or family input (Jolley and Shields 2009), to an approach that is collaborative and centered around the needs of the patient and family. In the old provider driven paradigm, systems and structures were largely crafted for the convenience of those who provide healthcare, whereas in family-centred care systems and structure enable and empower families to collaborate in care (Kuo et al. 2012).

In the general healthcare literature, patient and family-centred care is often referred to as one broad construct. This can be demonstrated through keyword searches of patient centred and family-centred care; regardless of which search term is used, the results are very similar, as illustrated by the following:

- "Patient and family centered is an approach to the planning, delivery, and evaluation of healthcare that is grounded in mutually beneficial partnerships among healthcare providers, patients, and families" (www.ipfcc.org/faq.html)
- "Patient centered care is about engaging the patient, the family, the care giver and the physician in the entire healthcare experience" (www.nrcpicker.com)
- "Patient centeredness refers to health care that establishes a partnership among practitioners, patients, and their families (when appropriate) to ensure that decisions respect patients' wants, needs, and preferences and that patients have the education and support they require to make decisions and participate in their own care" (IOM 2001, p. 7).

4.2.1 Paediatric Family-Centred Care

The notion of paediatric family-centred care emerged post World War II (Jolley and Shields 2009; MacKay 2009). Prior to this time, care in hospitals was provided from an expert model that often viewed parents as a "negative factor" in the care of hospitalized children and thereby relegated them to visitor or attendant status. Care decisions were made by care experts, patients and families were told what to do, and children with special needs were frequently institutionalized (Kuo et al. 2012). Limitations in visiting hours meant that, in some cases, children did not even see their parents during the entire admission (MacKay 2009). This approach to care was influenced by factors such as the industrial age where institutionalization thrived, a healthcare environment characterized by the "battle of science against infections diseases," and a lack of understanding of children's developmental, social, and psychological needs (Jolley and Shields 2009, p. 166). As a result, physical care needs, discipline, and asepsis were given greater attention than emotional needs. The extent to which the needs of parents were even identified is unclear and the discourse of cultural needs simply did not exist.

Post World War II, the social climate, particularly in countries impacted by the war, started to become increasingly interested in notions of psychological health. The phenomenon and impact of child-parent separation began to be studied and the negative effects of separation were highlighted by the work of leaders such as Spence, Robertson, Bowlby and others (see Jolly and Sheilds 2009 for a more thorough discussion). Based on evidence of the negative impact of separation and the advocacy efforts of parents, hospital policies were challenged and changed to include families in the care of children during hospital stays. The changes were slow and not without resistance. Over time, care models gradually evolved to reflect more parental participation and increasing partnership in care. Parents began to be viewed as participants in care who, with education and support from the nurse, could assume responsibility and control for aspects of care (Jolley and Shields 2009). System changes occurred through the evolution of policies favoring more liberal visiting hours, allowance for rooming in and sibling visits, and practices that brought hospital inpatient rounds back to the bedside, thereby allowing for increased opportunities for interaction and discussion between families and clinicians (Jolley and Shields 2009). These changes greatly facilitated family-centred care but are, in themselves, not sufficient. True collaboration requires commitment and valuing opportunities for family input and influence in care.

4.2.2 Domains and Principles of Paediatric Family-Centred Care

Paediatric family-centred care can be described as an approach to care delivery that is grounded in the values of respect, collaboration and shared decision-making (Institute for Family-Centered Care 2011). In a review of definitions and principles of family-centred care, Kuo et al. (2012) note that the term family-centred care is broad, non-specific and subject to wide interpretation with respect to implementation and measurement. As shown in Table 4.1, however, there is general consensus on core domains and key principles (Institute for Patient and Family-Centered Care 2011; Kuo et al. 2012).

From earlier discussion it is evident that family-centred care is essentially seen as an extension of patient-centred care, however there are conceptual differences that must be considered. Paediatric settings often prefer family-centred care as it is neither easy nor desirable to separate care of the child from the parents. For some, family-centred goes "beyond the patient clinician interaction by considering the needs of all family members, not just the child" (Kuo et al. 2012, p. 298). Other authors note "understanding and meeting children's needs in hospitals should always be the priority for both the patient's family and the healthcare providers" (Pettoello-Mantovani et al. 2009, p. 4) and indeed this is the basis for collaboration between family and healthcare providers. Paediatric family-centred care also reflects a view that as paediatric patients mature they should be encouraged to take a more active

Table 4.1 Domains and principles of paediatric family-centred care

Domain	Principles
Information sharing	The exchange of information is open, objective, unbiased, useful and affirming
Dignity and Respect	The working relationship is characterized by respect for diversity, cultural and linguistic traditions and values, care preferences, and builds on recognized strengths of the child and family
Participation	Patients and families are encouraged, empowered, and supported in participating in care and decision-making at the level they choose
Collaboration	Building on the principles noted above, collaboration requires trust and a willingness to negotiate in establishing the desired outcomes of care plans. Patients and families are encouraged to discover their own strengths, build confidence, and make choices about their health and well-being
Care in the context of family and community	Care decisions reflect the child within the context of family, home, and community. Family needs are recognized at all levels of the organization. There is flexibility in organizational policies, procedures, and services to adapt to cultural values and needs as needed

role in decision-making (Institute for Patient and Family-Centered Care 2011). Taken together these assertions reflect the following: (a) patient is at the center of care and patient needs drive care decisions, (b) patient needs are or should be understood the same way by healthcare providers and families, and (c) family needs are also important and family members may have distinct needs. On the surface these beliefs are commonly accepted principles, but are they universally applicable? Is there a bias in putting patient first and family second, especially when the relationship between the family and patient is such that the individual is considered a reflection of the family unit and not seen as being separate from the family? What happens when there is a lack of congruence or conflict between patient needs and family needs, or at least how best to meet patient needs? As will become evident later in this chapter, these questions become particularly important in care contexts characterized by cultural diversity where patient needs, priorities, and best interests may be viewed differently by families than their healthcare providers. The principles of paediatric family-centred care are laudable, but how easy or difficult is to apply them in an environment of cultural diversity? This question will be further highlighted and discussed later in the chapter.

4.3 Religion, Culture and Cultural Competence

There is little doubt that health and illness are inextricably linked to cultural issues. The impact of culture on health is significant and pervasive. Culture influences how illness is defined and experienced, what care and cure is sought, and who is consulted in the process (Srivastava 2007a). Although Canada has always been a land of immigrants, since the 1970's there has been increasing cultural diversification. Globalization and a shift in the source countries for immigrants means that whereas in the 1960's immigrants largely came from Europe, now the newcomers come more varied and different parts of the world, namely Asia, Middle East, Africa, and the Carribean. Current statistics indicate that nearly 20% of the Canadian population was born outside the country and this population is expected to reach 25% by 2013 (Statistics Canada 2008). These changing demographics mean that all healthcare providers must understand the influence and impact of culture on patients, families, and what constitues quality care. Increasing evidence of health inequities in selected populations and a recognition that lack of cultural competence leads to unsafe care has strengthened the need and urgency for appropriate care for all, including those who are relgious and culturally diverse. Integrating culture into care has become more than a nice thing to do, it is now considered a professional and quality imperative.

Culture is a difficult term to define. While sometimes defined narrowly in terms of race, ethnicity, religion, language, or country of origin, most scholars and practitioners view culture broadly as shared patterns of learned values, beliefs, and experiences of a group that provide a sense of identity and guide individuals, often unconsciously, in their thoughts, actions and decision-making (Srivastava 2008).

Cultural competence can be described as "a set of congruent behaviours, attitudes, and policies that come together in a system, agency, or among professionals and enables that system, agency, or those professionals to work effectively in cross-cultural situations" (Cross et al. 1989, p. iv). Culturally competent care can be described as "the ability to provide care with a client centred orientation, recognizing the significant impact of cultural values and beliefs as well as power and hierarchy often inherent in clinical interactions, particularly between clients from marginalized groups and healthcare organizations" (Registered Nurses Association of Ontario 2007, p. 70). Over the past decades there has been a marked increase in attention to the impact of culture and culural competence on care. The key messages from the literature on culture and cultural competence are highlighted in Table 4.2 (Srivastava 2008).

4.4 Religion

Although religion is situated within a particular cultural context and as such can be viewed as one dimension of an individual's or family's culture, in matters of health and illness, influence of religion can take on significant emphasis. Amongst healthcare professionals, religion is often a taboo topic. It can also be a political topic—religious freedom and rights are behind major conflicts in the world. Fowler and Reimer-Kirkham (2012) identify several reasons for recognizing and integrating religion into care. These reasons are highlighted in Table 4.3 (Fowler and Reimer-Kirkham 2012).

Table 4.2 Culture and cultural competence: key messages from the literature

Culture	Culture matters
	Culture of the healthcare provider and system matters as much as the culture of the client
	Culture is not a static concept and not a characteristic of certain individuals and families. Everyone has a culture
	Culture exists at the level of the individual, group, organization, and society
	Cultural issues are about patterns and about inclusion
Cultural competence	Cultural competence is a requisite skill for all healthcare providers
	Becoming culturally competent is a development process and an ongoing commitment
	Cultural competence requires awareness, knowledge, skills, and an ability to understand and navigate the dynamics of differences
	Cultural competence requires increased capacity within the individual healthcare providers as well as within the organization and system
	Many of the issues and strategies for cultural competence are also applicable to good patient- and family centred care; however they take on exaggerated importance when partnering with vulnerable families

Religions are imbued with moral authority and guide decision-making in simple and difficult situations. Such authority comes from four sources: religious traditions, sacred stories, reason, and religious experience (Fowler and Reimer-Kirkham 2012). The extent to which each source is used by a group or individual varies. Reasoning and decision-making in complex circumstance may be deductive or principle-based or it may be inductive and more contextual, recognizing the experiences of the moral community and its members as a valid source of knowledge and judgment. While some may rely more on experience and scripture, others draw more heavily on reason and/or tradition (Fowler and Reimer-Kirkham 2012). Conflict can arise when healthcare providers understand religious authority differently than the family and therefore may question the legitimacy of such authority.

4.5 Family-Centred Care and Cultural Competence: Synergies and Tensions

Although they have developed from different needs and frameworks, family-centred care and cultural competence are both approaches to care with many foundational similarities such as: (a) grounding in values of respect and collaboration and regard for patients and families as knowledgeable partners in care, (b) focus beyond the illness and disease to the experience and impact on patient and family, (c) call for a fundamental attitudinal change on the part of both clinicians and organizational systems and structures and (d) aim to improve healthcare quality through better health outcomes as well as increased patient and family satisfaction. Through principles of

Table 4.3 Reasons for recognizing religion in patient care

Many patients and families are religious
Religiosity is associated, generally positively, with health outcomes
Religious beliefs influence health decision-making
Religions offer positive coping strategies
Religious beliefs and practices may have health implications
Some patients and families want their care providers to support their religiosity
Professional standards and codes of ethics direct clinicians to identify and support client and family religious and spiritual needs.

respect and collaboration, family-centred care should reduce health disparities by changing the traditionally biased and potentially stereotypical approaches to clinical decision making (Slade 2007). Given these fundamental similarities one would expect to achieve the desired goal of high quality, holistic care for all through a focus on family-centred care without needing to look towards cultural competence? Although limited, evidence to date seems to indicate that while patient- and family-centred care approaches may have the potential for integrating the cultural context and addressing cultural needs, this potential is not being realized (Slade 2007; Guerrero et al. 2010; Montes and Halterman 2011). This gap is revealed in two key ways: (a) evidence of health disparities in family-centred care and (b) identified barriers to family-centred care.

4.5.1 Racial and Ethnic Disparities in Paediatric Family-Centred Care

The principles of paediatric family-centred care both explicitly and implicitly address the need for respecting cultural needs and providing individualized care tailored to the needs of the family. Thus it is logical to expect that family-centred care would be a significant step in reducing healthcare disparities (Slade 2007; Saha et al. 2008). There is, however, little empirical support that family-centred approach is sufficient. Using a large data set, Slade (2007) set out to examine the question "does patient-centred care affect racial disparities in health?" Her findings indicate that racial differences were not impacted by patient-centred care; however, literacy was a factor that needed to be considered when trying to improve health outcomes. In other words, although patient-centred care should account for needs associated with race, class, and literacy, this was not the case in the study, leading the author to conclude "…PCC [patient centred care] as a strategy for reducing health status disparities for blacks should be addressed cautiously, with skepticism…"(Slade 2007, p. 152).

Within the paediatric context, Croker et al. (2010) studied racial/ethnic and language disparities in family-centred care and its components for children with special health care needs. Parents were asked whether their doctor (1) spends enough time, (2) listens carefully, (3) provides care that is sensitive to the family's values and customs, (4) provides needed information, (5) helps the family feel like a partner in the child's care and (6) for parents who reported a non-English primary language spoken at home, whether interpreter services are received when needed. Parental response of usually or always to all 5 components of the family-centred scale and to the sixth component, if applicable, was categorized as having family-centred care. Findings indicate that significantly smaller proportions of Latino (47 %), African-American (53 %), multiracial (66 %), and other (58 %) children received family-centred care, compared with white children (72 %; $P < 0.001$). The other category included Asian, Native American, Alaskan Native and Pacific Islander families. These disparities persisted even after adjustment for factors of child health (condition severity and emotional, behavioral, and developmental needs), socioeconomic (poverty level, parental education, household composition, and residing in a metropolitan statistical area), and access (insurance type, usual source of care, and having a personal physician). The researchers also noted disparities for Latino and African-American children and children in households with a non-English primary language for the components of time spent with the provider and sensitivity to the family's values and customs. The authors conclude racial/ethnic and language disparities exist for children with special needs and initial efforts to address these disparities should focus on increasing provider time and cultural sensitivity (Coker et al. 2010).

Similarly, Montes and Halterman (2011) compared receipt of family centred care between parents of white and black children with autism spectrum disorder (ASD) and found that black parents were more likely to report not receiving family-centred care as measured by a four point likert scale in five dimensions: listening, time spent with the child, sensitive to values and customs, receiving needed specific information, and feeling like a partner in care. Responses of never or sometimes were coded as not receiving that component of family-centred care. Notably, with respect to time spent, sensitivity to values and customs, and feelings of partnership, black parents of a child with ASD were 2–5 times more likely than white parents not to have reported receiving family-centred care (Montes and Halterman 2011).

Guerrero et al. (2010) also studied the receipt of family-centred care among a general population of US children. Their findings too showed that families of Latino and black parents were less likely to report receiving family-centred care. Although, in this study, the difference in black and white families did not persist once adjustments were made for child characteristics that included health status, age, gender, and citizenship and socioeconomic factors, differences persisted amongst the Latino parents in areas of time spent and receiving adequate explanations.

It should be noted that the volume of research evidence in this area is sparse and largely US based. Nonetheless, findings are consistent and, in order to identify areas for further action and improvement, beg a deeper interrogation of processes within the family-centred care framework. It is particularly disconcerting that the gaps exist in the area of communication, an area considered to be essential for patient safety.

4.5.2 Barriers to Paediatric Family-Centred Care

A consistent theme in the literature on family-centred care is the difficulty in implementing paediatric family-centred care in ways that optimize its effectiveness. Carter (2008, p. 2092) notes that,

> despite the fact that family centered care has been welcomed with open arms by academics, practitioners, and educationalists alike, there is little clear evidence that it actually works… there is evidence that family centered care is problematic in some way: that difficulties need to be overcome; barriers need to be overthrown; relational and attitudinal issues that compromise the effectiveness of implementation need to be addressed.

Barriers exist within the individual practitioner (attitude and knowledge) and within the organization (expectations and resources). The most commonly identified barrier in the literature is that of understanding family centred care particularly in a context of cultural diversity and busy care environments (Shields et al. 2006; Pergert et al. 2007; MacKay 2009). Difficulties in caring for culturally diverse families include: not understanding the patient's or family's culture, fear of precipitating a negative event or outcome when lacking cultural knowledge, language barriers, non-traditional families including separation and divorce, contradictions and tensions between nurses and families, disagreeing with decisions made by patients and families, disagreements with families' religious and spiritual beliefs, and inability to fulfill patient or family wishes (Pergert et al. 2007; MacKay 2009). Challenges have also been identified with respect to communication and interactions including non-verbal communication, emotional expression, gender roles, truth-telling, racism and prejudice, and inequity in care as being areas of concern (Pergert et al. 2007). There is no doubt that cross-cultural interactions are complex and pose many challenges. Changing demographics of our society means that such complexities and difficulties will increase, not decrease. What does this mean to the future directions of family-centred care—is it, as Carter (2008) asks, an unachievable and un-implementable ideal? On the contrary, there is considerable evidence that paediatric family-centred care does result in improvements in care (American Academy of Paediatrics 2012), however the improvements may not be realized across all populations, particularly those that are racially or culturally diverse. It may be helpful to look at cultural competence literature as way to develop a deeper understanding of the barriers that exist in our current understanding and implementation of family-centred care.

4.6 Re-Examining Family-Centred Care

This section will review the principles of family centred care through the lens of cultural diversity, with the hope of uncovering unstated assumptions and identifying hidden truths. For ease of discussion, the domains of family-centred care identified in Table 3.1 will be used to organize this exploration. Before looking at the specific domains, some overarching issues must be noted. A key gap in family-centred care literature and discourse is lack of explicit acknowledgement of the dominant culture of the healthcare system and providers. Cultural diversity has been identified as a significant barrier to family-centred care. The question that is not asked loudly enough is whose culture is getting in the way? Is it only the culture of the family that needs to be considered? What about the culture of the healthcare provider(s), the organization, and the system? Literature on health disparities and cultural competence has clearly established that when it comes to the influence of culture on health outcomes, culture of the patient, provider and the healthcare system matter (Srivastava 2007a). In any healthcare interaction, enablers and barriers can exist in any or all of these components. Thus it is important to acknowledge and make visible the context in which family-centred care occurs—that is, values and assumptions embedded in the culture of the Western world, including Canada.

Although family-centred care challenges the values of the medical model, this is only one aspect of our healthcare culture. Another aspect to consider is that value of individualism as opposed to collectivism. The extent to which a culture values notions of individualism or collectivism varies across the world. Individualism and collectivism are often viewed as opposing points on a continuum and while individuals and societies can, and do, have both types of tendencies, one style tends to predominate. Individualistic societies regard the individual person as the central entity, and societies (families, work unit, community), are seen as a collection of individuals, not something that supersedes them. Individualism promotes the concept of equality for all and rights of the individual. By contrast, collectivist cultures regard the group as the primary unit (Debs-Ivall 2007). Recognizing that societies are made up of individuals, collectivism promotes the needs of the group and privileges that over the needs of the individual. The western world and countries such as Canada and the United States are seen as primarily individualistic societies, although groups within these countries, such as the Aboriginal peoples, may be more collective. Our healthcare culture also reflects the individualistic predomination with an emphasis on privacy, independence, and individual right to information and decision-making. Care decisions are based on best interest of the individual. Even when the decisions are made by others, they are to be based on expressed wishes or best interests of the individual not what the substitute decision maker believes is the right thing to do. Families that come from predominately collectivist cultures (for example Asia, South America, Africa), often struggle with this value and fail to comprehend why they may not have access to information, be the decision makers, how to separate the individual from the family and why its is not appropriate to make decisions that are deemed to be in the best interest of the family unit. Understanding and acknowledging these perspectives is critical to the establishment of trust and relationship based on respect.

Another fundamental difference in variability across cultures has to do with the how illness may be understood and responded to. In the western world, illness is generally regarded as a disruption—this may be a result of internal factors such as abnormal cell growth or physiologic damage (e.g., ischemia); or external factors such as infections or injury. In contrast, many see illness as disruption in balance between different chakras, meridians, or energy fields or an outcome of behavior, often moral behaviour. In the latter view, illness may be regarded as punishment of deeds and a child's illness may be deemed a result of parental or familial behaviour (Srivastava 2007a). Illness may also be a result of an intentional or unintentional curse or jinx that is placed upon the child, such as in the case of the evil-eye (Srivastava 2007a). Even in the Christian faith, variations may exist between the scientific view of illness and the religious views, ranging from no relationship to illness being viewed as sin by mankind[1]. Again, these are such fundamentally different concepts that tensions arising from them cannot be understood or addressed without explicit acknowledgement of contrasting views. The goal of such discussions would not be to reach agreement on the cause of the illness; rather the approach would be to focus on the mutually desired outcome. A framework that can be useful in such situations is that of L.E.A.R.N (listen to family viewpoint, explain your viewpoint, acknowledge differences, recommend options, and negotiate actions) (Srivastava 2007a).

4.6.1 *Dignity and Respect*

Values of dignity and respect are widely espoused by health providers but are often hard to translate into action. In order for the working relationship to be characterized by a respect for diversity, there first has to be a valuing of diversity that goes beyond superficial acknowledgement and persists even when the diverse view seems strange. This means health providers must be aware of their own ethnocentrism[2] and develop cultural humility, which requires a genuine understanding that our way of doing something is only one way, is likely to have limitations, and may, in fact, not be what is right for a particular patient and family.

Values of dignity and respect need to extent beyond the individual to their role in the family and care, and their wisdom, skills and expertise. Healthcare providers and family members need to trust each other's actions and motivations. Trust emerges from a sense of familiarity and cannot be readily assumed in cross-cultural interactions (Srivastava 2007a). Healthcare providers must recognize the need to earn and not assume trust. Family strengths need to be seen and understood through the eyes of the family, not just the providers'. It is not unusual for healthcare providers to see how particular family's worldview can be problematic without recognition of possible strengths associated with that worldview. All this goes beyond an attitudinal shift, to the development of skills that equip healthcare providers to recognize both their own biases and those that exist within the family. Recognition of such biases is the first step to managing the associated negative impact. Thomas Kuhn said, "[y]ou don't see something until you have the right metaphor to perceive it" (Gleick 1987, p. 262). Understanding culturally unfamiliar viewpoints requires

[1] For a discussion of explanations of illness associated with various religions see Taylor (2012).

[2] A belief that one's own cultural values, beliefs, and behaviours are the most superior, and best for everyone.

an understanding of the limitations of current metaphors or mental models and an ability to expand them through knowledge of other perspectives and worldviews. Utilizing tools and approaches that help elicit the patient's and the family's narrative can be very helpful in cross cultural interactions.

Another way to communicate respect is by making families feel welcomed with a non-judgmental attitude. Respect is felt when individuals feel their roles, actions, and views are understood and they do not feel judged or discriminated against on the basis of those views. Although organizational and social attitudes are shifting, discrimination and racism, albeit unintentional, continue to exist in our healthcare environments. Racial and ethnic minorities, Aboriginal families, and families from lower socio-economic groups are more likely to report experiences of discrimination and disrespect (Ngo-Metzger et al. 2006; Slade 2007; Saskatchewan Ministry of Health 2009). Without an explicit understanding of and attention to issues of provider and system bias, the principles of dignity and respect remain at risk of not being realized.

4.6.2 Information-Sharing

To be effective, information-sharing involves a reciprocal relationship. As providers, we have expertise to share and want to inform, but do we give sufficient thought as to what we are informed by? What do healthcare provides need to understand about the patient, family, and their culture in order to develop respectful relationships, foster collaboration, and identify what families see as useful and relevant information? Do we seek this information proactively, to include in our plan of care, or wait for it to emerge, generally in a conflict situation? What "family or cultural expertise" do we see as being integral to the care that needs to be provided? These questions have yet to be adequately explored in the literature.

Objective, unbiased, useful and affirming information-sharing is a laudable goal that cannot be achieved without a foundation of awareness of one's own biases, understanding of what is seen as useful and affirming by the family, and the ability to communicate effectively. Cultural diversity impacts on each of these pre-requisites. It is interesting to note that provider expertise is usually regarded as knowledge; however, family expertise is often referred to as beliefs, with less credibility and respect. Provider bias can be a significant challenge in cross-cultural communication, which is characterized by differences in communication styles (e.g., verbal and non-verbal), relevance and appropriateness of communication in different circumstances, as well as language (Srivastava 2007b).

Language is a significant barrier to-cross cultural communication that can greatly be reduced by interpreters. However, access to and making effective use of interpretation services remains an ongoing challenge. Although there is increasing recognition for the need and value of language support, in the Canadian heath care system there is an absence of legislated requirement for language support other than French and English in particular jurisdictions. Most often, it is the providers who determine whether or not language support is needed. Clinicians may be resistant to utilizing interpreters because of a lack of pervceived need for language support (i.e., believe we can get by), additional time required for interpretation, and a perceived loss of control since reliance on an interpreter means uncertainty as to whether the information exchange is accurate and complete (Srivastava 2007c; MacKay 2009). Underlying this reluctance seems to be a belief that because interpretation is less than perfect, it is not worthwhile and the effort required seems to overshadow potential benefits. If we were to look at this tension through the family lens, perhaps we would view interpretation as a way to strengthen patient and family voice, empowerment and inclusion, particularly for families who may already be feeling disenfranchised due to language barriers and unfamiliarity with the healthcare system. While not perfect, many fears and challenges associated with third party interpretation can be effectively addressed, provided that the healthcare provider and system are willing to make the investment of time, share power, give up some control, and develop cross-cultural communication skills.

Another area where the principles of information-sharing can become problematic is difference in values and beliefs with respect to who should have what information. This can take on particular challenge in situations where there is familial concern around disclosing a grave or poor prognosis. As noted earlier, the western healthcare system is based on values of individualism and thus information sharing and full disclosure are viewed as enabling and empowering behaviors to assist informed-decision making and allow patients to fully participate in care. Anything less than full disclosure is often seen as withholding information, deceitful, and unethical or unprofessional behaviour. Access to information is seen as basic right and anything that interferes with that can become a rights violation. In contrast, non-Western cultures may regard full disclosure as an imposition of truth that is disrespectful, disempowering, burdensome and even inhumane. There are many cultures who prefer a more "need to know" approach, are comfortable with less than full disclosure and fear that disclosure of poor prognosis may add additional burden to the patient and take away hope. Such value of less than full disclosure is common in many families with Aboriginal, Asian, Japanese, and African, and Italian (Srivastava 2007b) backgrounds and is often supported by religiosity. For example, religious views may state that only God decides when life ends, and that pronouncement of impending death, such as the case when families are told their loved one likely has XX days or months to live, are inappropriate. As well, simply speaking of death may hasten its arrival and thus should be avoided. Even when the health team has nothing to offer, interventions such as prayer and other religious rituals offer hope and comfort (Srivastava et al. 2012). It is therefore crucial that health providers take the time to fully understand the meaning and implications of limited or full disclosure and negotiate the communication approach with the family. Strategies for such negotiation include letting the family disclose the prognosis in their way, discussing with the family the benefits of such disclosure, and finally acknowledging one's own needs and expectations for professionalism through statements such as "if he patient asks me a direct question I will answer that truthfully". This challenge is not unique to the paediatric setting, although it is made more complex with the parent-child relationship, and even more so when a child is expected to gain understanding of their illness and have an increased role in decision-making as they mature. Even if a child is a pre-teen or teenager, a cultural view on the role of children, interdependence rather than independence, and authority and decision-making may limit or preclude the involvement of the child. Families may see it as their role to seek relevant information and make tough decisions for their child and not burden the child with this task, particularly when s/he is already experiencing distress and vulnerability due to illness (Debs-Ivall 2007; Pottinger et al. 2007; Seth 2010).

4.6.3 Participation

As is evident from the previous discussion, the domains of paediatric family-centred care are not mutually exclusive; rather, they influence each other. Respectful relationships are the foundation upon which to build the pillars of information-sharing, participation and decision-making. The degree to which familial participation in care is facilitated and welcomed is influenced by many factors, including how it is valued and understood by clinicians. Bruce et al. (2002) studied health professionals' perceptions and practices of family-centred care and found that not all principles of family-centred care were equally valued. In their study, clinicians highly valued the helping dimensions, such as providing emotional support and information, but identified parent-professional collaboration as the *least necessary* element of family-centred care. Other studies have identified that parental participation is more readily supported for routine tasks such as feeding, than for things that are traditionally in the healthcare provider's domain, such as giving medication, even when the parent normally gave those medications at home (Harrison 2010). Other factors that hinder participation include role expectations and lack of organizational supports such as adequate staffing, leading to time pressures for staff (Mackay 2009; Harrison 2010). Clinicians often regard themselves as care experts with legal and professional obligation to care for patients and thereby may feel the need to assess (and presumably evaluate) parental abilities before allowing participation (Harrison 2010). Such findings highlight the need for further attention to issues of power and control in the clinician–family relationship. It should be noted that these barriers exist across all types of families; it is not difficult to imagine, however, why they are more likely to surface negatively with families of children from culturally diverse, poor, or socially marginalized populations.

From the parents' perspective, participation in care is impacted by factors such as lack of adequate staffing, limited time for information-sharing and communication, the difficulty of forming rapport with the numerous care givers, and feelings of inadequacy related to their own abilities (Harrison 2010). In some instances, parental attempts to develop a partnership were either not recognized or misunderstood as questioning staff competence (Harrison 2010). Parental participation is also not without costs—financial costs due to time lost from work; additional expenses for travel, food, parking and care for siblings; and emotional costs in assuming the burden of care (Shields et al. 2006). Families who may be particularly challenged in participation include those least able to absorb the additional financial costs, those with limited emotional support, those who may be unfamiliar with the healthcare system, and those who may feel they are being judged or subject to discrimination. Families from culturally diverse communities and Aboriginal families experience increased vulnerability in all these areas.

4.6.4 Shared Decision-Making

Shared decision making through collaborative partnerships is an important and desirable goal for family-centred care. Challenges can arise, however, when the decision-making approaches of families are different those of the care providers. It is easy to involve others in decision-making as long as the decisions are what we think are right. When there are disagreements, it is the *family's* views that said to pose a problem. Shared decision-making requires a respect for the skills and expertise of all parties in the decision and a fundamental trust in each other's actions and motivations (Arango 2011). While clinicians rarely mistrust parental motivations, they often view them as stemming from misguided beliefs[3].

Family expertise with respect to knowledge of their context may be welcome when it is seen as augmenting the decisions recommended by the treatment team. However, when this expertise is in conflict, shared decision-making can be quickly abandoned and the situation reframed as the need to convince parents of the "right" decision. The "right" decision is one that focuses on the patient's best interest, almost to the exclusion of the interests of anyone else and family involvement can be problematic as it may be viewed as "muddling" the patient's decision making process (Ho 2008). This approach, embedded in many professional codes of ethics, reflects the Western values of individualism. For many families, the family is the smallest unit of identity and the value is placed on interdependence. Therefore, decisions are made in the best interest of the individual in the context of, not separate from, the family (Pottinger et al. 2007). In some families, decision-making is not based on ability (i.e., who has the most understanding or information), but rather on the role in the family. Variations also exist as to the kind of information and authority that is drawn upon for effective decision-making. While clinicians may privilege scientific evidence or their own experience, families may wish to draw on familial or religious authority. Ho (2008) argues that family and relational identity are in fact important to many patients' agency. Instead of labeling such disagreements as hostile and limiting, it is recommended that clinicians proactively engage with families to identify the decision-maker, explore family values and beliefs and listen to the family concerns and reasoning process (Pottinger et al. 2007; Ho 2008; Rodriguez-Osorio and Dominguez-Cherit 2008; Seth 2010). Such efforts can clarify expectations and misconceptions, and identify common goals upon which to anchor decisions.

---

[3] See previous discussion in truth telling as an example.

4.6.5 Collaboration

If respectful relationships can be viewed as the foundation for family-centred care principles, collaboration can be described as the roof, supported by the pillars of information-sharing, participation, and shared decision-making. Cross-cultural and religious issues that impact on the dynamics of these pillars ultimately impact the process and outcomes of collaboration. Thus far, the discussion has highlighted a need for healthcare providers' to challenge their own assumptions, navigate issues of power and hierarchy, and empower families to become equal partners in care. But do all families want to be equal partners in care? Some families may be very happy with provider-based decision-making, since they view the clinicians as the experts with the necessary knowledge and skill, and whose job it is to care for their child. Others may defer to provider decision-making because of a lack of confidence in their own abilities to communicate or navigate the healthcare system. Still others may wish to participate in decision-making but not in care, or vice versa. Variations may also exist as to when parents expect or want to collaborate.

Literature on parental views on collaboration is limited. In a review of family-centred care, Harrison (2010) describes collaboration as feeling prepared for discharge and knowing who to call after. A survey of parents of hospitalized children found that Hispanic families rated the importance of collaboration and support items lower than other racial groups (Harrison 2010). As well, parents who had been in the US less than 5 years ranked the importance of all three areas (respect, support, and collaboration) as lower (Harrison 2010). It is difficult to draw conclusions from such findings other than the need for further exploration of how culturally diverse families perceive and desire collaboration. These findings invite deeper reflection on the value of collaboration. Why is collaboration a good thing and does it lead to better health outcomes? If we believe that collaboration leads to better outcomes, then it behooves all clinicians to inform and enable families, without imposing upon those who may be reluctant to collaborate in the care of their child. Doing any less would be akin to settling for poorer health outcomes and thus a lower standard of care.

4.6.6 Care in the Context of Family and Community

To achieve the goal of paediatric family-centred care change and flexibility are needed at the individual as well as the organizational level, however, the latter lacks attention and articulation in the literature. Kuo et al. (2012) note that little attention is given to the community context or the system of care. Similarly, Bruce et al. (2002) note that the family-centred care element of healthcare delivery systems being flexible, accessible, and responsive to family needs was not highly valued by healthcare providers. This may be because paediatric family-centred care has primarily developed in the context of hospital care with a focus on immediate issues, thus providers are more focused on information-sharing and providing support. Other systemic barriers include things such as limited access to programs and services, poor educational support for clinicians, lack of conducive space, and a lack of recognition that paediatric family-centred care takes time on the part of the clinicians, thereby requiring adjustments in workload, staffing and reimbursement plans (Mackay 2009; Harrison 2010; Kuo et al. 2012).

Given the recognition that paediatric family-centred care is a fundamental paradigmatic shift in care, it is surprising that such little attention has been given to the changes necessary within organizations and health systems. It is worth reiterating that culture is not something that belongs only to patients and families—all individuals, including clinicians, are cultural beings; health organizations and systems are also cultural entities, and they in turn are located in the broader cultural context of society. This is where some parallels from the cultural competence field can be helpful. First, cultural competence recognizes that the "dynamics of difference" or issues of power imbalance, bias and discrimination, and exclusion exist at all levels: in family—team relationships, family—healthcare organization relationships, and family—society relationships (Fung et al. 2012). Systemic approaches that address issues at all these are necessary to achieve the kind of transformational change that is required to make family-centred care a reality. Issues of access and organizational supports such as language support strategies, support for teams, and space for families are all important issues that require an organizational strategy. Given that hospitals play only one part in the episode of care, more attention is needed for transitions and community support, so children and families receive the care that they need in the context in which they live.

4.7 From Awareness to Application: Bridging the Gaps

This chapter started out with the aim of exploring family-centred care through the lens of cultural diversity, in order to illuminate the synergies, tensions and gaps that impact care. Key points that have emerged in the discussion can be summarized as follows:

- Paediatric family-centred care and cultural competence are synergistic approaches to care. They both recognize a need to shift from provider-based care to a care environment that recognizes the strength of families and the critical role they play in health and well-being.
- The tensions and challenges that exist in implementing family-centred care are intensified when working with culturally diverse families.
- The primary gap that emerges when paediatric family-centred care is viewed through the lens of cultural diversity is a lack of recognition of the impact of "own" culture—that is, the culture of the providers and the culture of the healthcare system.
- Gaps also exist in the discussion of family-centred care from the systems perspective. For example, family-centred care recognizes the power issues with the

clinician-family relationship, but does not adequately address power, hierarchy and exclusion at the broader levels of the organization. There is limited literature on strategies needed at the organization level to support family-centred care.

So where to go from here? Given the synergies and the gaps noted above, one approach to integrating religion, culture and family-centred care is to further develop paediatric family-centred care through the approach of cultural competence. Cultural competence literature teaches us that to see change in practice, attitudinal change is not enough; rather, clinicians need knowledge and skill to elicit and understand family needs, and to negotiate the dynamics of difference that arise within a diverse environment (Srivastava 2008). Elsewhere, I have argued for an approach to care that takes the ABC's of cultural competence (affective, behavioral, and cognitive domains) and adds the elements of D (dynamics of difference) and E (recognition of the practice environment and a commitment to equity) (Srivastava 2008). It may be helpful to extend the ABC(DE) framework to ABCDE(F), with the F for family-centred care. In this way, the focus on family-centred care is further supported through cultural competence. The aim would be to emphasize development of inclusionary practices at the individual, team, and organizational levels.

Paediatric family-centred care literature highlights the importance of negotiation with respect to roles, decision-making, and care expectations (Shields et al. 2006). Negotiation is a critical skill in cross-cultural interaction, but is not the only mode of collaboration. Accommodation, validation, and reframing are also strategies that can be effectively utilized in cross cultural interactions.[4] Cultural care validation refers to actions and decisions that help patients and families acknowledge and retain values that are of most significance to them in a given situation. Healthcare providers are encouraged to explicate the values and practices that are important to the family and, where possible, work with these values and practices as a foundation for mutual goal-setting (Srivastava 2008). It is important to note that explicating and acknowledging a family's values does not constitute an endorsing of those values or a loss of alternate perspectives. Acknowledging and validating these important care values can have extremely positive effects with respect to the relationship and also signals to the families that cultural diversity is respected and not discriminated against.

Culture care accommodation refers to actions that healthcare providers can take to accommodate family needs and values. Negotiation is helpful when accommodation is not possible. It is important that healthcare providers avoid seeing the options as "either/or," and explore the "this and… " approach instead. In this way, confrontation and animosity can be avoided in favour of ongoing engagement and collaborative problem-solving. Cultural care reframing helps individuals to reorder, change, and modify their views and approaches to discover new possi-

[4] These strategies were first presented by Madeline Leininger (1978) as part of the Culture Care Theory of Universality and Diversity which outlines three modes of decision-making: Culture care preservation, culture care accommodation and negotiation, and culture care repatterning. The discussion here is an adaptation of Leininger's work as developed by Srivastava (2007a).

bilities and ways of achieving health goals. Reframing is about seeing something differently, trying out new behaviours. In this way things that were previously considered undesirable can be seen to have some value. Reframing should not be confused with imposition of viewpoints. Sharing alternative ways of understanding behaviour and discovering new patterns and meanings leads to increased options that can be chosen for meaningful health outcomes. For example, Western society is focused on the individual and the value of taking care of oneself. In many cultures, the comparable value is to care for others, and putting oneself first may even be regarded as selfish behaviour. Healthcare providers working in this context must acknowledge the value of taking care of others but offer the alternative explanation that being able to take care of oneself and being healthy is a prerequisite to being able to care for others. To draw on an airline analogy, flight safety instructions advise that in case of emergency when oxygen is required adults are told to first put on their own masks first and then assist children or others needing assistance (Srivastava 2007a). It should be noted that reframing applies equally to clinicians as it does to patients and families. Sometimes it is our understanding that needs to deepen or shift in order to fully understand and collaborate with family values and preferences.

4.8 Conclusion

Paediatric family-centred care is an innovative approach to care that can and does lead to improved health outcomes and quality care. However, gaps continue to exist in the realization of this potential, particularly where issues of cultural diversity is involved. Although family-centred care and cultural competence are very synergistic approaches to care, one cannot assume that individuals and organizations that are family-centred are, thereby, culturally competent (National Center for Cultural Competence 2007). In this chapter, I have argued that the paradigm shift of collaboration and partnership desired by paediatric family-centred care can be greatly facilitated through a more deliberate understanding and integration of cultural competence in the paediatric family-centred care agenda. This will not only begin to address the perception of "cultural diversity as barrier" to effective collaboration, but also allow for a deeper understanding of why clinicians and care environments continue to experience difficulties in understanding and supporting paediatric family-centred care. Cultural competence literature is clear that achieving excellent care for all requires a systemic approach. Frameworks for cultural competence highlight that transformation in care requires awareness and attitudinal shifts, but also that such shifts must be supported by new skills, tools and resources. Reframing our conceptual understanding of family-centred care and its associated competencies through the lens of cultural competence offers considerable promise. This integration will lead to a shift from the current state where religion and culture are regarded as foundational aspects of collaboration and care, not as barriers to care. Only then will the promise of paediatric family-centred care be fully realized.

References

American Academy of Paediatrics. 2012. Patient and family centered care and the paediatrician's role. *Paediatrics* 129:394–404. doi:10.1542/peds.2011-3084, ISSN: 1098-4275.

Arango, P. 2011. Family-centered care. *Academic Paediatrics* 11:97–99.

Bruce, B., N. Letourneau, J. Ritchie, et al. 2002. A multisite study of health professionals' perceptions and practices of family-centered care. *Journal of Family Nursing* 8:408–429. doi:10.1177/107484002237515.

Carter, B. 2008. Commentary on Shields L, Pratt J, Hunter J (2006) Family centred care: A review of qualitative studies. *Journal of Clinical Nursing* 5: 1317-1323, 17 (15): 2091–2093.

Coker, T. R., M. A. Rodriguez, and G. Flores. 2010. Family-centered care for us children with special health care needs: Who gets it and why? *Paediatrics* 125:1159. doi:10.1542/peds.2009-1994.

Cross, T., D. K. Bazron, and M. Issacs. 1989. Toward a culturally competent system of care. Washington, D.C.: Georgetown University Child Development Center. http://www.mhsoac.ca.gov/meetings/docs/Meetings/2010/June/CLCC_Tab_4_Towards_Culturally_Competent_System.pdf. Accessed 16 Oct 2012.

Debs-Ivall, S. 2007. Caring for diverse families. In *The healthcare professional's guide to clinical cultural competence,* ed. R. H. Srivatava, 144–171. Toronto: Mosby.

Fowler, M., and S. Reimer-Kirkham. 2012. Religious ethics, religious social ethics, and nursing. In *Religion, religious ethics, and nursing,* eds. M. Fowler, S. Reimer-Kirkham, R. Sawatzky, et al., 27–60. New York: Springer.

Fung, K., H. T. Lo, R. Srivastava, et al. 2012. Organizational cultural competence consultation to a mental health institution. *Transcultural Psychiatry* 49:165–184. doi:10.1177/1363461512439740.

Gleick, J. 1987. *Chaos: Making a new science*. London: Viking.

Guerrero, A. D., J. Chen, M. Inkelas, et al. 2010. Racial and ethnic disparities in paediatric experiences of family-centered care. *Medical Care* 48:388–393.

Harrison, T. M. 2010. Family centered paediatric nursing care: State of the science. *Journal of Paediatric Nursing* 25:335–343. doi:10.1016/j.pedn.2009.01.006.

Ho, A. 2008. Relational autonomy or undue pressure? Family's role in medical decision-making. *Scandinavian Journal of Caring Science* 22:128–135.

Institute for Patient and Family-Centered Care. 2011. Advancing the practice of patient- and family-centered care in primary care and other ambulatory settings: How to get started. http://www.ipfcc.org/pdf/GettingStarted-AmbulatoryCare.pdf. Accessed 12 Oct 2012.

Institute of Medicine (IOM). 2001. Envisioning a National Health Care Quality Report. National Academy of Sciences.

Jolley, J., and L. Sheilds. 2009. The evolution of family-centered care. *Journal of Paediatric Nursing* 24:164–170. doi:10.1016/j.pedn.

Kuo, D. J., A. J. Houtrow, P. Arango, et al. 2012. Family-centered care: Current applications and future directions in paediatric health care. *Maternal Child Health Journal* 16:297–305. doi:10.1007/s10995-011-0751-7.

Leininger, M. 1978. Transcultural nursing: Concepts, theories, and practices. New York: John Wiley and Sons.

MacKay, L. J. 2009. Exploring family-centered care among paediatric oncology nurses. Unpublished thesis University of Lethbridge. https://www.uleth.ca/dspace/bitstream/handle/10133/2483/mackay,%20lyndsay.pdf?sequence=1. Accessed 19 Sept 2012.

Montes, G., and J. S. Halterman. 2011. White-black disparities in family-centered care among children with autism in the united states: Evidence from the ns-cshcn 2005-2006. *Academic Paediatrics* 11:297–304.

National Center for Cultural Competence. 2007. A guide for advancing family-centered and culturally and linguistically competent care. http://nccc.georgetown.edu/documents/fcclcguide.pdf. Accessed 12 Oct 2012.

Ngo-Metzger, Q., J. Telfair, D. H. Sorkin, et al. 2006. Cultural competency and quality of care: Obtaining the patient's perspective. Commonwealth fund. http://www.commonwealthfund.org/Publications/Fund-Reports/2006/Oct/Cultural-Competency-and-Quality-of-Care-Obtaining-the-Patients-Perspective.aspx. Accessed 20 Sept 2012.

Pergert, P., S. Ekblad, K. Enskar, et al. 2007. Obstacles to transcultural caring relationships: Experiences of health care staff in paediatric oncology. *Journal of Paediatric Oncology Nursing* 24 (6): 314–328.

Pettoello-Mantovani, M., A. Campanozzi, L. Maiuri, et al. 2009. Family-oriented and family-centered care in paediatrics. *Italian Journal of Paediatrics* 35:12. doi:10.1186/1824-7288-35-12.

Pottinger, A., A. Perivolaris, and D. Howe. 2007. The end of life. In *The healthcare professional's guide to clinical cultural competence,* ed. R. H. Srivastava, 227–246. Toronto: Mosby.

Registered Nurses Association of Ontario (RNAO). 2007. *Embracing diversity: Developing cultural competence*. Toronto: Author. www.rnao.org/Page.asp?PageID=122&ContentID=1200.

Rodriguez-Osorio, C. A., and G. Dominguez-Cherit. 2008. Medical decision making: Paternalism versus patient-centered (autonomous) care. *Current Opinion in Critical Care* 14:708–713.

Saha, S., M. C. Beach, and L. A. Cooper. 2008. Patient centeredness, cultural competence, and health care quality. *Journal of the National Medical Association* 100:1275–1285.

Saskatchewan Ministry of Health. 2009. The need for more effective patient- and family-centred care. http://www.health.gov.sk.ca/adx/aspx/adxGetMedia.aspx?DocID=21252c45-e338-4811-9c16-9296ce7e2075&MediaID=3458&Filename=detailed-research-report-patient-first-web.pdf&l=English. Accessed 2 Sept 2012.

Seth, T. 2010. Communication to paediatric cancer patients and their families: A cultural perspective. *Indian Journal of Palliative Care* 16 (1): 26–29. doi:10.4103/0973-1075.63131.

Shields, L., J. Pratt, and J. Hunter. 2006. Family centred care: A review of qualitative studies. *Journal of Clinical Nursing* 15:1317–1323. doi:10.1111/j.1365-2702.2006.01433.

Slade, C. 2007. Does patient-centered care affect racial disparities in health? Public management and policy dissertations. Paper 24. http://digitalarchive.gsu.edu/pmap_diss. Accessed 2 Sept 2012.

Srivastava, R. 2007a. Culture care framework II: Culture knowledge, resources, and bridging the gap. In *The healthcare professional's guide to clinical cultural competence,* ed. R. H. Srivastava, 75–100. Toronto: Mosby.

Srivastava, R. 2007b. Cross cultural communication. In *The healthcare professional's guide to clinical cultural competence,* ed. R. Srivastava, 101–124. Toronto: Mosby.

Srivastava, R. 2007c. Working effectively with interpreters. In *The healthcare professional's guide to clinical cultural competence,* ed. R. Srivastava, 125–143. Toronto: Mosby.

Srivastava, R. 2008. The ABC(DE) of cultural competence in clinical care. *Ethnicities and inequities in health and social care* 1 (1): 27–33.

Srivastava, R., B. Srivastva, and R. Srivastava. 2012. Hinduism and nursing. In *Religion, religious ethics, and nursing,* eds. M. Fowler, M. Reimer-Kirkham, S. Sawatzky, et al., 173–196. New York: Springer.

Statistics Canada. 2008. Canada's ethnocultural mosaic: 2006 census. Ottawa: Ministry of Industry. Catalogue no. 97-562-X.

Taylor, E. J. ed. 2012. Religion a clinical guide for nurses. New York: Springer.

Chapter 5
Ethical and Legal Issues in Patient and Family-Centred Care: Canadian First Nations, Métis and Inuit People

Julija Kelečević

> *"Well I believe that the child, each and every child, is a gift from the Creator and it's your responsibility to take care of that gift. It's your responsibility not to mislead them. I believe in telling a child their birth story, that's what I do, because that sets them into the family....And at nighttime is usually when they really enjoy the story, and you don't change it, it's THEIR story. They'll correct you if you miss something. They know their story, that's how you know they are listening. You do those things for their security, so they know they are wanted and they belong to a family.*
>
> *Ojibwe Elder, Freda MacDonald*

5.1 Aboriginal People in Canada

Aboriginal people[1] are the original people in Canada. Based on linguistics, physical anthropology and archaeological evidence, it has been suggested that the first Aboriginal people migrated from Asia to the Americas some 13,000 years ago (Kirk and Szathmary 1985).[2] Although there are some commonalities among Canadian Aboriginal people, they do not belong to a uniform group. Each Aboriginal group has a distinctive history, including specific approaches to association and collaboration with groups from neighbouring territories. Distinct histories, both pre- and post-contact with the European settlers, resulted in great cultural diversity, with

[1] The term "Aboriginal people" is used throughout the text to collectively describe all original people in Canada and their descendants. It also includes the term "Indian" as defined in the *Indian Act,* 1985. The term "Indigenous" refers to any Indigenous people throughout the world.

[2] This is a scientific hypothesis that may be in conflict with Aboriginal epistemology about the origin of the Aboriginal people.

J. Kelečević (✉)
Alberta Health Services, Alberta, Canada
e-mail: julija.kelecevic@albertahealthservices.ca

R. Zlotnik Shaul (ed.), *Paediatric Patient and Family-Centred Care: Ethical and Legal Issues,* International Library of Ethics, Law, and the New Medicine 57,
DOI 10.1007/978-1-4939-0323-8_5, © Springer Science+Business Media Dordrecht 2014

over 50 Aboriginal groups recognized in Canada (Dickason and McNab 2008). Each of the groups is characterized by the use of its own language (often with a distinct dialect), presence of a specific system of kinship relationships, and continuation of traditional land base. According to Statistics Canada (2006) census data, the number of people who identify themselves as Aboriginal (First Nations People, Métis or Inuit)[3] is 1,172,790. This figure represents an absolute rise in the number of Aboriginal people, as well as a proportional increase of Aboriginal people with respect to Canada's entire population. According to Statistics Canada (2008), this figure still does not accurately describe the Aboriginal population. Some discrepancy between the number of self-identified Aboriginal Canadians on the 2006 census and the number of status Indians[4] still exists (MacIntosh 2011). There is evidence of incomplete enumeration in some Aboriginal reserves and settlements (Statistics Canada 2008), as well as among homeless Aboriginal people (Hanselmann 2001).[5] Further, it is also unclear how many Aboriginal people adopted outside of Aboriginal communities and not repatriated with their birth families are missed on census data.

Aboriginal people are less likely to live in urban areas; however, the 2006 Census data show that more Aboriginal people reside in large cities and smaller urban centres than ever before. There was an increase from 50 % in 1996 to 54 % in 2006 (Statistics Canada 2008). Evidence of improvement of living conditions for Aboriginal people is somewhat modest, with a slight decline in the number of people who reside in over-crowded households. The number who noted the need for major household repairs remained the same in 1996 and 2006 (Statistics Canada 2008). The Aboriginal population is younger than the non-Aboriginal population, with a median age of 27, as compared to 40 (Statistic Canada 2008). According to Statistic Canada (2008), there is a variation of this age variable among distinct Aboriginal groups, with the median age of 22 for the Inuit, 25 for the First Nations People, and 30 for the Métis. Although the number of Aboriginal seniors doubled from 1996 to 2006, seniors represent 5 % of the Aboriginal population, compared to 13 % for non-Aboriginal Canadians.

Demographic characteristics of Aboriginal children are different from those of non-Aboriginal children. Aboriginal children are a growing proportion of all Canadian children, particularly in urban areas (Statistics Canada 2008). As compared to their non-Aboriginal counterparts, Aboriginal children under 5 years of age are more likely to grow up in a larger family with younger parents, in a household led by a single parent, in multi-generational households, and in families with three or

[3] These data are collected under the category of "Ethnic origin" on the Canada Census 2006 questionnaire. The available options to highlight Aboriginal ancestry are North American Indian, Métis or Inuit according to Statistics Canada.

[4] The criteria for being registered as a status Indian through the federal Indian Register are fairly specific, according to the *Indian Act*. Some of self-identified Aboriginal people do not fulfil such criteria, including Métis. Other may lost the status or opt out from being registered.

[5] For more comprehensive discussion of whether a systemic under-counting of Aboriginal people is a result of continuous colonizing and discriminatory policies in Canada, see, for example, Hanselmann (2001) and Potvin (2005).

more children (Statistics Canada 2008; Statistics Canada 2009; The First Nations Information Governance Centre 2011). Also, more people are involved in the raising of Aboriginal children, including members of extended family and other community members (Statistics Canada 2009). Aboriginal children are also twice as likely to live in a grandparent household as compared to non-Aboriginal children (Statistics Canada 2009). Depending on their geographical location and ancestry, between 42 and 57 % of young Aboriginal children live in low-income households and in over-crowded dwellings (Statistics Canada 2009). Aboriginal parents reported a high degree of satisfaction with the community supports from family and friends to assist them in raising children (Statistics Canada 2009). The survey results also showed a lack of specific community activities that promote traditional values and provide cultural activities for young Aboriginal children. Except for Inuit children, small proportions of Aboriginal children were in care arrangements that promote traditional values and customs (Statistics Canada 2009). According to the Aboriginal Children's Survey (Statistic Canada 2009), Aboriginal mothers are on average younger than non-Aboriginal mothers. Further, young Aboriginal mothers are more likely to be heads of single-parent households as compared to mothers from the general population. Approximately one in four Aboriginal children has a mother who is between the ages of 15 and 24 (Statistics Canada 2009). The presented demographic differences between Aboriginal and non-Aboriginal adults and children will be taken into account when discussing any healthcare related topic, including ethical and legal issues in patient and family-centred care specific to Canadian Aboriginal people.

5.2 Aboriginal-Non-Aboriginal Relationships

"No Canadian acquainted with the policies of domination and assimilation wonders why Aboriginal people distrust the good intentions of non-Aboriginal people and their governments today." (Royal Commission on Aboriginal Peoples 1996, p. 1). The relationship between Aboriginal people and the descendants of European migrants to Canada is a result of several centuries of coexistence between different cultures with distinct underlying value systems. The pre-contact Aboriginal people in Canada were hunters, fishers and gathers, who lived in small and highly mobile social groups (Waldram et al. 2007). The communities relied on oral tradition to share and preserve knowledge about the territories they lived on, including the seasonality of food resources, healing practices and spiritual ceremonies. The contact with European explorers was far from beneficial to Aboriginal people. Although initial economic benefits resulting from trade was recorded on both sides, long-lasting negative consequences on Aboriginal identity, cultural continuity and health status remain (Waldram et al. 2007). The construction of Aboriginal self-identity contrasts considerably to European political concepts of state, nation and ownership. The Aboriginal people did not subscribe to European legal models that rule relationships among people and between peoples and the land they inhabit.

The British and, later, Canadian governments signed land "treaties" with different Aboriginal groups, which directly impacted Aboriginal self-identity (McIntosh 2011). Legislated initiatives were accompanied by more subtle cultural and social pressures. The most notable pressures of legislated initiatives include the creation of residential schools, suppression of Aboriginal languages, adoption of Aboriginal children into non-Aboriginal families, relocation of Aboriginal groups and encouraged abandonment of traditional Aboriginal ways of life in favour of economies based on domestication and industrialization.

It is almost incredible that the last residential school closed in Saskatchewan in 1996 (Health Canada 2013). Miller (1996) and Grant (1996) provide a comprehensive history of the residential school system in Canada, examining its impact on both Aboriginal people and the rest of Canadian society. The goal of residential schools (run either by state or church) was simple: to assimilate and acculturate future generations of Aboriginal children; "'civilize' and 'Christianize'" (Truth and Reconciliation Commission of Canada 2012, p. 26). Aboriginal children were taken from their families, often forcibly, and were relocated to schools distant from their communities. The geographical remoteness resulted in cultural isolation for Aboriginal children. The children were forbidden to communicate in their mother tongues and were discourage to engage in any spiritual practice. When a young girl saw the Shingwauk school in Sault Ste. Marie, Ontario for the first time, she observed, "Nothing could ever go wrong in such beautiful surroundings." (Willis 1973, p. 136). However, the conditions in residential schools were harsh. The children lived in an unfamiliar environment without family support, were forced to learn a new language, and often went hungry and without adequate healthcare (Truth and Reconciliation Commission of Canada 2012). The deplorable conditions in residential schools were made worse by the physical, emotional and sexual abuse of Aboriginal children, creating spaces "where a lot of children's prayers did not get answered" (Knockwood 2001, p. 27). The impact of residential schools was both immediate and long-lasting. It truly represents an assault on Aboriginal culture and values, specifically targeting Aboriginal children. It led to the disintegration of Aboriginal families and loss of parenting skills. The experience of residential schools affected educational and employment histories, the use of social and healthcare services, interactions with the Canadian legal system, and the healthcare status of Aboriginal people (Truth and Reconciliation Commission of Canada 2012).

June 11, 2008 marks an important step forward towards reconciliation, as it was when Prime Minister Harper offered a full apology on behalf of all Canadians for the residential school system. However, many other assimilation and acculturation initiatives still need to be fully understood and addressed. In 1965, the Federal-Provincial Child Welfare Agreement granted the Province of Ontario the authority and power to oversee welfare services under the *Ontario Child Welfare Act* for status Indians on Reserve. It is unclear how many Aboriginal children were taken away from their parents and their communities from the 1960s to 1980s and fostered or adopted out to non-Aboriginal families. Aboriginal children were removed by Children's Aid Societies (CAS) based on criteria of "the best interests of a child," "parental inadequacies," and "community disorganization," although no measurable standards were described. It does seem that there was lack of congruence between the norms of raising children in Euro-Canadian households and the Aboriginal family values of permissiveness, sharing and discipline. Although the design of such government policies might be guided by the best interests of children in mind, the implementation of the policies perpetuated continuous systemic assimilation of Aboriginal children. The term "Sixties Scoop" was coined by Johnston (1983) to describe such practices. He examined the causes behind disproportionate numbers of Aboriginal children being in care of child welfare agencies. He also outlined some of the potential solutions for fixing this problem, including the development of Aboriginal-controlled and -operated child welfare services. Although some of Johnston's data are outdated, the conclusions of his report are similar to the findings of Trocmé et al. (2004), who examined the existing overrepresentation of Aboriginal children in the Canadian child welfare system, in which Aboriginal children are twice as likely as non-Aboriginal children to be placed in care. The inter-generational oral tradition of knowledge-sharing has been interrupted by the systemic relocation of Aboriginal children out of their birth communities. There is terrific cultural loss related to lifestyle, healing practices and spiritual ceremonies within Aboriginal communities.

5.2.1 State-Aboriginal People Legal Relationships

Mcintosh (2011) summarizes the legal foundations of the state-Aboriginal relationships as applicable to healthcare coverage and healthcare access. The responsibility of how healthcare is delivered to Aboriginal people, and by whom, differs from that of the general population in Canada. According to the *Constitution Act,* 1867 provinces have jurisdiction over matters related to health, except in matters related to Indians and Indian lands,[6] which falls under federal jurisdiction. Section 91(24) of the *Constitution Act,* 1867 has not been directly litigated, and McIntosh (2011) questions whether federal authorities could relinquish legal responsibilities related to Aboriginal health to provincial or territorial jurisdictions. The relationships between the state and Aboriginal people's access to healthcare was further regulated (and complicated) by the creation of the *Indian Act*, in 1951. The federal government has created criteria for who can be registered in this category. The criteria has been modified several times, and the most controversial question is whether this represents an attempt to limit federal responsibility towards people who are registered, including the state's duty to oversee health-related matters for Aboriginal people. The two mentioned Acts do not represent all the legislation that guide these relationships. It remains difficult to isolate direct impact they had on healthcare delivery and, consequently, the healthcare status of Aboriginal people.

[6] The terms "Indian" and "Indian lands" are taken directly from Section 91(24) of the *Constitution Act,* 1867.

5.3 Health Status of Aboriginal Children

Aboriginal children on average experience worse health than their non-Aboriginal peers. Unfortunately, this is a universal finding for Indigenous children throughout the world (Smylie and Adomako 2009). The *Indigenous Children's Health Report* specifies data for the health status of Aboriginal children in Canada (McShane et al. 2009). Key disparities in very early age identified in the report include higher infant mortality (twice the rate in First Nations People and four times higher among the Inuit population compared to the general Canadian population), higher rates for Sudden Infant Death Syndrome (varying from 3 to 12 times higher in Aboriginal populations compared to the rest of Canadians), and overall higher incidences of premature births compared to the general Canadian population(McShane et al. 2009). For some 30 % of Aboriginal children, food security contributes to health issues, especially in remote or Northern communities where nutritious food is both rare and expensive. Approximately 36 % of Aboriginal children are obese, compared to 8 % of non-Aboriginal children(McShane et al. 2009). The rate of Aboriginal children between ages of 6 and 14 years who cannot participate in physical activities due to debilitating health conditions is double than that of their non-Aboriginal peers(McShane et al. 2009). First Nations and Inuit children living on reserves are disproportionally affected by respiratory infections compared to the general Canadian population, with double the rates (The First Nations Information Governance Centre 2011). Also, there are twice as many Aboriginal children living on reserves affected by infectious Hepatitis A compared to the general population (Jin and Martin 2003). Although the challenges related to the mental health of Aboriginal children are evident, some positive findings were described in recent reports. The percentage of youth who have thought about suicide or attempted suicide declined from 21.1 to 16.5 % between 2002 and 2008 (The First Nations Information Governance Centre 2011). A similar trend has been reported for suicide attempts, where the percentage decreased from 9.6 to 5.9 %. Approximately, six out of ten First Nations youth age 15–17 reported that they have never used non-prescription cannabis, while one of ten respondents reported daily or almost daily use. About two-fifths of youth responded that they consumed an alcoholic beverage in the year prior to the survey. Approximately 20 % of First Nations youth smoke daily.

In self-reported surveys, it has been recorded that the majority of children are in "excellent," "very good" or "good" health (Statistics Canada 2009). Similarly, youth described their health to be "very good" (34.7 %) or "excellent" (30.1 %) (The First Nations Information Governance Centre 2011). The discrepancy in the self-reported evaluation of health status and epidemiological data may be the result of how Aboriginal people understand health and illness. If they follow traditional Aboriginal models of wellness, the presence of physiological disorder may not be interpreted as an illness and therefore may not be self-reported (Adelson 2007). The statistical data show that Aboriginal children in Canada are under a disproportionate burden of different diseases compared to their non-Aboriginal peers. The reasons for such differences are multifaceted and are deeply rooted in the historical relationship between Aboriginal people and the state, as well as Aboriginal and non-Aboriginal people in general. The health disparities of Aboriginal children are associated with historical, environmental, political and socio-economic factors, and "those who are the poorest and the most disempowered are the sickest and the least likely to be able to change or remove themselves from their immediate circumstances" (Adelson 2005, p. S58).

5.4 Bioethics and Aboriginal Values

Theoretical and empirical literature on healthcare ethics focusing on Aboriginal people in Canada is scarce. More strides have been achieved in the field of research ethics. In light of their cultural histories, Aboriginal communities aspire to "generate and disseminate knowledge for and about themselves" (Brant Castellano 2004, p. 113). To facilitate the preferred approach of Aboriginal people to research, *CIHR Guidelines for Health Research Involving Aboriginal People* (CIHR 2008) was created. The document is an example of thoughtful and meaningful collaboration of stakeholders from both Aboriginal and research communities. The principles described in the *CIHR Guidelines* were later incorporated into the 2nd Edition of the *Tri-Council Policy Statement: Ethical Conduct for Research Involving Humans* (2010) to ensure the Canadian research community's commitment to conduct culturally safe and mutually beneficial research with Aboriginal people.

Both Ellerby et al. (2000) and Kotalik (2009) try to unpack the relationship between the traditional Aboriginal values-based and the western principles- and rights-based approaches in healthcare decision-making. Kotalik (2009, p. 124) stated that "the Aboriginal population as a minority group shares many basic values with the majority of Canadians steeped in European moral tradition." He compared traditional values of Anishnabe people with moral virtues describe in bioethics literature. He discovered "a remarkable degree of congruence, suggesting that the character traits of a good person, traits that an Aboriginal patient would appreciate to find in an Aboriginal traditional healer, are very much those that the European philosophical tradition considers as important to cultivate in physicians and other health professionals" (Kotalik 2009, p. 124). Brant (1990) described similar findings. She discussed behavioural norms that promote group unity among Aboriginal people. These norms included emotional restraint, noncompetitiveness, non-interference and sharing. For example, the concept of non-interference behaviours relates to western concept of respect, "discouraging coercion of any kind, be it physical, verbal, or psychological" (Brant 1990, p. 537). However, Ellerby et al. (1998, p. 846) added that "in their approach to ethical decision-making, Aboriginal cultures differ from religious and cultural groups that draw on Scripture and textual foundations for their ethical beliefs and practices." Aboriginal people make health-related decisions not based only on individual values, but within the context of their families and communities (Garvey et al. 2004). They also use narratives to assist them in decision-making related to healthcare. These narratives are not uniform and cannot be reduced to a single Aboriginal story (Garvey et al. 2004).

5.4.1 K'aila's Story

Parents from different cultural, religious and spiritual backgrounds sometimes encounter situations in which the suggested care plan for their children challenges their beliefs and values. The case of K'aila is one example where parents' values clashed with healthcare recommendations about what kind of treatment would be in their son's best interest (Paulette 1993). K'aila was a baby boy born in Alberta to an Aboriginal family. He was diagnosed with abnormal liver function at 3 months of age. The only potential treatment for K'aila was a liver transplant, as recommended by the attending paediatrician. The parents considered the proposed care plan within the context of their spiritual beliefs and decided to refuse it. They considered this to be the best decision for their son, as accepting the liver transplant would "bring with it the spirit of the other person" (Coward and Hartrick 2000, p. 262). The paediatrician believed that K'aila's parents made an inappropriate decision, limiting K'aila's future autonomous choices. The paediatrician reported the parents to the Alberta Social Department. Facing a difficult situation and potential legal proceedings, the parents relocated to Saskatchewan. There, another application to the Department of Social Services in Saskatchewan was made, resulting in a court application for K'aila to be taken into custody and undergo liver transplantation. The court upheld the parents' decision rejecting the application; however, the court's opinion was based on the uncertainty of the long-term success of the offered treatment, rather than the parents' spiritual beliefs that the treatment could harm K'aila. K'iala returned to Alberta with his parents, where he died at the age of 11 months. When "he was taken gently from [the family], [the family] held him in [their] arms and talked and sang to him" (Paulette 1993, p. 17). Even if the paediatrician had belonged to the same spiritual tradition as K'aila's parents, there is no guarantee that everyone would have agreed what the best course of action would have been in this difficult situation. K'aila's case is often used in education not only to illustrate how his parents reached a decision informed by their spiritual beliefs, but also "how culture and ethical practice is intimately connected with the structure of social relations that produce forms of oppression and dependency" (Coward and Hartrick 2000, p. 264).

5.4.2 Clinical Ethicist's Story[7]

Sharing stories is central to Aboriginal traditions. It seems fitting to share my experience working as a clinical ethicist, describing a consultation involving a pae-

[7] The identifying information has been modified to maintain the confidentiality of the patient, his family, and his community. Prior to Adam's discharge, I received their permission to share the story.

diatric Aboriginal patient. Several years ago, I received a call from a mental health in-patient unit with respect to an adolescent. The charge nurse asked me to meet Adam, the 14-year old patient, and his family. The nurse provided only some basic information about Adam's diagnosis, without clearly articulating what was the ethics concern. I went to the unit and reviewed Adam's chart.

Three days prior to the consultation request, Adam was found wondering through the streets in the early afternoon, talking to himself and occasionally yelling. After Adam broke a window, somebody called the police. When police and ambulance arrived, they observed that Adam was disheveled and his hand was bleeding. When paramedics tried to examine him, Adam became further agitated, and was brought to the hospital. For first six hours, he was not able to reply to any questions, and nobody knew even his name. That evening Adam's parents reported him missing and finally somebody made a connection that the young man admitted earlier was Adam. He was treated for the cut on his hand and was examined by an adolescent psychiatrist, who admitted him under diagnosis of first psychotic episode. In progress notes, a nurse wrote that Adam's parents expressed that they did not believe that Adam suffered from mental illness. Notes from a family meeting also contained similar statement, including the parents' request for decreasing the dose of the newly prescribed antipsychotic medications with which he was being treated. This triggered a request for a clinical ethics consultation. When I met Adam's parents, they disclosed that they belong to an Inuit cultural group. This was only my second time interacting with an Aboriginal family within the context of a clinical ethics consultation, and my first time that the patient was a minor. Though I knew, in theory, that people's understandings of mental health and mental illness are culturally determined, I knew nothing about how this particular family perceived mental illness and whether their cultural affiliation impacted their opinion. I had also just a basic knowledge about the history of Aboriginal people in Canada, having moved here as an adult, completing most of my education abroad. I met Adam's family in a case conference room. They were accompanied by an elderly woman. I introduced myself, and explained my role and the purpose of clinical ethics consultations. Adam's parents turned towards the elderly woman and translated what I just shared. They spoke a language I had never heard before. It seemed to me that they spoke much longer in their mother tongue than I did in English. They all took turns speaking. The periods of talking were interrupted by periods of silence. Every time I observed silence, I expected that the conversation would continue in English. At one point, Adam's mother explained that the elderly woman was Adam's grand-aunt. "Are you in hurry? If you have other appointments today, we can meet tomorrow when you have time," the mother was worried. "We don't want to cause you any problems, for instance if you have to meet other patients and families who need you," the father said. I had nothing else scheduled and assured them that I would be able to spend as much time with them as they needed. Next I asked how much the parents knew about Adam's illness, the suggested medications, and in particular any risks associated with the therapy. I also asked whether they knew what potential consequences Adam would suffer if they decided against antipsychotic medications. The mother translated my questions to the grand-aunt, and she looked a bit confused.

The grand-aunt talked to the parents for a while and the father translated back to English that she did not understand what "mental illness" meant. I repeated what I had read in Adam's chart, including his diagnosis. I also asked if Adam was able to communicate in English. "Of course, he goes to high school here. That's the reason we moved to the city. Our community had no high school and we had cousins here," Adam's mother replied. The father continued to translate my questions to the grand-aunt. The father described the grand-aunt's role: "She is what you would call a specialist; a specialist in our community. She knows what is good for our community and the people who live there. She will know what is best for Adam and for us as a family. She is a healer." He also shared that their native language had no equivalent to term "mental illness." Now, it was my turn to be confused. How could we continue the discussion if the central concept could not be translated? I was in need of a specialist, someone who had lived experience similar to Adam and his family.

This was first of many encounters I had with Adam and his family. I met with an Aboriginal facilitator, who introduced me to some basic concepts of the family's cultural background. The facilitator explained to me that, just as any other cultural group, Aboriginal people varied in how strictly they adhered to traditional values. "No different than Chinese, Kurds or Italians," the facilitator said, "but keep in mind that Adam's family may not trust you easily. Ask them if they are worried that you will take Adam away and put him in foster care. Let them tell you about his childhood, the place he was born in, how he likes the city." I wanted the Aboriginal facilitator, my specialist, at the next meeting with Adam and his family. "Meet with them one more time alone and see how it goes," the facilitator suggested. The following week, I spent several days with Adam, his parents and his grand-aunt. Adam felt a bit better. "The pills seem to clear some of fog in my head. I am still confused, I am not sure why I am still here, but I think I am getting better," Adam shared. The grand-aunt suggested that we take a stroll, "to feel the ground under our feet," the mother translated. It was a sunny day and Adam could leave the unit accompanied, so we went for a walk. I told them that I could not assure them that we would have a private conversation while walking, as other people may be around. We walked for an hour and talked, Adam's parents translating back and forth. We didn't talk about Adam's illness that time. They talked about their small community, told me a story about their neighbours' new baby, and described what kind of traditional meals grand-aunt made. They shared how everything in life is interconnected, that we all were part of a whole. "Is relatedness one of core values for this family?" I was wondering quietly. Aloud, I shared, "I have never had a family meeting outdoors; it's simply not done."

I continued meeting with Adam and his family during his hospital stay. I learnt how they believe that mental health is the result of the balance of body, mind, emotion and spirit. Later I read a similar statement in Mitchell's paper (2005) that, within the framework of the Medicine Wheel, mental illness was not believed to be some identifiable pathological disorder, but rather the manifestation of an existing imbalance of four aspects of health. In Aboriginal models of wellness, healthy individuals contain balanced physical, emotional, mental and spiritual characteristics. Also, the balance extends to, is related to, and is influenced by, persons' family members, members of their community, and the land they inhabit (Vukic et al. 2011).

The grand-aunt spoke once about healing techniques practiced in their community. She spoke about spiritual ceremonies, as well as use of traditional herbs. She explained that her role is to communicate with the spirit world in order to assist Adam's recovery. Kirmayer et al. (2009) described some of the obstacles in fully integrating different approaches. I wrote a consultation report, that I shared with Adam and his parents. The recommendation was to create a comprehensive care plan that will include both western therapeutic modalities and Aboriginal traditional healing practices. The recommendation was similar to what Brant (1996) described as an integrated approach to treating mental illness among Aboriginal people. For me, the application of the patient and family-centred (PFCC) model did not mean that health care professionals ought to simply follow the wishes of Adam and his family. The team had a primary obligation to provide the best possible care for Adam based the existing standards of care. The health care team did not replace their professional values with the cultural values of Adam's family; rather, they collaborated with Adam and his family to create mutually acceptable and safe discharge plan. The parents asked to meet one last time, so they could ask questions about the report. I expected to meet with Adam, his mother, father, and perhaps grand-aunt. When I entered the small conference room 3 h later, I was surprised to see eight more people sitting around the table. It was crowded; the space was not designed for large gatherings. I knew Adam's nurse and social worker, and I recognized some of the new people, having seen some of them around the hospital visiting Adam. After two and half hours spent with them, I realized that the meeting was not about who had authority over Adam's illness and what would be the best option for Adam's care after he was discharged. The meeting was a sharing circle, as required in their community when a member goes through significant life change. Adam was to be discharged from the hospital, his life would continue to transform, and this was an opportunity for everyone involved in his care to meet one last time. It was a moment filled with a sense of mutual respect.

5.5 How Does Patient and Family-Centred Care Fit with Aboriginal Traditions?

An implementation of PFCC in the paediatric setting may assist in answering this question. It is generally accepted that the implementation of PFCC should lead to better quality of care and consequently better outcomes for the paediatric patient. The body of literature about PFCC continues to grow. However, only a small segment of literature speaks about Aboriginal engagement in PFCC, and impact that PFCC application may have on changes in Aboriginal children's health status and existing health disparities.

To examine whether or how the PFCC model could be applied to Aboriginal paediatric patients and families, it is necessary to acknowledge past troubling and current slow-healing relationships between Aboriginal and non-Aboriginal Canadians.

Some of the challenges of implementing PFCC are a result of continuous difficulties that Aboriginal people face when accessing healthcare.

Some of the difficulties in the paediatric setting commence even before children are born. Couchie and Sanderson (2007) reviewed current policies, in which relocation of all pregnant women between 36 and 37 gestational weeks from remote and northern communities to larger centres was recommended. Although the recommendations were evidence-based, they were in contrast to communal values and traditions, especially among Inuit groups. Unsafe and impractical deliveries in remote and northern communities were raised as the main reasons for creation of policies. Couchie and Sanderson (2007) questioned whether such guidelines interfere with autonomous decision-making of Aboriginal women regarding their reproductive rights. The practice of relocated delivery has been recognized as harmful not only to mothers and newborns, but to their families and communities. As a result of the engagement of Aboriginal stakeholders, new guidelines that foster more clear communication between Aboriginal communities and healthcare professionals replaced outdated policies (The Society of Obstetricians and Gynaecologists of Canada 2010). The guidelines promote the return of birth to rural and remote Aboriginal communities, promoting self-determination among pregnant women.

5.5.1 Parents' Choice

The differences between parents' and healthcare providers' values in the paediatric setting will affect all existing relationships, including child-parent relationships. Parents who make different decisions from those suggested by a healthcare team could be deemed as "bad" parents, as described by Paulette (1993). Healthcare providers have a legal avenue to address any disagreements about parents' choices by reporting parents to child welfare agencies. As previously discussed, it could be understood that distrust among Aboriginal parents towards child protection agencies may exist. Their multi-generational negative experiences are Aboriginal parents' realities in which they both have to exercise their values, beliefs and traditions, and provide a voice for their children. Parents following traditional Aboriginal teachings may choose to include community members with specific expertise on the unique way of living in their community. The parents may even defer decisions about their child's care to others. The PFCC model of care can facilitate a variety of parents' choices, respecting that parents maintain self-governance by either making decisions of deferring decision to whomever they consider to be an expert in their child's well-being (Fiester 2011).

5.5.2 Stories of Success

Sioux Lookout MenoYa Win Health Centre developed a model of cultural safety to address tensions between their Aboriginal patients and healthcare providers

(Walker et al. 2009). They described a model of healthcare based on the Anishnabe philosophy that includes the use of traditional natural products and foods, and life processes. The goal was to come up with an organizational framework that is comfortable for both Aboriginal and non-Aboriginal stakeholders, that will result in *Menoyawin*(health and wellness) of the patients through practices of *Odabiidamageg* (governance and leadership, including representatives from the Elders Council), *Wiichi'iwewin* (patient, resident and client supports, including linguistic and cultural interpretations), *Andaw'iwewin* (traditional healing practices, incorporating smudging and sweet lodge ceremonies, traditional birth practices, etc.), *Mashkiki*(use of traditional medicines while monitoring potential interactions with mainstream medications), and *Miichim* (traditional eating practices).

In Shields et al.'s (2006) description of PFCC model in a paediatric setting, the relationships in healthcare extend beyond the individual patient and healthcare provider to include all family members engaged in the child's care. Each family member is considered to be an individual recipient of care and the care plan is tailored towards the family, rather than just the individual child. Hospitalized Aboriginal children are often accompanied by members of extended family or their birth community. The engagement of a larger number of people in a child's care is not commonly addressed in hospital policies, including visiting policies. In addition, enforcing a legally driven concept of privacy of health information may be challenging for healthcare providers when it is not easily discernible who should have such information.

The PFCC models may address some of the concerns of everyone involved in a child's care. Although the definitions of PFCC vary, the underpinnings of the model include active partnership, mutual respect and exchange of knowledge. Applying the PFCC models have led to observable changes in paediatric hospitals, including easier navigation of healthcare and the development of respectful and trusting relationships among all stakeholders (Howitt 2011).

Also, the implementation of PFCC has resulted in the development of hospital spaces that are more conducive to parents' and other family members' involvement in a child's care. An example of is the Hospital for Sick Children, where children's experience of architectural space have been examined (Adams et al. 2011). The researchers concluded that "hospital design [was]… driven by effective cultural rather than medical models" (Adams et al. 2011, p. 666). MenoYa Win Health Centre's (2013) initiative illustrates the importance of incorporating both traditional and modern senses of aesthetics and function. The new hospital was intentionally built to incorporate the Aboriginal philosophy of holistic healing, honoring Aboriginal values and providing non-medicalized spaces for patients and their families. The hospital opens to all four cardinal directions, following the structure of the medicine wheel, as a commitment to uphold the Aboriginal teachings related to health and well-being. All examples provide evidence of the benefit of PFCC to patient care outcomes. It is an illustration of the value of respectful healthcare spaces that address specific cultural needs.

PFCC also provides space for narrative. Aboriginal tradition is based on sharing stories, and the art of narrative is somewhat lost in high-stress and fast-paced healthcare practice. Family-centred rounding provides an opportunity for Aboriginal parents to share their child's preferences, hopes and dreams about their care.

Muething et al. (2007) describe the difference between family-centred and traditional bedside rounds. Patients and their families have an opportunity to decide to what extent they want to participate. The main ethical concerns were about confidentiality and privacy of shared information. Some healthcare providers questioned the efficacy of family-centred rounds, as they may last longer and could potentially disrupt workflow and negatively affect other patients. Having in mind historical relationships between Aboriginal people and people in positions in power, family-centred rounds seem to provide a safer place for sharing stories[8] and further building trusting relationships.

5.6 Future Considerations

Similar to any other specific patient population,[9] healthcare professionals must keep in mind that Aboriginal people's beliefs, values, customs and traditions are not uniform. It is vital to embrace the diversity of Aboriginal values and belief providing care for Aboriginal patients. Aboriginal patients who uphold more traditional ways of life may have different sets of values than healthcare professionals providing care for them. The provision of healthcare is often coloured by stoicism, the clear delineation between personal and professional, and the formation of boundaries between the providers and patients. Many questions remain unexplored. Do healthcare providers have an extraordinary duty to care for Aboriginal children in light on existing health disparities? Should healthcare providers accommodate Aboriginal parents' and children's values that are contrary to clinical values? This is the central problem faced by advocates of PFCC, regardless of which cultures are at play.

Many Aboriginal healing practices have disappeared as a result of acculturation and assimilation. Under the infamous Potlatch Law, Aboriginal spiritual activities were outlawed, including many healing ceremonies (Waldram 1997). Those practices that survived to date have been practiced in secrecy by a small number of traditional healers. Knowledge about ceremonies and healing practices are sacred and only a few people in any Aboriginal community are privy to it (Waldram 1997). The evidence of benefit and safety of alternative Aboriginal therapies are missing in scientific literature. If health care professionals recognize that health and healing are culturally constructed, potential benefits and risks of a treatment ought to

[8] The family-centred rounds may be understood as sharing circles, as used in Aboriginal traditions. Children, their parents and other important people involved in childcare can share the child's story of illness and recovery. Similarly, healthcare providers may consider this as an opportunity to communicate complicated diagnoses and care plans in a form of narrative that is more easily understood by non-healthcare providers.

[9] Specific patient populations represent groups of people who share clear distinguishable criteria. Based on such criteria, ethically competent practitioners ought to take into account how such characteristics may alter their approach to care. Much of literature deals with challenges in providing ethical and culturally safe care for patients from diverse ethnics groups. However, the identifiable criteria can extend beyond ethnicity, including, for example, levels of ability/disability, sexual orientation or gender.

include relevant cultural information. As health care professionals are not experts in Aboriginal healing practices, the risks and benefits should be communicated by traditional practitioners.

I would argue that health care professionals have a duty to not only acknowledge, but also understand the predicament of Aboriginal people in Canada in order to provide appropriate care. Historically contentious relationships between Aboriginal and non-Aboriginal people in Canada call for this. Without such effort, the application of PFCC model for pediatric patients would not be possible as safety of a treatment would always outweigh benefits defined in spiritual terms.

Fiester's recommended (2012, p. 24) that "patient-centered care can obligate us only to provide the options we actually have to offer", but this would be true only if alternatives provided by patients and families are included. Would the outcome for K'aila have been any different if PFCC concepts had been applied? There is no evidence that the final result would have been any different; however, the experience that K'aila and his family endured during months of his illness may have resulted in a different story for K'aila and his family.

5.7 Conclusion

One of the fundamental values in Aboriginal cultures is the special place that children occupy. They are seen as gifts not only to parents, but also to their communities. They are seen as a true future of Aboriginal tradition and failure to protect is the greatest shame in an Aboriginal family. Engaging Aboriginal young patients and their families in all stages of PFCC should result in mutually trusting and respectful relationships between Aboriginal people and healthcare providers, perhaps easing some of the historically tense relationships. Further, the ethical imperative to provide care that is consistent with patients' wishes can be operationalized through the PFCC models. This is especially important for groups that do not prescribe to strict individualistic autonomous decision-making. Although Aboriginal people do not represent the only group in Canadian society that follow more of a communitarian approach in constructing the concepts of a good life, and right and wrong actions, the strained relationships between Aboriginal and non-Aboriginal Canadians deserve close and continuous examination of questions of power, ethics, law and culture.

References

Adams, A., D. Theodore, E. Goldenberg, et al. 2010. Kids in the atrium: Comparing architectural intentions and children's experiences in a pediatric hospital lobby. *Social Science & Medicine* 70:658–657.

Adelson, N. 2005. The embodiment of inequity: Health disparities in aboriginal Canada. *Canadian Journal of Public Health* 96 (supp2): S45–S61.

Adelson, N. 2007. Biomedical approach a poor fit with aboriginal views on health and healing. *Canadian Psychiatry Aujourd'hui* 3 (1). http://publications.cpa-apc.org/browse/documents/142. Accessed 5 Jan 2013.

Brant, C. 1990. Native ethics and rules of behaviour. *Canadian Journal of Psychiatry* 35:534–539.

Brant, C. 1996. Native issues. In *Images in psychiatry: Canada*, ed. Q. Rae-Grant, 91–94. Washington, D.C.: American Psychiatric Press, Inc.

Brant Castellano, M. 2004. Ethics of aboriginal research. *Journal of Aboriginal Health* 4:98–114.

Canadian Institutes of Health Research. 2008. *CIHR guidelines for health research involving aboriginal people*. Ottawa: Canadian Institutes of Health Research.

Canadian Institutes of Health Research, Natural Sciences and Engineering Research Council of Canada, and Social Sciences and Humanities Research Council of Canada. 2010. Tri-Council Policy Statement: Ethical Conduct for Research Involving Humans. Canadian Institutes of Health Research, Natural Sciences and Engineering Research Council of Canada, and Social Sciences and Humanities Research Council of Canada, Ottawa, ON.

Constitution Act. 1867.

Couchie, C., and S. Sanderson. 2007. A report on best practices for returning birth to rural and remote aboriginal communities. *Journal of Obstetrics Gynaecology Canada* 29 (3): 250–254.

Coward, H., and G. Hartrick. 2000. Perspectives on health and cultural pluralism: Ethics in medical education. *Clinical and Investigative Medicine* 23 (4): 261–265.

Dickason, O., and D. T. McNab. 2008. *Canada's first nations: A history of founding peoples from earliest times*. Don Mills: Oxford University Press.

Ellerby, J. H., J. McKenzi, S. McKay, et al. 2001. Bioethics for clinicians: 18. aboriginal cultures. *Canadian Medical Association Journal* 163 (7): 845–850.

Ells, C., M. R. Hunt, and J. Chambers-Evans. 2011. Relational autonomy as an essential component of patient-centered care. *International Journal of Feminist Approaches to Bioethics* 4 (2): 79–101.

Fiester, A. 2012. What "patient-centered care" requires in serious cultural conflict. *Academic Medicine: Journal of the Association of American Medical Colleges* 87 (1): 20–24.

Garvey, G., P. Towney, J. R. McPhee, et al. 2004. Is there an aboriginal bioethic? *Journal of Medical Ethics* 30 (6): 570–575.

Grant, A. 1996. *No end of grief: Indian residential schools in Canada*. Winnipeg: Pemmican Publications, Inc.

Hanselmann, C. 2001. *Urban aboriginal people in Western Canada: Realities in policies*. Calgary: Canada West Foundation.

Health Canada. 2013. Indian residential schools. http://www.hc-sc.gc.ca/fniah-spnia/services/indiresident/index-eng.php. Accessed 5 Jan 2013.

Howitt, M. J. 2011. The family care coordinator paving the way to seamless care. *Journal of Pediatric Oncology Nursing: Official Journal of the Association of Pediatric Oncology Nurses* 28 (2): 107–113.

Indian Act, R.S.C. 1951.

Jin, A., and D. Martin. 2003. Hepatitis A among residents of first nations reserves in British Columbia, 1991–1996. *Canadian Journal of Public Health* 94 (3): 176–179.

Johnston, P. 1983. *Native children and the child welfare system*. Toronto: Canadian Council on Social Development.

Kirk, R., and E. Szathmary, eds. 1985. Out of Asia: Peopling of the Pacific and Americas. *The Journal of Pacific History* (Inc. Canberra).

Knockwood, I. 2001. *Out of the depths: The experiences of Mi'kmawhildren at the Indian residential school at Shubenacadie*. Roseway: Fernwood.

Kotalik, J. 2009. Aboriginal health care in Canada—Bioethics perspective. *Formosan Journal of Medical Humanities* 10:19–135.

Lavergne, C. C., S. S. Dufour, N. N. Trocme, et al. 2009. Visible minority, aboriginal, and caucasian children investigated by Canadian protective services. *Child Welfare* 87 (2): 59–76.

McIntosh, C. 2011. Indigenous peoples and health law and policy: Responsibilities and obligations. In *Canadian health law and policy*. 4th ed., eds. J. Downie, T. Caulfield, and C. M. Flood, 575–614. Markham: LexisNexis, Inc.

McShane, K., J. Smylie, and P. Adomako. 2009. Health of first nations, inuit, and Métis children in Canada. In *Indigenous children's health report: Health assessment in action*, eds. J. Smylie and P. Adomako, 11–65. Toronto: Centre for Research on Inner City Health and the Keenan Research Centre in the Li KaShing Knowledge Institute.

MenoYa Win Health Centre's. 2013. A commitment to health and well-being. http://www.slmhc.on.ca/history. Accessed 5 Jan 2013.

Miller, J. R. 1996. *Shingwauk's vision: A history of Native residential schools*. Toronto: The University of Toronto Press.

Mitchell, T. 2005. Healing the generations: Post-traumatic stress and the healthstatus of aboriginal populations in Canada. *Journal of Aboriginal Health* 2 (1): 14–23.

Muething, S. E., U. R. Kotagal, P. J. Schoettker, et al. 2007. Family-centered bedside rounds: A new approach to patient care and teaching. *Pediatrics* 119 (4): 829–832.

Ontario Child Welfare Act. 1965.

Paulette, L. 1993. A choice for K'aila. *Humane Medicine* 9 (1): 13–17.

Potvin, M. 2005. The role of statistics on ethnic origin and 'race' in Canadian anti-discrimination policy. *International Social Science Journal* 57 (183): 27–42.

Prime Minister of Canada. 2008. Prime Minister Harper offers full apology on behalf of Canadians for the Indian residential schools system. http://www.pm.gc.ca/eng/media.asp?id=2149. Accessed 22 Oct 2012.

Royal Commission on Aboriginal Peoples. 1996. People to people, nation to nation: Highlights from the report of the Royal Commission on Aboriginal Peoples. http://www.aadnc-aandc.gc.ca/eng/1100100014597/1100100014637. Accessed 2 Sep 2012.

Shields, L., J. Pratt, and J. Hunter. 2006. Family centred care: A review of qualitative studies. *Journal of Clinical Nursing* 15 (10): 1317–1323.

Smylie, J., and P. Adomako, eds. 2009. Indigenous children's health report: Health assessment inaction. www.crich.ca. Accessed 22 Oct 2012.

Statistics Canada. 2007. 2006 Census dictionary. Statistics Canada Catalogue no. 92-566-XWE. http://www12.statcan.ca/english/census06/reference/dictionary/index.cfm. Accessed 23 Sep 2012.

Statistics Canada. 2008. Aboriginal peoples in Canada in 2006: Inuit, Métis and first nations, 2006 census. Statistics Canada Catalogue no. 97-558-XIE. http://www12.statcan.ca/census-recensement/2006/as-sa/97-558/pdf/97-558-XIE2006001.pdf. Accessed 3 Aug 2012.

Statistics Canada. 2009. Aboriginal children's survey. Statistics Canada Catalogue no. 89-634-X. http://www.statcan.gc.ca/pub/89-634-x/89-634-x2008001-eng.pdf. Accessed 21 Aug 2012.

The First Nations Information Governance Centre. 2011. RHSRHS phase 2 (2008/10) preliminary results. The First Nations Information Governance Centre, Ottawa, ON.

The Society of Obstetricians and Gynaecologists of Canada. 2010. Returning birth to aboriginal, rural, and remote communities. *Journal of Obstetrics and Gynaecology Canada: JOGC* 32 (12): 1186–1188.

The Truth and Reconciliation Commission of Canada. 2012. They came for the children: Canada, aboriginal peoples, and residential school. The Truth and Reconciliation Commission of Canada. Winnipeg, MB.

Trocmé, N., D. Knoke, and C. Blackstock. 2004. Pathways to the overrepresentation of aboriginal children in Canada's child welfare system. *Social Service Review* 78 (4): 577–600.

Vukic, A., D. Gregory, R. Martin-Misener, et al. 2011. Aboriginal and Western conceptions of mental health and illness. *Pimatisiwin: A Journal of Aboriginal and Indigenous Community Health* 9 (1): 65–86.

Waldram, J. B. 1997. *The way of the pipe: Aboriginal spirituality and symbolic healing in Canadian prisons*. Toronto: University of Toronto Press.

Waldram, J. B., D. A. Herring, and T. K. Young. 2007. *Aboriginal health in Canada: Historical, cultural and epidemiological perspectives*. Toronto: The University of Toronto Press.

Walker, R., H. Cromarty, L. Kelly, et al. 2009. Achieving cultural safety in aboriginal health services: Implementation of a crosscultural safety model in a hospital setting. *Diversity in Health and Care* 6 (1): 11–22.

Willis, J. 1973. *Geniesh: An Indian girlhood*. Toronto: New Press.

Chapter 6
Accounting for the Family in Law: An Impartial but not Impersonal Point of View

Lee A. Chapman

6.1 Introduction

> Convinced that the family, as the fundamental group of society and the natural environment for the growth and well-being of all its members and particularly children, should be afforded the necessary protection and assistance so that it can fully assume its responsibilities within the community. Recognizing that the child, for the full and harmonious development of his or her personality, should grow up in a family environment, in an atmosphere of happiness, love and understanding, ...

While this Preamble to the U.N. *Convention on the Rights of the Child* (1989) is perhaps offered more as an aspirational goal, the idea that the family is the fundamental unit of society is universally recognized. There may be no global consensus on what constitutes a family, but state and international laws both recognize and support the rights of individuals to come together to form families, and impose obligations based on familial membership. Generally, international laws do not seek to prescribe what should comprise a family unit, but allow people to self identify, recognizing various forms of legitimate social arrangements based on a variety of beliefs and values.

Family-centred care similarly recognizes and supports the family as central to the life of the child and, therefore, a key component to optimizing health. The ethos of family-centred care includes respect for the family's core values, and recognizes that families present with many different cultural and religious backgrounds. Family-centred care principles encompass collaboration, including information sharing, something which helps build trust and contributes to partnerships between patients, caregivers and health care providers. However, not all families present as a cohesive unit; nor do they have a single point of view. Inevitably this leads to tensions and difficulties in legal representation.[1]

The first hurdle for the clinician and lawyer is identifying who is a member of the family. There is no universally agreed on definition. Individuals self identify as

[1] Unless otherwise stated legal references are according to the laws of Ontario, Canada.

L. A. Chapman (✉)
PBLO (Pro Bono Law Ontario), SickKids, Toronto, Ontario, Canada
e-mail: lee_ann@pblo.org

R. Zlotnik Shaul (ed.), *Paediatric Patient and Family-Centred Care: Ethical and Legal Issues,* International Library of Ethics, Law, and the New Medicine 57,
DOI 10.1007/978-1-4939-0323-8_6,

belonging to a family unit, as well as identifying the other people they call "family". Each parent[2], and each child[3], may have a different notion of who constitutes their family unit. Is it only the biological or adoptive parents and their children or, as a remarried parent might insist, does it extend to the step parent who shares a significant portion of child care responsibilities and financial support? What if a child considers the paternal grandmother, who provides comparable child care and emotional support, to be an essential member of his family, but for whom the biological mother is a source of tension and even antagonism? Further, children may consider that they do not have just one family; in cases where separations have occurred, and families re-formed, they may locate themselves in two or more families. Though separated families may go through a formal legal separation process, custody is not necessarily assigned to one parent; both parents may be allotted joint custody or decision-making.

Even members of intact families with agreed membership do not necessarily share the same beliefs and aspirations. Nor will they necessarily process information, weigh risks, and come to the same conclusions with respect to their child's healthcare. How then does one determine the family's position on health care decisions with respect to the incapable child?

Further, parents of a child not capable of making health care decisions, even when in agreement, are not free to make unrestricted decisions. As surrogate or substitute decision-makers, their choices are circumscribed in accordance with the best interests of the child. Parents are accustomed to exercising autonomy when making decisions about their children, and the family's broader interests may be the basis for that decision-making. Therefore, the greater scrutiny given health care decisions, based on an externally imposed legal standard, can leave some parents with a sense that there has been an unwarranted intrusion into their private realm, and in extreme cases they may feel that their role as primary caregivers and substitute decision-maker has been usurped. Parents may assume that a legal system which obligates them to care for and protect their children, will also bestow upon them a reciprocal right to determine what is in the best interests of their child within the context of their own family.

Moreover, when parents are in agreement on a course of treatment, and it is in the best interests of the child, further complications and conflicts in legal representation of the family arise where parents and capable children have differing views with respect to health care decisions. The young person is assumed capable unless otherwise determined and has the legal right to make independent health care decisions which may not align with those of the parent (*Health Care Consent Act (HCCA),* 1996. S.O., s. 4(2)). Though parents might come to terms with the evolving capacities of their children, they may also expect that their own opinions and broader family interests will be taken into account with respect to health care decisions for their still dependent children.

Further, having made the healthcare decision, the capable young person may then wish to keep the health information from her parents and exercise attenuated

[2] The term "parent(s)" will be used as a proxy for all legal guardians of children.

[3] For the purposes of this chapter the word "child" denotes the relationship within the family. It is recognized that many patients in a paediatric setting who are older may not identify with the term "child"; the use of the term "child" should not be read as pejorative.

rights to privacy (*Personal Health Information Protection Act (PHIPA)*. S.O. 2004, s. 23 (1) and (3)). Parents who are legally obligated to provide for and protect their children, may find it incoherent that the law would not require the disclosure of health information to them, if not by a co-existing or shared right, then at the least in order to facilitate their supportive role in care, and most particularly when it may be in the child's best interest.

Family-centred care ethos endeavours to ensure each family member's wishes are taken into account in order to come to some kind of shared position. However, can the law sensibly account for the notion of the family in this way? Is there a legal position for the family as a whole, such that it can represent an impersonal and coherent point of view, abstracted from the individual opinions and rights of each of its constituent members? Understanding the legal status of the family and family members, beginning with a brief historical perspective, may aid in locating family-centred care in law.

6.2 Legal Status: Historical Overview

Historically, the family unit not only provided a sense of identity based on kinship, but also provided protection. In western European cultures, the family was inclusive of all members of a household, even servants, and at times it was extended to include the retinue of the master of the house. Belonging to a family unit brought with it obligations of both protection and loyalty (van Caenegem 1992). However, as feudal communal structures broke down, and further changes were brought about by industrialization, the size of the household was reduced. Constituent members of the family unit also changed, and membership was less defined by loyalty or protection and more on the basis of blood ties and marriage. By the eighteenth century the definition of family narrowed to closely related members who lived under one roof, the "domestic family".

Under English Common Law as reported in the eighteenth century, the prescribed family was narrowly defined as a man and woman who were married, and their legitimate children. Further, the man alone had full rights and legal identity; married women were non-persons.

> By marriage, the husband and wife are one person in law: that is, the very being or legal existence of the woman is suspended during the marriage, or at least is incorporated and consolidated into that of the husband: under whose wing, protection, and cover, she performs everything;[4]

As non persons women had virtually no legal power or rights with respect to their children. Sir William Blackstone, in his Commentaries on the Laws of England (1765–1769), discussed the rights of parents with respect to children and succinctly refers only to "[t]he legal power of a father,—for a mother, as such, is entitled to no power, but only to reverence and respect."

---

[4] *Femme-covert-* whereby a woman's legal existence was incorporated to that of the husband on marriage. Sir William Blackstone, Commentaries on the Laws of England Chapter: XV Husband and Wife.

In Canada, progress in both case law and legislation during the twentieth century brought women full and equal rights to vote[5] and recognition as persons under the law (*Edwards v. Canada (Attorney General)* [1930]). Women gradually obtained equal property rights and legal status in other realms, including the family, such that both parents were equally entitled to custody of the child.[6] Women also gained equal rights to maintain their aboriginal status in marriage and to pass on those rights to their children (*Bill C. 31,* 1985, amended the *Indian Act,* R.S.C.1985). Further change broadened the definition of the spouse from the traditional marriage between a man and a woman to include same sex marriages and unmarried persons who met certain conditions.[7]

Historically, children also lacked legal status and were treated as or akin to property. In Book V of Aristotle's *Nicomachean Ethics* the child was considered an extension of the father until they reached an age of independence and, accordingly, there could be no conduct towards them that would be unjust (or just). In Roman law the doctrine of *Patria potestas* (power of the father) gave absolute power to the father over his children's property and their lives, including the right to banish them and originally even the right to prescribe death. Any legal rights the child had were acquired through the father (Smith 1875).

Children made the progression to a more independent legal status through a series of legal and social reforms. Blackstone's Commentaries (1765–1769) chronicle the obligations that parents owed to children under law and by nature.[8] The obligations toward legitimate children included maintenance, protection, and education. The duties of the parent to provide, protect, and educate their children was not expressed in terms of the rights of children, even though the obligation of one typically corresponds to a right in the other. Rather, they were fiduciary duties assigned to a parent, in particular a father, and owed according to his station or role as parent. Further, though parental duty was an articulation of natural law, the obligation was acquired by societal status, and the benefits of the obligation extended beyond the individual child to the broader society. While the law provided that children should be maintained and protected, as individuals they had no legal status with respect to property, contracting or marriage, and their interests were subordinate to those of their parents or legal guardians. Until children reached the age of majority, legal decision-making was left to the father with respect to property and person:

> The power of parents over their children is derived from the former consideration, their duty: this authority being given them, partly to enable the parent more effectually to perform his duty, and partly as a recompense for his care and trouble in the faithful discharge of it. (Blackstone)

---

[5] 24 May 1918 all female citizens over 21 became eligible to vote in federal elections in Canada.

[6] *Family Law Reform Act,* 1978, S.O. 1978, c. 2. Ontario introduced a limited equal property division rule for married couples; extended. With respect to custody of children, *Children's Law Reform Act* R.S.O. 1990, c. C.12, s. 20 (1).

[7] Two persons who had co-habited for a period of three years or who had lived together in a relationship of some permanence if they were the natural or adoptive parents of a child. *Family Law Act,* R.S.O. 1990, c. F.3. s. 29.

[8] It is a principle of law, that there is an obligation on every man to provide for those descended from his loins; Blackstone (1893).

Children who were *filius nullius* (son of no man) had few if any rights and Blackstone comments that they were entitled to maintenance only if the mother identified the father and paternity could not be disproved. Most Canadian provinces, including Ontario, have repealed laws which made distinctions based on legitimacy. Expanded definitions in family law have further changed the definition of a child to include not only biological or adopted children, but also "a person whom the parent has demonstrated a settled intention to treat as a child of his or her family" (*Family Law Reform Act,* 1978, S.O.), a definition which is shared by several other common law jurisdictions. Child welfare laws in Ontario mandate that parents must care for, provide for, and supervise and protect, their child until the age of 16 years (*Child and Family Services Act.* R.S.O. 1990, s. 37(2)). While the age at which a person gains full status as an adult (age of majority) in Ontario is 18 years, parents may be financially responsible for financially supporting their children to age 18 and beyond in some circumstances, with exceptions for children who withdraw from parental control at 16 years of age (*Family Law Act,* R.S.O. 1990. s. 31.).

The laws regarding children have not only changed with respect to protection and entitlements, but also increasingly reflect their independent legal status. The obligations of both parents and governments to not only act in a child's best interests, but also recognize a child's right to participate to varying degrees in decisions that affect them, are now overarching legal principles recognized in a variety of contexts.[9] A child's right to make automous health care decisions based on capacity, not age, is evidence of legal and social progress of children over the last century.

However, given the continuing legal obligation of a parent to provide care and support, parents may reasonably expect a defined role for the family in decisions concerning their dependent children. At the very least they may expect the law to be consistent and easily discernible with regard to their right to participate in decisions affecting the health and welfare of their children. Seemingly incoherent or indeterminate laws with respect to the rights of children, and obligations and rights of parents and the family, can exacerbate tensions when there is disagreement over treatment. The following scenarios illustrate the resulting problems in legal representation in the family-centred care context.

6.3 Three Scenarios—Locating the Law in Family-centred Care[10]

6.3.1 Scenario 1: When Parents Disagree

Tia is 4 years old. Doctors are recommending surgery to minimize scarring to her face brought on by a vicious dog attack. The mother wishes the surgery to go ahead as soon

[9] For example custody decisions; health care decisions; special education.

[10] Please note that none of these scenarios represent actual cases at SickKids or any other case that we are aware of; similarly the names are fictional representations.

> as possible to minimize the psychological damage, including possible bullying at school which may occur as a result of her significant facial scars. The father believes his child has spent enough time in the hospital receiving treatment as a result of the attack, including an emergency operation. He does not want the surgery to go ahead at this time, but wants her to return to home and school as soon as possible.

Each parent has coherent reasons for arriving at their position on how their daughter's health care should proceed. Further, each parent believes that they have assessed what is in the best interest of the child in coming to their decision. The surgery is not life saving, nor in this scenario can the medical team say with certainty that delay will be a significant impediment to best results. Family-centred care ensures that each parent is listened to, and that their opinions based on their beliefs and values are respected, but can it inform the legal position on consent?

While the *HCCA* 1996, s. 20 (1), presumes capacity, when someone is found not to have capacity then there is a listed hierarchy of substitute decision-makers:

Consent on Incapable Person's Behalf

List of persons who may give or refuse consent

20 (1) If a person is incapable with respect to a treatment, consent may be given or refused on his or her behalf by a person described in one of the following paragraphs:

1. The incapable person's guardian of the person, if the guardian has authority to give or refuse consent to the treatment.
2. The incapable person's attorney for personal care, if the power of attorney confers authority to give or refuse consent to the treatment.
3. The incapable person's representative appointed by the Board under section 33, if the representative has authority to give or refuse consent to the treatment.
4. The incapable person's spouse or partner.
5. A child or parent of the incapable person, or a children's aid society or other person who is lawfully entitled to give or refuse consent to the treatment in the place of the parent. [emphasis added] This paragraph does not include a parent who has only a right of access. If a children's aid society or other person is lawfully entitled to give or refuse consent to the treatment in the place of the parent, this paragraph does not include the parent.
6. A parent of the incapable person who has only a right of access.
7. A brother or sister of the incapable person.
8. Any other relative of the incapable person.

The list is exhaustive, and it is important to note that each category refers to individualized decision-making. While a parent can make decisions on behalf of the incapable child, there is no category for collective or "family" decision-making. The law does not allow for the aggregation of each parent's or (more broadly) each family member's position so as to arrive at the family's overall legal position for the purposes of decision-making. If, as in this case, the mother wishes to have the surgery done tomorrow, and father wishes to wait a year, there is no "averaging" of their positions such that the legally right answer would be to have the surgery in 6 months. While in this case averaging might appear fair and reasonable as a way to come to the family's overall position, imagine two parents who disagree on circumcising their child, each claiming religious reasons for his or her decision; obviously averaging to a half circumcision is not an option.

Rather, family members are encouraged to work together toward the common goal of optimizing their child's health. However, in all cases they may not see a common path for achieving that end. Many factors outside the scope of the health care regime may influence their inability to come to a shared opinion or even work together in an attempt to find commonality. In extreme cases there may also be legal impediments to co-operative decision-making, such as restraining orders in cases of domestic violence. While clinicians and legal institutions may try to help the parents reach a consensus, if they cannot, then it is a zero sum game. There is no "family" right to decision-making under healthcare legislation, nor is there a legal way of arriving at a common family position when parties disagree. The surgery either will, or will not go ahead at the present time, and the parent whose wishes are not adopted may believe their inherent rights as a parent, and as a member of the family unit, have been diminished.

Unfortunately, without a legal accounting for the family as a unit, the legal system can push families further apart as each family member becomes invested in defending one position as right, or at least, more beneficial for the child. While the law may allow either parent to make the decision, in cases where there is disagreement and there is no clear and overriding best interest the procedure may not go ahead. Parents may be encouraged to seek a "tie-breaker" through the family courts, or an application may be made to the Superior Court of Justice to appoint a guardian of the person to make the health care decision, in some cases the public trustee and official guardian (*Substitute Decisions Act,* 1992, S.O. *s. 55*).

In cases where families have fragmented and previously sought clarification of decision-making in the family court system, there may be a single voice for decision-making, that of the custodial parent. However, this parent is not empowered to represent a family position; rather, there is merely an affirmation from the court that a shared point of view is unlikely or impossible and the court, either by mutual agreement or by order, affords one parent the role of decision-maker. While these parents are obligated to act in the best interests of the child, they may not have to consider the wishes of other family members, including the other parent. The courts may award full custody for a variety of reasons, for example, when parents have a history of conflict such that common ground on any issue is highly unlikely, or when the court may have reason to believe one parent will not act in the best interests of the child, or because of necessity due to distance, etc.. In cases where courts award joint custody, inability of the parents to come to an agreement may result in further litigation and turmoil, something which may not only impact the parents, but also the children, who may suffer increased stress and behavioural problems as a result of the continuing conflict (Emery 1982).

Families who are not intact can even be further polarized by the court process. For intact families, resorting either to family court or a "consent and capacity" hearing to determine the decision-maker, can have a lasting impact on the family, exacerbating any existing marital discord and, even worse, locating the child at the centre of the conflict.

6.3.2 Scenario 2: Parental Decisions and the Best Interests of the Child

> Sasha is a 10 year old boy with medically intractable epilepsy. Multiple drugs have been tried over the years but the numbers and severity of seizures have not decreased. Clinicians are concerned with Sasha's quality of life as well as further, possibly permanent, cerebral degeneration. Doctors have located the seizure focus and are recommending removing a lesion located in his brain. Sasha's parents are active members in a small religious sect that is strongly opposed to the removal of any tissues from a living person, and will not consent to surgery. The family believes that surgery has serious implications for their son in the afterlife, and will also result in being isolated from their extended family, and other community members. Sasha wishes to go along with his parent's decision.

Children are no longer viewed legally as mere extensions of their parents. Accordingly, parents do not have unfettered power to make decisions, but must do so in a in a manner consistent with their child's status as a rights holder with a distinct legal identity. As the Supreme Court of Canada case *B.(R.) v. Children's Aid* [1995] at 432 confirmed: "The rights enumerated in the Charter are individual rights to which children are clearly entitled in their relationships with the state and all persons—regardless of their status as strangers, friends, relatives, guardians or parents"

When children are too young to exercise their rights independently, domestic and international law requires parents to act in the best interests of the child. In Canadian law the furtherance and protection of the best interests of the child take priority over the wishes or desires of parents.

In Ontario, an expansive directive is found in the *Children's Law Reform Act* R.S.O. 1990, s. 20 (2).with respect to obligations flowing from custody of a child:

> Rights and responsibilities: "A person entitled to custody of a child has the rights and responsibilities of a parent in respect of the person of the child and must exercise those rights and responsibilities in the best interests of the child."

Further, the "best interests of the child" is a universal principle, which not only applies to parents and legal guardians but to government bodies as well as to the judiciary. Article 3 of the comprehensive directive of the U.N. *Convention on the Rights of the Child* states:

> 1. In all actions concerning children, whether undertaken by public or private social welfare institutions, courts of law, administrative authorities or legislative bodies, the best interests of the child shall be a primary consideration.

Despite the obligation to act in the best interests of the child, as a general matter there is no government intrusion into the family sphere except in cases of abuse or neglect, or where there are health or safety concerns. It is presumed that determination of the best interests of the child lies with parents, and that unnecessary interference with the parent-child relationship may seriously affect the child's well being. The Supreme Court, in *New Brunswick Minister of Health and Community Services v. J.G.,* [1999] 46, has affirmed that in severing those ties "Not only is the parent's right to security of the person [under the *Canadian Charter of Rights and Freedoms*] at stake, the child's is as well."

In addition to instances of child abuse and neglect, when parents are not making medical decisions in the best interest of their child, clinicians and other concerned parties may call on child protection legislation which provides for apprehension of children in cases where parents do not provide consent to medical treatment necessary to the best interests of the child (*Child and Family Services Act*, R.S.O. 1990, s. 37 2(e)), or seek a substitute decision-maker through an administrative tribunal. Further, the courts have *parens patriae*[11] jurisdiction with respect to minors, which allows them to override the decisions of parents and other legal guardians in the best interest of the child.

Even a clear direction to act in the best interests of the child does not necessarily provide guidance on how to realize best interests. As the Canadian Coalition for the Rights of the Child (CCRC) points out, "[t]he amorphous nature of the BIC is both a source of strength and weakness. Its subjectivity allows it to be responsive to the situation of an individual child and to evolving knowledge about child development" (CCRC 2009, p. 8). However because it is vague, it gives courts little guidance on the kinds of factors that ought to be taken into account and, thereby, allows the possibility of subjective determinations sometimes based on personal views. There is no single legislative test nor definition and it is, therefore, open to infinite interpretations. Even with a seemingly exhaustive list of the constituents of best interest, such as is found below in the *Child and Family Services Act*, R.S.O. 1990, s. 37 (3), the list typically does not provide guidance on which factors are particularly relevant nor what weight is to be given to each factor, but leaves this up to the person authorized under the *Act*. Not surprisingly, in such circumstances evaluations of best interests are neither unambiguous nor consistently applied.

(3) Where a person is directed in this Part to make an order or determination in the best interests of a child, the person shall take into consideration those of the following circumstances of the case that he or she considers relevant:

1. The child's physical, mental and emotional needs, and the appropriate care or treatment to meet those needs.
2. The child's physical, mental and emotional level of development.
3. The child's cultural background.
4. The religious faith, if any, in which the child is being raised.
5. The importance for the child's development of a positive relationship with a parent and a secure place as a member of a family.
6. The child's relationships and emotional ties to a parent, sibling, relative, other member of the child's extended family or member of the child's community.
7. The importance of continuity in the child's care and the possible effect on the child of disruption of that continuity.
8. The merits of a plan for the child's care proposed by a society, including a proposal that the child be placed for adoption or adopted, compared with the merits of the child remaining with or returning to a parent.

[11] Latin "parent of the country" invoked in common law jurisdictions- role of the state as guardians of persons with a disability such as children; mentally incompetent.

9. The child's views and wishes, if they can be reasonably ascertained.
10. The effects on the child of delay in the disposition of the case.
11. The risk that the child may suffer harm through being removed from, kept away from, returned to or allowed to remain in the care of a parent.
12. The degree of risk, if any, that justified the finding that the child is in need of protection.
13. Any other relevant circumstance.

The application of the "best interests" standard in custody and access decisions has, at different times, both allowed and denied parents the right to indoctrinate children in the parent's religion. It has been found in the best interests of children to be allowed to move to a distant country from Canada with one parent, and also been found not to be in the best interests of the child to move from one city to another with a parent. Given the number of factors and the seemingly inconsistent results of applying the standard, it is not surprising that there is not a clear understanding of the principle.

In the preceding scenario the parents believed their decision not to consent to surgery was consistent with what was "best" for their child. Most parents are not familiar with the legal concept of "best interests". When they do become acquainted with the term, they may presume best interests is specific to the family in which the child resides and to the values and beliefs the family espouses and, in varying degrees, the law would support such a contextual interpretation. Further, as parents, they expect broad discretion in determining which factors constitute best interests. They assume a subjective standard and, in the context of their family, religious beliefs are paramount to all other considerations. They may also believe that a decision for their child which is consistent with how they would decide the matter for themselves constitutes making a best interests decision.

Under the *HCCA* 1996, s. 21 (2), best interests determinations on behalf of all incapable persons is more constricted, and must be made in accordance with a number of factors listed below:

In deciding what the incapable person's best interests are, the person who gives or refuses consent on his or her behalf shall take into consideration,

a. The values and beliefs that the person knows the incapable person held when capable and believes he or she would still act on if capable;
b. Any wishes expressed by the incapable person with respect to the treatment that are not required to be followed under paragraph 1 of subsection (1); and

1. Whether the treatment is likely to,
   - Improve the incapable person's condition or well-being,
   - Prevent the incapable person's condition or well-being from deteriorating, or
   - Reduce the extent to which, or the rate at which, the incapable person's condition or well-being is likely to deteriorate.

2. Whether the incapable person's condition or well-being is likely to improve, remain the same or deteriorate without the treatment.

3. Whether the benefit the incapable person is expected to obtain from the treatment outweighs the risk of harm to him or her.
4. Whether a less restrictive or less intrusive treatment would be as beneficial as the treatment that is proposed.

The focus of the best interest inquiry is the impact on the health of the person if treatment is given or withheld; accordingly there is considerable deference to the opinions of medical providers. There is also an acknowledgment that the person's autonomy is being infringed without consent, and therefore that the least intrusive and least restrictive means to achieve the benefit should be undertaken, an in the case of someone who has previously had capacity, to take into account their values and beliefs (*HCCA,* 1996, 21(1)(2)). There are various conflicting opinions and theories over what constitutes best interests, in particular when it comes to decision-making on behalf of those who lack capacity. Depending on one's understanding of the underlying principles of best interests, the outcomes are very different. Take, for example, the parent who wants their young child A to donate a kidney to a sibling B who will otherwise face increasingly poor health and quality of life and, perhaps, a premature death.

If one's view is that best interests is not just about welfare but also a stand in, or "place keeper", for full autonomy and rights, one must preserve as many options as possible for the child until they are able to make their own decisions. Therefore, one cannot use the child instrumentally, disregarding their autonomy, even if it is to benefit a member of their own family, as would be the case in the kidney donation example. Furthermore, as the child moves closer to full capacity, their wishes are increasingly taken into account, giving the child scope to influence the outcomes to varying degrees. Finally, when they reach capacity they are allowed to make unencumbered decisions. Accordingly the decision is not static, but may be re-evaluated as the child's decision-making ability increases or matures. Encouraging their involvement in the decision process to the appropriate degree benefits the child by nurturing their capacity to become a rational decision-maker in the future.

This autonomy enhancing value is reflected in the Supreme Court of Canada's interpretation of best interest in *A.C. v. Manitoba (Director of Child and Family Services)* [2009] at 87. The Court, in considering its own obligations to act in the best interests of the child under *parens patriae,* noted the increasing importance of the child's wishes as they mature: "The more a court is satisfied that a child is capable of making a mature, independent decision on his or her own behalf, the greater the weight that will be given to his or her views when a court is exercising its discretion." There is no authority for the parent, nor other substitute decision-maker to take into account the broader family's wishes, nor consideration for the impact of the decision on any other member of the child's family, when making a health care decision for the incapable child.

Another interpretation of best interests equates it to a general welfare principle, which permits the decision-maker to substitute their own opinion, through a kind of cost benefit analysis of what is beneficial for the child, without commensurate considerations being given to rights and future autonomy. So, for example, a parent could consent to have one child's kidney donated to a sibling B. The parent decision-maker assumes child A benefits by improving or even saving B's life, and that this outweighs the cost to child A in losing one kidney and undergoing an otherwise unnecessary and invasive surgical procedure. Making this kind of determination either disregards the future autonomy of child A or, at the least, presumes a good deal of subjective information about the kind of person child A will become, and the sort of pain or medical risk that the child would be willing to undertake for a sibling. It also presumes the child to have a future close relationship to sibling B. Rather than enhance autonomy and preserve potentialities, it significantly diminishes or even extinguishes vital choices A will have in the future, for example, if A wanted to donate a kidney to their future child who might have need of the donation.

There is another position which gives more latitude to the family's general welfare when determining the best interests of the child, and yet also claims to pay appropriate attention to the rights of the child and his or her future potentialities. As such, a procedure which would be moderately intrusive to the child and not impede their future choices, but have a profound benefit to the family as a whole, is allowed. Examples of this are marrow donations from one child to save the life of a sibling, which is permissible under law if certain conditions are met. Marrow is replaceable; so the donor child's future ability to donate if needed is not impeded and, therefore, future autonomy is preserved. The benefits in this case are not only to the recipient in helping to preserve the life of a sibling, but to the family as a whole, including the donor, and are deemed greater than the costs of what is a more short term invasive procedure. Taking in this context, a best interest calculation acknowledges that autonomous decisions may include the impact of our decisions on the significant people in our lives, in particular, family members. These so called "relational interests" are often proposed as a constituent part of best interest calculations, and this is an improvement on understanding the latter calculations in purely individualistic terms.

The more contentious situation is when a procedure that might be a slight benefit to the child, but a large burden on the family, is not pursued. To withhold any beneficial treatment that would improve the child's health is difficult to reconcile with best interests. Mere inconvenience can never be a justification for medical decision-making where there is any negative impact on the health of the child. However, families make decisions in the context of their lives, and externalities beyond the consideration of the individual child, may fetter optimizing choices. Where families live, their financial conditions, the available family support, and other issues, may all impact choices. To the extent that it does not have a significant impact on the health of the child, there are accounts, beyond relational interests, as to why the impact on others may be part of a best interest computation. For example, Herring and Foster (2012) argue that best interests should not only reflect relational interests (e.g. the child's interests relating to family members), but also the virtue and benefit to the child of promoting social obligations and a more general altruism when making decisions on their behalf. According to Foster and Herring, altruism and obligations are both aspects of well being in individuals and, as such, should be incorporated into a calculation of best interests.

However understood, legally, the interests of the family do not come at the cost of the child's dignity interests and must be minimally intrusive to present and future autonomy. So, while a parent may consent to male circumcision in order for the boy to be a fully participating member in the family's religion and culture, female circumcision, which has been found discriminatory in purpose, without health benefits and results in diminished future autonomy of young women, (World Health Organization 2013), is legally prohibited. Tensions inevitably arise when families do not see the legal distinctions between these two cases as being based on preserving rights and autonomy of the child; they view the law as preferring one religion or culture over another, rather than taking an equal accounting of both positions.

Thus, the family who, because of religious convictions, refuses surgery which will enhance their child's current health and ability to partake in normal activities, and lessen the chances of brain damage that could impede future potential, will likely find the law does not uphold their right to deny surgery. They are unlikely to find this a satisfactory outcome based on their stated family values and beliefs. Further, they might suggest that the wishes of their 10 year old should be taken into account. However, the courts have expressed deep concerns about the ability of younger children to develop independent and stable decisions while living with, and dependent on, their parents. There is concern that these children have not yet freely developed their own beliefs and values, but are merely reflecting those of the family and other affliations. (*A.C. v. Manitoba (Director of Child and Family Services)* [2009] at 96).

6.3.3 Scenario 3: Mature Minor's Rights in Disagreement with Parents

Gila is an intelligent 14 year-old. She has been undergoing dialysis for the last 3 years. At this point the medical team and her parents believe that it is in her best interest to proceed with a kidney transplant now that they have found a family donor. Gila has decided that she does not want to proceed with the transplant; the reasons are complex, but include fear of surgery, a concern about how her life will change post transplant, and also concerns for the family donor. Further, she does not want her parents to know the reasons, and has told the team to respect her privacy and not convey the information to them.

In 1985 the "mature minor" principle for healthcare consent was first articulated in the U.K. case *Gillick v. West Norfolk* and Wisbech Area Health Authority, [1985] at 409. A mother challenged the local health authority's right to prescribe birth control to minors without parental consent. The landmark case was decided by the House of Lords: "Provided the patient, whether a boy or a girl, is capable of understanding what is proposed, and of expressing his or her own wishes, I see no good reason for holding that he or she lacks the capacity to express them validly and effectively and to authorise the medical man to make the examination or give the treatment which he advises."

However, despite the precedent, British courts and other legal authorities have not been consistent in upholding the right of competent minors to make health care decisions. In particular, the Court of Appeal confirmed in *Re R (a Minor) Re*

(Wardship Consent to Treatment) [1992], that it will not necessarily prevent overriding the competent child's wishes in situations where the child's life is threatened. In such cases, the court may exercise its *parens patriae* jurisdiction to authorize treatment based on an assessment of what is most conducive to the child's welfare. This leads to the seemingly incoherent position that a person may have capacity to consent to treatment, but not capacity to refuse treatment, if it is not in their best interests. Further, parents may retain some decision-making authority even with respect to minor children with capacity. However, the more recent UK case *R. (Axon) v. Secretary of State for Health,*[2006] confirmed the privacy rights attached to a *Gillick*-competent minor's health care, and the ongoing application of *Gillick.*

The principle that age is not determinative of capacity is also referred to as the "mature minor" or the "emancipated minor" doctrine in American law where an individual may apply to the court to be emancipated from the general prohibition against consent based on minority. In addition, certain "mature minor" laws allow non-emancipated minors to get medical treatment for specific medical conditions without their parents' or legal guardian's consent and, in some cases, without their knowledge. The right includes not only the right to consent to treatment, but also the right to refuse treatment in some states. In 1989, the Illinois Supreme Court ruled that a minor patient (17 year old) should be permitted to refuse medical treatment necessary to save her life (*re E.G.,* 549 N.E.2d 322 (1989)).

Article 12 of the U.N. *Convention on the Rights of the Child* affords the child both voice, in all matters affecting them, and agency, in accordance with maturity:

1. States Parties shall assure to the child who is capable of forming his or her own views the right to express those views freely in all matters affecting the child, the views of the child being given due weight in accordance with the age and maturity of the child.
2. For this purpose, the child shall in particular be provided the opportunity to be heard in any judicial and administrative proceedings affecting the child, either directly, or through a representative or an appropriate body, in a manner consistent with the procedural rules of national law.

Canadian domestic laws are a patchwork where rights of children and adolescents are concerned, and families may find the inconsistencies incomprehensible. Under civil laws young people under the age of 18 are *non sui jurius*, something which does not allow them full social and civil rights. In order to participate they require a litigation guardian who is accountable to the court and to act in the best interests of the child. Laws do not allow voting until the age of 18, or driving before 14–16 years (depending on the province). Children have no standing on custody and access hearings in Ontario courts; the minor who wishes to be heard in court may apply for intervener status to independently voice his or her concerns, or the court may appoint a lawyer to represent his or her interests (Wilson 1994).

However, among the rights that children do have are freedom of speech and association, the right to due process, and the right to an education, including special education services in school. Under Ontario child welfare law, they have the right to protect their privacy with respect to counseling records at the age of 12, and those with capacity have the right to consent or refuse health treatment as well as the right to protect the confidentiality of their personal health information. Under Ontario law, people over the age of 16 may also sign an advance health care directive and appoint an attorney for personal care. *(HCCA,* 1996, s. 26; *Substitute Decisions Act,* 1992, s. 43.)

While the *Child and Family Services Act* 1990 does provide the state power to apprehend a child where the parent refuses to consent to necessary medical treatment, it does not grant powers of consent to treatment that are greater than a parent would have; therefore it does not override the capable child's authority to make their own treatment decisions.

In Canada the right to consent to treatment also includes the corresponding right to refuse treatment. In 2009, in *A.C. v. Manitoba*, the Supreme Court of Canada was asked whether the Charter's guarantees of liberty, equality and religious freedom oblige the state to respect the decisions of children under age 16 to refuse medical treatment in circumstances when those decisions could cost the children their lives or seriously damage their health. The finding was that children *may* make life and death decisions about their medical treatment. In the majority opinion, Justice Rosalie Abella wrote at 187:

> If, after a careful and sophisticated analysis of the young person's ability to exercise mature, independent judgment, the court is persuaded that the necessary level of maturity exists, it seems to me necessarily to follow that the adolescent's views ought to be respected. Such an approach clarifies that in the context of medical treatment, young people under 16 should be permitted to attempt to demonstrate that their views about a particular medical treatment decision reflect a sufficient degree of independence of thought and maturity.

The absence of a declared age when one reaches sufficient maturity and independence of thought to make health care decisions, leads to uncertainty for families with respect to understanding legal rights of their children. Not only is there no age, but capacity is not global; it is assessed relative to a specific decision, at a particular time, and in a particular context *(HCCA,* 1996, s. 15). Accordingly, a 5 year old may have the legal right to decide whether to have an injection or oral medication if the choice is treatment neutral, but not the right to refuse the medication altogether.

The health care provider must determine if the person is able to understand the nature of the treatment and the consequences of consenting or refusing it, and must also determine if the person is capable of independent decision-making. However, there is no single objective test for assessing if one meets the elements of the test required for capacity. Further, a finding of capacity must allow for seemingly irrational decisions, or "bad" choices. In the above scenario the health care providers may take extra care to ensure the patient understands the consequences of refusing or delaying transplant, and try to assuage her fears. However, even though they do not agree with her decision, if she is found capable, her choice not to proceed with the transplant will be the correct outcome based on law, despite her age. If she has capacity to make the decision, her right to consent will extend to her right to privacy. In this case it includes her wish not to release personal health information to her family.

Even parents who acknowledge and respect their child's right to make decisions may nevertheless regard the complementary privacy rights as incomprehensible and impracticable to supportive family care. The *Personal Health Information Protection Act* 2004, 23 (2)i provides that the young person must consent to the release of health care information if it is in respect of "treatment within the meaning of the *Health Care Consent Act,* 1996, about which the child has made a decision on his or her own in accordance with that Act (*HCCA* 2007, s. 5(2)). Parents may find duties of care hindered by a minor child's right to keep personal health information from them. For example, parents can point to the tendency of their children to forget medication or ignore dietary restrictions, and assert that without appropriate shared health information they cannot know if their children are following the prescribed treatment. Further, if they are unaware of what that treatment consists of, they will be unable to support the treatment at home. Thus, they find that their roles as caregivers are incompatible with their dependent child's legal right to privacy, and may not see the logical connection between capacity and privacy rights. However, both capacity and privacy are constitutive of the legal acknowledgement of the same notion of autonomy.

In the U.K. there is a requirement (under the *Fraser Guidelines*[12] with respect to contraception) to try to persuade the minor to inform their parents of their health care procedures. There is no such role for the health provider under the Canadian health law regime; autonomy is upheld and respected in this situation as much as for any other person. There are no residual rights for decision-making, or access to information for the family, with respect to the "mature" or capable minor in the health care context: "…proof of capacity entitles the "mature minor" the right to exercise personal autonomy free from parental or judicial control" (*A.C. v. Manitoba* 2009, 175). However, in the words of one social worker: "I try to encourage whole family involvement in care for all kinds of therapeutic reasons, but often there are reasons why [capable patients] don't want their parents/caregivers to be involved or aware, and of course I respect those."[13]

6.4 Conclusion

Most families will not have had personal or direct contact with the legal system until a fundamental problem or dispute arises. Generally, families function unfettered by the law, making a myriad of daily decisions based on promoting what is best for the family, e.g., where to live, where to worship if at all, and where to attend school. Families are often about collective interests, however, and legal decisions are most often concerned with upholding individual rights, and in this distinction there lies

[12] The Fraser guidelines refer to the guidelines set out by Lord Fraser in his judgment of the *Gillick* case in the which apply specifically to contraceptive advice.

[13] Thanks to many insightful comments by social workers at the Hospital for Sick Children and in particular, Lysa Toye.

a constant tension. In the healthcare context the tensions are heightened because of the engagement of our most fundamental rights—bodily integrity, dignity, and autonomy. Unequivocally, by their very nature, these rights are securely attached to the individual. Accordingly, when the decision is exercised by a substitute decision-maker, such as a parent, the decisions will receive the utmost scrutiny. Further adding to the tension is the requirement that a legal substitute decision-maker is a singular voice, accountable in law, not an aggregation of positions. Thus the family as a unit does not have legal status as substitute decision-maker.

Historically, the family did have one legal position, one point of view, but it was not the impersonal family point of view; it was that of the male head of the household who was legally empowered to speak for all other members of the family. As the only one who could exercise most legal rights, and who had sole decision-making over minor children, the family was essentially reduced to a single opinion in law. However, we no longer impose a single voice for the family, nor is there legal authority or a process to aggregate the wishes of individuals within the family so that we arrive at one impersonal family view. While the law may equally regard each person according to their status, there is no impersonal or aggregate status in which the family's position as a singularity is regarded as legally decisive. The law may have equal for regard family members, but only as individuals; it seeks an impartial, but not an impersonal point of view. As such, there will continue to be inevitable tensions between legal representation and realizing the collaborative goals of family-centred care.

References

Cases

A.C. v. Manitoba (Director of Child and Family Services) [2009] 2 S.C.R. 181.
B.(R.) v. Children's Aid of Metropolitan Toronto [1995] 1 S.C.R. 315.
B.(S.G.) v. L (S.J.), (2010) O.R. (3) 796 (C.A.).
Edwards v. Canada (Attorney General) [1930] A.C. 124.
Gillick v. West Norfolk *and Wisbech Area Health Authority*, [1985] 3 All E.R. 402.
New Brunswick Minister of Health and Community Services v. J.G., [1999] 3 S.C.R. 46.
R.(Axon) v. Secretary of State for Health, [2006] EWHC 37.
re E.G., 549 N.E.2d 322 (1989).
Re R(a Minor) Re (Wardship Consent to Treatment) [1992] 3 WLR 592.
Young v. Young, [1993] 4 S.C.R. 3.

Legislation

Bill C. 31, 1985, amended the Indian Act, R.S.C., 1985, c. 1–5.
Child and Family Services Act, R.S.O. 1990, c. C.11, s. 37 (3); 2006, c. 5.
Children's Law Reform Act R.S.O. 1990, c. C.12.
Family Law Act, R.S.O. 1990. c. F. 3, s. 31.
Family Law Reform Act, 1978, S.O. 1978, c. 2.
Health Care Consent Act, 1996. S.O. c. 2, Schedule A, ss. 4(2).
Personal Health Information Protection Act, 2004, S.O. 2004, Sched. A.
Substitute Decisions Act, 1992, S.O. 1992, c 30.
UN General Assembly, Convention on the Rights of the Child, 20 November 1989, United Nations, Treaty Series, vol. 1577, p. 3.

Articles and Books

Aristotle. *Nicomachean Ethics Book V*. (W.D. Ross trans.) http://classics.mit.edu/Aristotle/nicomachaen.mb.txt. Accessed 15 Oct 2012.

Blackstone, Sir William. 1765–1769. Commentaries on the laws of England in four books. Notes selected from the editions of Archibold, Christian, Coleridge, Chitty, Stewart, Kerr, and others, Field's analysis, and additional notes, and a life of the author by George Sharswood. In *Two volumes* (1893). Philadelphia: J.B. Lippincott Co.

Canadian Coalition for the Rights of the Child. 2009. Best interests of the child report. Toronto. http://rightsofchildren.ca/wp-content/uploads/bic-report-eng-web.pdf.

Eekelaar, J. 1994. The interests of the child and the child's wishes: The role of dynamic self-determinism. *International Journal of Law, Policy and the Family* 8:42–61.

Emery, Robert E. Sept 1982. Interparental conflict and the children of discord and divorce. *Psychology Bulletin* 92 (2): 310–330.

Herring, J. J. W., and Charles Foster. 2012. Welfare means relationality, virtue and altruism. *Legal Studies* 32 (3): 480–498.

Smith, William. 1875. *A dictionary of Greek and Roman antiquities*, 873–874. London: John Murray.

van Caenegem, R. C. 1992. *An historical introduction to private law (trans. Johnston D.E.L.)*. Cambridge: Cambridge University Press.

World Health Organization. 2013. Fact Sheet No. 241: Female genital mutilation.

Wilson, Jeffrey. 1994. *Wilson on children and the law*. Toronto: LexisNexis Canada.

Chapter 7
Legal Considerations in Paediatric Patient and Family-Centred Healthcare

Joan M. Gilmour

7.1 Introduction

Patient-centred care and patient and family-centred care have come to the forefront in paediatric practice. While both are evolving concepts, as detailed elsewhere in this volume, there is wide agreement on the broad outlines of such an approach. In particular, the intent to increase parental and patient participation in the care provided is clear. Under that broad umbrella, there are different emphases on the extent and nature of parental involvement, for instance, in decision-making, in the physical and technical requirements of care-giving, and in other aspects of meeting a child's healthcare needs.

This model of caregiving is both enabled and limited by law. Law shapes the environment in which healthcare is provided, conditioning what can and cannot be done. As such, it supports yet also sets bounds on implementation of this model. Consequently, it is important for all involved to understand the legal framework in which parents, patients, healthcare providers and health facilities operate. This chapter examines several issues that frequently arise: parental and healthcare providers' authority and responsibilities, children's best interests, decision-making by mature minors, the role of state intervention, and responsibilities and potential liability when care is shared or delegated.

7.1.1 *Parental Authority and Responsibility*

Law puts the patient and his or her parents at the centre of decision-making. This strongly reinforces paediatric patient and family-centred care. Healthcare providers are legally obligated to obtain consent before treatment (*Health Care Consent Act* s. 4; *Malette v Shulman* 1990). In the case of adults, the patient's consent is sufficient,

J. M. Gilmour (✉)
Osgoode Hall Law School, Toronto, Canada
e-mail: JGilmour@osgoode.yorku.ca

R. Zlotnik Shaul (ed.), *Paediatric Patient and Family-Centred Care: Ethical and Legal Issues*, International Library of Ethics, Law, and the New Medicine 57,
DOI 10.1007/978-1-4939-0323-8_7,

unless he or she is incapable of making such a decision. The situation differs with children, who may not be able to give legally valid consent. In order to do so, the patient must be competent (sometimes referred to as "decisionally capable"), meaning that he or she can both understand and appreciate the nature and consequences of the recommended treatment, alternative treatments, and non-treatment (*Health Care Consent Act*; *Fleming v Reid* 1991). Additionally, the patient must be able to make a decision that is voluntary (*Re T.D.D.* 1999). Many minors, especially those who are younger, cannot meet these tests. In that case, a substitute decision-maker (SDM), usually the child's parent or guardian, has the legal authority to consent to or refuse treatment (*Health Care Consent Act* s. 20). Parents' power and authority in decision-making about their child has been affirmed by the Supreme Court of Canada, which noted: "…the parental interest in bringing up, nurturing and caring for a child, *including medical care* and moral upbringing, is an individual interest of fundamental importance to our society…" (emphasis added) (*B.(R.) v Children's Aid Society of Metropolitan Toronto* 1994, p. 371).

The Court has made it clear that parental rights are given effect in law to enable parents to meet the responsibilities they have for their children. It has also confirmed that judicial recognition of parental rights is "…based on the presumption that parents act in the best interests of their child" *B.(R.) v Children's Aid Society of Metropolitan Toronto* 1994, p. 371). While the law allows parents and guardians considerable discretion in deciding about treatment, their authority in deciding where their child's best interests lie is subject to limits.

7.1.2 Best Interests

It is clear that the "best interests" parents are to act on when deciding whether to consent to or refuse treatment are those of their child, not anyone else (*Director (Child, Youth and Family Enhancement Act) v D.L.* 2012, para 34). Thus, in *E v Eve.*, an application seeking judicial authorization for the non-therapeutic sterilization of a developmentally disabled young woman, one argument advanced by the applicant (Eve's mother, who was her SDM) was that because of her own advancing age, it would be difficult for her to assume responsibility for raising a baby should Eve become pregnant and have a child (*E. v Eve* 1986). In rejecting this argument and the mother's application, the Supreme Court of Canada noted that decisions about best interests must be made for the benefit and protection of the decisionally incapable individual, not others. This case serves as a reminder that in implementing a family-centred care model, in which each member of the family "…may be considered a care recipient, with care planned around the whole family, not just the individual child" (this volume, literature review, p. 5), the best interests of the patient must remain paramount. Most often, the patient's best interests will be advanced by attention to the care requirements of his or her family as a whole. But the interests of the family or parents will not always coincide with those of the child. It is the child patient's best interests that must guide both healthcare providers and parents in planning treatment and care (*E v Eve* 1986, *Syl Apps Secure Treatment Centre v B.D.* 2007).

However, what will be in a patient's "best interests" is not always self-evident or uncontested. Views will differ, affected by value systems, religious and other beliefs, cultural background, perceptions of the risk and benefit of treatment alternatives, and other considerations (*Scardoni v Hawryluck* 2004; Gilmour et al. 2011a). In some jurisdictions, healthcare consent legislation specifies factors SDMs should take into consideration. In Ontario, for instance, these include relative burdens and benefits of treatment and alternatives, whether treatment would improve the person's condition or slow deterioration, congruence with the patient's values and beliefs, and whether less intrusive treatment would be as beneficial (*Health Care Consent Act*, s. 21). Even when there is no detailed legislative guidance, courts undertake a similar assessment when determining best interests (*Sweiss v Alberta Health Services* 2009; *E v Eve* 1986; *May v Alberta Health Services* 2010). It is evident from the decided cases that expert medical opinion weighs heavily in judicial assessments of children's best interests, although courts also recognize the importance of and act on other values as well (*Saskatchewan (Minister of Social Services) v P.(F.)* 1990; *Couture-Jacquet v Montreal Children's Hospital* 1986; *Chmiliar v Chmiliar* 2001; Gilmour et al. 2011c).

People may disagree about how to weigh these factors, and consequently, about what treatment decisions are indicated. For instance, the Ontario Consent and Capacity Board concluded it was in the best interests of EJG, an 8-month-old infant who was left in a persistent vegetative state following significant oxygen deprivation prior to birth, and who suffered numerous associated sequelae, including quadriplegia, recurrent respiratory arrests, chronic lung disease and other infections and injuries, to accept a treatment plan proposed by his physicians that involved discontinuing mechanical breathing support, not attempting resuscitation in the event of respiratory failure, and limiting his treatment to comfort care. The Board held that the focus in determining his best interests must be on both the expected effects on EJG of the treatment plan and any alternatives, as well as on his well-being, since he was too young to have values and beliefs of his own. It concluded his parents had been wrong in basing their decisions solely on their own values and beliefs. In a decision upheld on appeal, the Board ordered his parents to consent to the treatment plan proposed, or be replaced as his substitute decision-makers (Gilmour 2011; *In the Matter of EJG* 2007). However, in *Scardoni v Hawryluck*, the court overturned a decision of the Ontario Consent and Capacity Board that an elderly patient's adult daughters, whom she had appointed as her substitute decision-makers, were not acting in her best interests when they insisted on continuing life support for their seriously ill and debilitated mother against medical advice. The court held that the Board had erred in focusing too heavily on the onerous effects of continued treatment on the patient, and failed to give proper consideration to all the statutory factors to be considered in assessing her best interests, in particular, the patient's values and belief system (2004). As is apparent, these can be very difficult decisions, particularly when the benefits of treatment are far from certain and the burdens and risks are significant.

Most decisions parents make about their child's healthcare are not made in such extreme circumstances, or with such grave consequences. Recognizing this, the law allows parents considerable latitude. As LaForest J noted in *B.(R.)*:

> If one considers the multitude of decisions parents make daily, it is clear that in practice, state interference in order to balance the rights of parents and children will arise only in exceptional cases. In fact, we must accept that parents can, at times, make decisions contrary to their children's wishes—and rights—as long as they do not exceed the threshold dictated by public policy, in its broad conception (*B.(R.) v Children's Aid Society of Metropolitan Toronto* 1995, para 86).

In everyday life, parents are not held to a standard of perfection. There will be times when the decisions they make will not be in the best interests of their child, but nonetheless, do not contravene the "threshold dictated by public policy." For instance, while courts accept that routine childhood immunization is generally in a child's best interests, it has been held that parental refusal did not provide sufficient grounds for state intervention, absent evidence of a greater need to protect either the child or the community from vaccine-preventable infectious diseases (*Newfoundland (Director of Child Welfare) v CRB* 1995; Gilmour et al. 2011b). Thus, when the risks of foregoing treatment are less serious or less immediate, parents' treatment refusal is more likely to be allowed to stand by healthcare providers, child protection authorities, and courts. But when will parents' decisions cross the line? The next section of this chapter examines limits on parental authority, as well as healthcare provider and state responsibilities to intervene.

7.1.3 Limits; Intervention Principles

Parents have a legal and ethical duty to ensure their minor children are provided with needed healthcare (*B.(R.) v Children's Aid Society of Metropolitan Toronto* 1995). When illness seriously threatens a child's health and the parents will not obtain or permit available effective treatment, then the child's healthcare providers have a legal obligation to notify child welfare authorities (*Child and Family Service Act* 1990; Gilmour et al. 2011c). If a child is found in need of protection, legislation in all Canadian jurisdictions authorizes measures to ensure that needed medical care is provided. For instance, in *B.(R.)*, because of their religious beliefs, Jehovah's Witness parents objected to blood transfusions that their infant daughter's physicians considered were required to address medical conditions posing a serious risk to her health (1995). The Supreme Court of Canada upheld the constitutionality of the child protection proceedings, and confirmed that she should receive the recommended medical treatment. The parents' refusal of treatment had fallen below the "socially acceptable threshold" and justified state intervention (*B.(R.) v Children's Aid Society of Metropolitan Toronto* 1995, p. 373).

Nor can parents choose treatment that seriously endangers a child's life or health contrary to medical advice. In *Re T.D.D.*, the court concluded it was in the best interests of a 13-year-old boy with bone cancer to undergo medically recommended chemotherapy and amputation of the affected leg despite objections by the boy and his parents, who wanted to rely on alternative therapies and prayer to treat a recurrence of his cancer (ultimately, the cancer was too advanced to proceed) (*Re T.D.D.* 1999).

7.1.4 Informed Consent

Parental or patient consent prior to treatment must be informed. This means that patients, or if decisionally incapable, their substitute decision-makers, must be given material information about the treatment proposed, its risks and benefits, and alternative courses of action, including non-treatment–in other words, the information that a reasonable person in similar circumstances would want to help them decide (*Reibl v Hughes* 1980; *Health Care Consent Act*, s. 11). This information-sharing, required by law, should also reinforce paediatric patient and family-centred care.

Parents must sometimes make tremendously difficult decisions about their child's best interests, especially when their son or daughter is very ill, the treatment proposed is risky, and its prospects are limited or uncertain. They want to act responsibly, to do their best for their child. They need the support of the treating healthcare providers, who should enhance parents' decision-making abilities by providing the best information available about their child's condition and prognosis, the treatment recommended and the alternatives, in understandable terms (Gilmour et al. 2011c; Gilmour et al. 2011e). From the perspective of paediatric patient and family-centred care, clear communication with parents about their child's condition, prospects and treatment can be an important way to attend to their care needs, as well as their child's. Indeed, information-sharing is one of the core components of the guidance principles for paediatric patient and family-centred care identified by the American Academy of Pediatrics, as discussed in Chap. 1 of this collection.

7.1.5 Mature Minors

The requirement to obtain informed consent to treatment grows out of the strong protection that law accords people's rights to bodily integrity and autonomy (*Fleming v Reid* 1991). While many children will not be able to give a legally valid consent because they cannot meet the legal tests for decisional capacity and independence, some young people can satisfy both these requirements (Gilmour et al. 2011e). They are able to understand and appreciate the nature and consequences of the recommended treatment, alternative treatments, and non-treatment, and can make a voluntary decision (*Health Care Consent Act; A.C. v Manitoba (Director of Child and Family Services)* 2009). Absent additional statutory or common law constraints, minors who meet these tests can make their own decisions about treatment (*Van Mol v Ashmore* 1999). At common law, this is referred to as the mature minor doctrine. If they are not able to decide for themselves, parents or guardians will make healthcare decisions on their behalf, as described previously. Decisional capacity can be especially difficult to assess in adolescents. As I have noted elsewhere, "[a]chieving the necessary maturity and understanding to make decisions about treatment is an incremental process that takes place over time, but progress is not necessarily constant, uniform or without setbacks. Nor do all adolescents mature at the same rate. Development is affected by the individual's environment as well as personal characteristics"(Gilmour 2002, p. 210). Refusing treatment needed to preserve life or prevent a serious risk to health requires greater appreciation of the consequences than does refusing more minor treatment (*A.C. v Manitoba (Director of Child and Family Services*) 2009). This is true of decisions that are difficult and complex as well.

There had been conflicting views on whether provincial child welfare legislation displaced the mature minor doctrine, such that if a minor was deemed in need of protection pursuant to that legislation, then child welfare authorities could be authorized to consent to treatment that the minor's physicians considered necessary, even when the minor met the tests for decisional capacity and refused treatment. This interpretation of child welfare legislation was rejected by the Supreme Court of Canada in *A.C. v Manitoba (Director of Child and Family Services)* (2009). A.C. was almost 15 years old when she was admitted to hospital because of gastrointestinal bleeding caused by Crohn's Disease. Her physicians concluded blood transfusions were necessary to prevent serious risk to her health, and perhaps her life. She and her parents were members of the Jehovah's Witness faith, and strongly opposed transfusions on religious grounds. A.C. and her parents appealed the lower court decisions authorizing treatment to the Supreme Court of Canada, arguing a breach of her rights under the *Canadian Charter of Rights and Freedoms* (1982). The Court held that if a young person is sufficiently mature and independent, her decisions about treatment must be taken into account, regardless of her age. It noted that, because child protection authorities are only involved in a limited class of case, those with the most serious consequences when treatment is refused, then "the ineffability inherent in the concept of 'maturity'...justifies the state's retaining an overarching power to determine whether allowing the child to exercise his or her autonomy in a given situation actually accords with his or her best interests," but that generally, "[i]f...the court is persuaded that the necessary level of maturity exists, it seems to me necessarily to follow that the adolescent's views ought to be respected" (*A.C.* 2009, paras. 86, 87). The Court ultimately upheld the lower court's decision authorizing blood transfusions because A.C.'s capacity to decide about treatment had never actually been judicially determined in the initial proceeding. While the implications of this decision for mature minors' decision-making power in situations not involving intervention by child welfare authorities have yet to be fully developed, the Court's strong support for minors' decision-making even when the consequences of refusing treatment will be grave make it likely that mature minors' decisions about treatment will prevail in less serious circumstances as well. The decision in *A.C.* makes it clear that healthcare providers and parents must go beyond involving mature minors in decision-making to sharing, and in some instances ceding, power and authority. This will have implications for paediatric patient and family-centred care.

Even when a minor is not decisionally capable, it is respectful to include him or her in discussions and decisions about treatment in accordance with his or her abilities. As Gilmour et al. note, doing so can "…show the patient that the physician is mindful of his or her independence and respectful of his or her ability to take part in developing the management plan"(Gilmour et al. 2011e, pp. S164–165). Both the young person's assent and sustained dissent or resistance to treatment, although not controlling when a minor is decisionally incapable, should nonetheless be given serious consideration by healthcare providers and parents.

7.2 Responsibility and Potential Liability When Care is Shared or Delegated

The import of paediatric patient and family-centred care goes well beyond increased attention to how decisions are made about treatment. It may entail delegating responsibility for aspects of the child's care while at a health facility or at home to parents or other family members. It also envisages a much more collaborative care model, expanding the healthcare "team" to include lay members—parents, families, and patients themselves. The decision-making framework outlined previously does not speak to how care is actually delivered. Changing roles and responsibilities require attention to ensuring that quality of care and patient safety are maintained, and concerns about potential liability addressed. Paediatric patient and family-centred care may also raise questions of professional responsibility, since physicians, nurses, and other allied health professionals must practice in ways that maintain the standards of practice of their professions.

Judicial decisions not only determine liability for harm caused by deficient care, but also set legal standards that practitioners and health facilities must meet. For that reason, I turn to examine the law governing the legal liability of healthcare providers and institutions, to explain the duties that they owe to patients, and how changes in the way care is delivered may affect their legal obligations. I focus on negligence (sometimes referred to as "malpractice"), because that is the type of claim most frequently advanced in lawsuits about healthcare.

To establish liability for negligence, a plaintiff must prove on a balance of probabilities that:

1. The defendant owed him or her a duty of care;
2. The defendant breached the standard of care established by law;
3. The defendant's breach caused injury or loss to the plaintiff; and
4. The plaintiff's damages are not too remote to be recoverable in law (Klar 2012; Gilmour 2009).

Duty of Care It is well established that health professionals owe their patients a duty of care, as do hospitals and other health facilities. In the healthcare context, issues may arise with respect to the scope of that duty, but not its existence vis a vis their patients. With paediatric patient and family-centred care, there may be questions about whether the health professional owes a duty of care to others as well—for instance, a patient's parents whom a nurse may have instructed about some aspect of their child's treatment. If the instructions were wrong and the child is injured as a result, then the nurse may have breached a duty of care that she owes not only to the patient, but also to his parents, who relied on her for accurate information. Foxcroft suggests that in such a situation, a duty of care would also be owed to others who may foreseeably be harmed by the health practitioner's errors (Foxcroft 2010). However, as explained previously, if there is potential for conflict with the duty of care that health professionals owe to their paediatric patient, the duty to the patient will prevail (*Syl Apps Secure Treatment Centre v B.D.* 2007).

Standard of Care As in negligence law generally, healthcare professionals must act so as not to cause an unreasonable risk of harm to others. In determining the standard of care to be met, they are held to the standard of a reasonably competent member of their profession. Health professionals "must bring to [the] task a reasonable degree of skill and knowledge and must exercise a reasonable degree of care. He is bound to exercise that degree of care and skill which could reasonably be expected of a normal, prudent practitioner of the same experience and standing…" (*Crits v Sylvester* 1956, para 31). As a general rule in Canada, expert evidence of compliance with generally approved practice on questions of treatment and care is conclusive evidence of absence of negligence (*ter Neuzen v Korn* 1995; Gilmour 2009). If the common practice is divided, a practice is acceptable if followed by at least a respectable minority of competent practitioners in the same field (*Lapointe v Hôpital Le Gardeur* 1992). Professional judgment prevails in determining the standard of care, except in very limited circumstances. However, if a practice is "'fraught with obvious risks' such that anyone is capable of finding it negligent without the necessity of judging matters requiring diagnostic or clinical expertise," a court can find an approved practice, and the defendant who followed it, negligent (*ter Neuzen v Korn* 1995, p. 17; Picard and Robertson 2007, p. 356).

Delegated or Shared Care When responsibility for a child's care is delegated or shared, several issues may arise with respect to the standard of care. First, is it reasonable to shift responsibility for care in this way, whether to the patient, parents or family members? Depending on the particular type of health service, there may be regulatory or professional limits on the types of care that can be delegated, and to whom (*Regulated Health Professions Act* 1991). Has the health professional given the person selected to perform particular tasks adequate instruction to do so (Picard and Robertson 2007)? Is that individual able to perform those functions reliably and safely (Foxcroft 2010)? Is he or she willing to do so? Have the care plan and the various participants' roles been clearly communicated to all involved, and sufficiently documented? Are appropriate arrangements in place to ensure proper supervision and monitoring, where required (Foxcroft 2010)? Are the individuals to whom tasks have been delegated provided with sufficient resources and supports to enable them to act safely? Have any additional material risks in changing how care is delivered and by whom been properly disclosed to patients or their parents or guardians, and consent obtained (Foxcroft 2010; *Ciarlariello v Schachter* 1993)?

Even this short list gives an indication of the range of issues that may arise when care is shared or delegated. This is not meant to discourage the practice; paediatric patient and family-centred care can be better care, with better results for patients. But those adopting this model must be aware of the obligation to ensure that they, and all those involved in the patient's care, are able to meet the applicable standard of care. Managing the risks associated with shared or delegated care requires carefully thinking through how implementing paediatric patient and family-centred care affects the legal responsibilities owed.

Causation In a lawsuit for negligence, a plaintiff must prove that the defendant's wrongdoing caused him or her injury. The general, but not exclusive test for establishing that the defendant's breach of the standard of care caused harm to the plaintiff is the "but for" test, which requires the plaintiff to show that "but for" the negligence of the defendant, the injury would not have occurred. In a medical malpractice context, the plaintiff may assert that he or she was injured by the defendant's substandard or deficient care (*Wilson v Swanson* 1956), or that if the defendant had properly informed the patient or parents of the risks of the treatment, no consent to treatment would have been given (*Reibl v Hughes* 1980). Causation can be difficult for a plaintiff to establish in a medical malpractice case, particularly given the risks often inherent in treatment, however skillfully performed, and the debilitating effects of the patient's illness.

Damages Damage awards in negligence cases are compensatory. Once liability has been established, damages are meant to compensate the plaintiff for all losses incurred—that is, to return him to the position he would have been in if the injury had not occurred, insofar as money can do so.

Health Facilities Hospitals and other health facilities also owe a legal duty to provide reasonable care for patients. A hospital may be directly liable to a patient for its own negligence. It may also be vicariously liable for the negligence of those for whom it is legally responsible, such as its employees and those with apparent authority to act on its behalf (Gilmour 2009; Gilmour et al. 2011d). The most common duties hospitals owe patients are: "(1) to select competent staff and monitor their continued competence; (2) to provide proper instruction and supervision; (3) to provide proper facilities and equipment; and (4) to establish systems necessary for the safe operation of the hospital" (Picard and Robertson 2007, p. 460). It is apparent on reviewing these duties that the same types of questions and issues that arise with respect to the potential liability of health professionals for shared or delegated care can affect hospital and health facility liability as well.

7.3 Conclusion

This chapter has traced some of the ways in which law affirms and reinforces the centrality of parents' role in their child's healthcare. It has also explained how their authority is constrained by legal limits. Legal frameworks governing healthcare were not initially developed with paediatric patient and family-centred care in mind, but they can certainly accommodate, and even encourage its development. As this model of caregiving is implemented more widely, it will be important to ensure that provisions are in place so that all those involved can provide safe, quality care.

References

A.C. v Manitoba (Director of Child and Family Services). 2009. 2 S.C.R. 181, 2009 SCC 30.

B.(R.) v Children's Aid Society of Metropolitan Toronto. 1995. 1 S.C.R. 315.

Canadian Charter of Rights and Freedoms. 1982. Part I of the Constitution Act, 1982, being Sch. B to the Canada Act 1982 (U.K.), c.11.

Child and Family Services Act, R.S.O. 1990. c. C–11.

Chmiliar v Chmiliar. 2001. A.J. No. 838, 2001 ABQB 525.

Ciarlariello v Schachter. 1993. S.C.J. No. 46, [1993] 2 S.C.R. 119.

Couture-Jacquet v Montreal Children's Hospital. 1986. 28 D.L.R. (4th) 22 (Que. C.A.).

Crits v Sylvester. 1956. 1 D.L.R. (2d) 502 (Ont. C.A.), aff'd (1956), 5 D.L.R. (2d) 601 (S.C.C.).

Director (Child, Youth and Family Enhancement Act) v D.L. 2012. ABCA 275.

E. v Eve. 1986. 2 S.C.R. 388.

Fleming v Reid. 1991. O.J. No. 1083, (1991) 4 O.R. (3d) 74 (Ont. C.A.).

Foxcroft, L. 2010. Professional and legal issues. In *Child and family-centred health care*. 2nd ed., eds. L. Smith and V. Coleman, 232–254. London: Palgrave Macmillan.

Gilmour, J. 2002. Children, adolescents and health care. In *Canadian health law and policy*. 2nd ed., eds. J. Downie, T. Caulfield, and C. Flood, 205–250. Markham: Butterworths.

Gilmour, J. 2009. Duty of care and standard of care. In *Commission of inquiry on hormone receptor testing*. vol. 2, ed. M. Cameron (Comm'r), 35–64. St. John's Nfld: Government of Newfoundland and Labrador.

Gilmour, J. 2011. Death, dying and decision-making about end of life care. In *Canadian health law and policy*. 4th ed., eds. J. Downie, T. Caulfield, and C. Flood, 385–436. Markham: Lexis Nexis.

Gilmour, J., C. Harrison, L. Asadi, M. Cohen, and S. Vohra. 2011a. Childhood immunization: When physicians and parents disagree. *Pediatrics* 128 (suppl 4): S167–174.

Gilmour, J., C. Harrison, L. Asadi, M. Cohen, S. Aung, and S. Vohra. 2011b. Considering complementary and alternative medicine alternatives in cases of life-threatening illness: Applying the best interests test. *Pediatrics* 128 (suppl 4): S175–180.

Gilmour, J., C. Harrison, L. Asadi, M. Cohen, and S. Vohra. 2011c. Hospitals and complementary and alternative medicine: Managing responsibilities, risk and potential liability. *Pediatrics* 128 (suppl 4): S193–200.

Gilmour, J., C. Harrison, L. Asadi, M. Cohen, and S. Vohra. 2011d. Informed consent: Advising patients and parents about complementary and alternative medicine therapies. *Pediatrics* 128 (suppl 4): S187–192.

Gilmour, J., C. Harrison, L. Asadi, M. Cohen, and S. Vohra. 2011e. Treating teens: Considerations when adolescents want to use complementary and alternative medicine. *Pediatrics* 128 (suppl 4): S161–166.

Health Care Consent Act. 1996. Being Sch. 2 to the Advocacy, Consent and Substitute Decisions Statute Law Amendment Act, S.O. 1996, c.2.

In the Matter of EJG. Sept. 30, 2007. Ont. Consent and Capacity Board, HA-07-2284, HA-07–2285.

Klar, L. 2012. *Tort law*. 5th ed. Toronto: Thomson Carswell.

Lapointe v Hôpital LeGardeur. 1992. 90 D.L.R. (4th) 7 (S.C.C.).

Malette v Shulman. 1990. 67 D.L.R. (4th) 321 (Ont. C.A.).

May v Alberta Health Services. 2010. ABQB 213.
Newfoundland (Director of Child Welfare) v CRB. 1995. NJ No. 389 (SC).
Picard, E., and G. Robertson. 2007. *Legal liability of doctors and hospitals in Canada*. 4th ed. Toronto: Thomson Carswell.
Re T.D.D. 1999. S.J. No. 143 (Q.B.).
Regulated Health Professions Act, S.O. 1991. c.18.
Reibl v Hughes. 1980. 2 S.C.R. 880.
Saskatchewan (Minister of Social Services) v P.(F.). 1990. 69 D.L.R. (4th) 134 (Sask. Prov. Ct).
Scardoni v Hawryluck. 2004. O.J. No. 300, (2004) 69 O.R. (3d) 700 (S.C.).
Sweiss v Alberta Health Services. 2009. A.J. No. 1303, 2009 ABQB 691.
Syl Apps Secure Treatment Centre v B.D. 2007. SCC 38, [2007] 3 S.C.R. 83.
ter Neuzen v Korn. 1995. 10 W.W.R. 1 (S.C.C.).
Van Mol v Ashmore. 1999. 168 D.L.R. (4th) 637 (B.C.C.A.).
Wilson v Swanson. 1956. 5 D.L.R. (2d) 113 (S.C.C.).

Part III
Implementation and Lived Experience

Chapter 8
Paediatric Patient-Centred Care at SickKids: Advancing Child & Family-Centred Care

Krista Keilty, David B. Nicholas, Karen Sappleton and Karima Karmali

This chapter presents an overview of how C&FCC is currently implemented at SickKids. It describes in brief some historical influences and portrays some exemplary practices led by staff and families at SickKids. The use of case studies is employed to help illustrate some of the persistent tensions and challenges in delivering C&FCC in this specialty paediatric hospital that serves children and families with highly complex and diverse needs. The chapter ends with a description of some of the considerations and contemporary strategies planned for continuing to advance C&FCC at SickKids.

8.1 Brief History of C&FCC at SickKids

Opened in 1875 in Toronto by Elizabeth McMaster to serve children with illness, SickKids has a rich history that, in its early years, followed the historical convention related to the role of children and family in health care. Of foundational and historical note, the first children's hospitals opened in Europe in the nineteenth century (Jolley and Shields 2009), and SickKids was the first children's hospital to open in Canada (Young 1992).

While SickKids brings a long-term commitment to the health & wellbeing of children and families, its early history reflects an era when C&FCC was not a prior-

K. Keilty (✉) · K. Sappleton · K. Karmali
The Hospital for Sick Children, Toronto, ON, Canada
e-mail: krista.keilty@sickkids.ca

K. Sappleton
e-mail: karen.sappleton@sickkids.ca

K. Karmali
e-mail: karima.karmali@sickids.ca

D. B. Nicholas
University of Calgary, Calgary, AB, Canada
e-mail: nicholas@ucalgary.ca

R. Zlotnik Shaul (ed.), *Paediatric Patient and Family-Centred Care: Ethical and Legal Issues*, International Library of Ethics, Law, and the New Medicine 57,
DOI 10.1007/978-1-4939-0323-8_8, © Springer Science+Business Media Dordrecht 2014

ity in paediatric health care. It is described that SickKids, "was reluctant to adopt innovations in psychosocial aspects of care and, in particular, was slow to allow families' greater access to their children" (Young 1992, p. 1422). Early experience suggests inequalities that existed due to a family's socio-economic status. It is reported that parents with lower socioeconomic status had less access to their children than those from middle-class families during the inter-war period (Young 1992).

Perspectives shifted with the emergence of research, theory and advocacy. While some health care providers, prior to 1960, strongly opposed the presence of parents in the hospital, the work of John Bowlby and James Robertson (Alsop-Shields and Mohay 2001) on attachment and separation was taught by the Hospital for Sick Children School of Nursing instructors (Young 1992). These teachings eventually led to attitudinal shifts by SickKids staff and trainees who learned that parents' presence had the potential for improved child development. Over the span of a number of decades, owing to the influences of child development research and child and family advocates, paediatric care teams became increasingly committed to considering the experiences and perspectives of children and families within the planning, delivery and evaluation of health care.

Today, as a paediatric academic health sciences centre partnered with the University of Toronto, SickKids provides multi-level care to children and their families, meeting the needs of families locally, nationally and internationally. Primary and secondary care is provided for the population of children and families in metropolitan Toronto, Canada, with tertiary and quaternary care being provided to children from across the country and from around the world. Led by SickKids International, the hospital also engages in global opportunities to enhance child health and build system capacity through their Global Child Health Program. The facilities and activity level at SickKids are remarkable and continuing to grow. Annually, in just under 300 beds, the atrium and adjacent facilities host approximately 100,000 inpatients days; in over 100 clinics, there are more than 200,000 ambulatory visits; and in the emergency department (ED) there are more than 60,000 visits (Hospital for Sick Children 2011/2012). A new Research and Learning Tower opened in 2013 that has brought SickKids researchers under one roof and enabled enhanced collaboration and innovation in child health. In total, over 10,000 staff, physicians, trainees and volunteers work together at SickKids to approach care from the premise that families are the most important partners in the successful care of a child (Hospital for Sick Children 2010/2011).

At present, C&FCC is firmly positioned as a priority at SickKids. Led in recent years by key organizational leaders committed to excellence in child health and C&FCC, and in keeping with recent policy thrusts that include the directive to be patient-centred, SickKids has renewed its commitment to the continuous advancement of C&FCC. Systems and supports have been put in place to ensure that the hospital builds on its strong base and leads in the development and implementation of care practices informed by C&FCC. To this end, SickKids has established a Centre for Innovation and Excellence in Child & Family-Centred Care (the Centre). The Centre will identify and implement best practices in C&FCC, with the aim of enhancing care not only through clinical practice, but also by advancing education and research. A key aim of the Centre is to test and evaluate new ideas that can bring advancements in C&FCC. The Centre is building structures and processes to lend support to the implementation and evaluation of C&FCC practices. It will also conduct primary research in developing and testing promising practices examining the impact of C&FCC on patients and families.

8.2 C&FCC Practices at SickKids

Many recent initiatives serve to demonstrate SickKids' success in adopting and leading in the delivery of exemplary C&FCC. In this section of the chapter, a number of these leading practices are described. They have been organized under a model of C&FCC as originally described by the Institute of Patient and Family Centered Care (IPFCC 2010) and since revised for application at SickKids with the addition of the statement, '*Communication with children in a way that it is easy for them to understand and be understood by others*' (adapted, Keilty et al. 2012). The model of C&FCC describes four core components, outlined below, which further guide the organization of this chapter:

- Dignity and Respect—Health care providers listen to and respect the decisions of patients and their families, attempting to incorporate their values and beliefs into the rendering of care including planning and delivery.
- Information Sharing and Communication—Health care providers communicate with patients and their families in a clear, timely and unbiased fashion. Patients and families receive timely, complete and accurate information in order to effectively participate in care and decision-making. Communication with children should be in a way that it is easy for them to understand, and be understood by, others (adapted, Keilty et al. 2012).
- Participation—Patients and families are encouraged and supported to participate in care and decision-making to the extent that they choose.
- Collaboration—Patients and families collaborate with health care providers to advance and evaluate policy and program development, implementation and evaluation; health care facility design; and professional education (IPFCC 2010).

Each of these components is a priority in the delivery of comprehensive C&FCC at SickKids. This approach broadly recognizes both the child and family as central in the child's care, with the child and family duly recognized as key 'systems' to be 'engaged with,' rather than entities for whom health care practitioners may 'do for' in the delivery of paediatric care. Families, not the health care team, are the constant in the child's life; hence their perceptions and information are critical in complementing the role of the health care team. Moreover, these commitments to C&FCC at SickKids have increasingly been linked to the literature that describes the desired outcomes of C&FCC on quality (Davidson et al. 2007; National Consensus Project for Quality Palliative Care 2009) and patient safety (Institute of Medicine, Committee on Quality of Health Care in America 2001; O'Malley et al. 2008; Muething et al. 2007; Rathert and May 2007; Reid Ponte and Peterson 2008).

8.3 Recent Innovations at SickKids

Over the last 20 years, SickKids has built a strong foundation to support C&FCC. The hospital has undertaken many initiatives that have resulted in C&FCC becoming a core philosophy and part of an organizational culture whereby the centrality of the family in the life of the child is well understood, and partnership with the family is fundamental to the care delivery process. The integration of the family perspective has increasingly been recognized as vital to the development of all hospital processes, and is tied to strategy and performance management, along with patient care outcomes.

This next section highlights examples of more recent innovation in C&FCC that have served to further the understanding and evolution of C&FCC at SickKids.

8.3.1 Respect and Dignity through Cultural Competence

Toronto is one of the most culturally diverse cities in the world, with over 190 languages spoken in the Greater Toronto Area. Canada welcomes more than 200,000 immigrants annually, with 31 % choosing Toronto as their home (Citizenship and Immigration Canada 2011). This diversity is reflected in the population of children and families served at SickKids.

There is growing evidence that the quality of care and patient safety can be compromised when healthcare providers do not respond appropriately to language barriers and cultural factors (Gany et al. 2010; Institute of Medicine, Committee on Quality of Health Care in America 2001). Guerrero et al. (2010) describe that while parents generally reported positive experiences of C&FCC, disparities existed across racial and ethnic groups. C&FCC inherently needs to be equitable and recognized as important for all patients and families, given the emphasis that C&FCC models place on respect and dignity for patient and family values and beliefs. At SickKids, the delivery of culturally competent care is regarded as integral to C&FCC. A culturally competent health care system has been defined as one that, "…acknowledges and incorporates—at all levels—the importance of culture, assessment of cross-cultural relations, vigilance toward the dynamics that result from cultural differences, expansion of cultural knowledge, and adaptation of services to meet culturally unique needs." (Betancourt et al. 2003, p. 294).

In 2009, SickKids embarked on a multi-year initiative to systematically advance cultural competence at both the clinical and organizational levels. A number of initiatives that focused on enhancing the experience of patients and families new to Canada were implemented, with a particular focus on comprehensive cultural competence education for clinicians and other healthcare staff. About 2,000 clinicians participated in this education, which aimed to increase understanding of (1) health inequities faced by new immigrants, (2) personal biases and its impact on patient- and family-healthcare provider relationships, (3) the relationship between cultural competence, C&FCC and patient safety, (4) differences in health beliefs, parenting practices and expressions of pain and grief across cultures, and (5) cross-cultural communication, including the importance of working with interpreters to help address language barriers. Clinicians participated in role-plays and simulations with standardized cases, and were equipped with tools and resources they could utilize in their day-to-day practice. And many made written commitments regarding changing their practice by utilizing specific interventions to deliver culturally competent care.

SickKids has further translated key patient/family education materials and other documents into nine commonly spoken languages, to facilitate the communication of important information to families with limited or no English proficiency. These materials have been placed on an open access SickKids-sponsored health information website for broad distribution (www.aboutkidshealth.ca). These initiatives cumulatively have resulted in significant changes in clinician practice, including an increase in the utilization of interpreter services and translated documents. These changes in practice, in keeping with the core component of dignity and respect for all children and families at SickKids, have also had a clear and significant positive impact on patient and family satisfaction scores related to the delivery of culturally sensitive care by staff (Karmali et al. 2011).

8.3.1.1 Information-Sharing Through the Use of Complex Care Plans

Information-sharing is an essential component of C&FCC. It consists of integrated steps: first, listening to the perceptions of children and families; and second, timely and unbiased communication to patients and their families. Information-sharing is vital to effective care. This includes learning among children and families, and engagement in optimal health care decision-making.

An important initiative in support of quality information-sharing and communication with children and families at SickKids has been the widespread development of family-mediated, written care plans for children in the Complex Care Program (Adams et al. 2010). This care plan is jointly developed by a paediatric nurse practitioner in conjunction with the family, and it contains diagnostic and care details related to the child with special health care needs. Reviewed and updated regularly, the care plan offers an accessible resource for timely and comprehensive information-sharing and communication between the family and care providers as needed, including those in the community that are outside the immediate circle of health care (e.g., emergency room, school, etc.). The plan, importantly, is held by the family and can be changed as indicated by the context of the child's health needs. This comprehensive plan outlines detailed information about the diagnoses and care requirements, allowing the family to limit the frequency with which they must repeat their child's complex medical history and complex treatment regimes. Accordingly, rather than having to share their health-related 'story' at length with each new care provider, the family has an accessible tool that can be used by the family as needed and desired. Of importance, the family is a partner in the development of this communication tool, hence both the information and its transmission are mutually shared in ways that respect family preferences and foster egalitarianism, comprehensiveness and rapid access to needed information, which in turn, promote patient safety as well as seamlessness and quality of care.

8.3.2 Participating in Supportive Transitions

The transition from paediatric to adult health care is a time of stress and opportunity for adolescents with special health care needs and their families. With over 90 % of youth with special health care needs surviving into their young adulthood due to the significant advancements in treatments and medical technology, there is substantial need for youth and their families to participate in preparing for the adult system (Betz and Smith 2011; Pai and Schwartz 2011). The goal of transition is to engage youth with a coordinated, uninterrupted and developmentally appropriate transfer to the adult system (Canadian Paediatric Society 2007), but this is often complicated by the already complex medical, social and psychological needs that adolescents experience as they navigate healthcare and other facets of their developing selves.

In keeping with government funding and the hospital's mandate, SickKids requires youth to graduate to an adult system by 18 years of age. Patients, families and health care providers have perceived multiple barriers to a successful transition, including practice differences between paediatric and adult systems, and difficulty in finding a knowledgeable and experienced adult provider (Shaw and DeLaet 2010). Such barriers can create a sense of stress and fear of the unknown, and the youth and families benefit from being active participants and receiving support from the health care systems, in facilitating a successful process as they transition from adolescence to adulthood in the context of their developmental and medical needs.

Housed under the Division of Adolescent Medicine, the 'Good 2 Go Program' at SickKids, established in 2006, provides direct transition support to patients and families, as well as education and consultation with other hospital programs. It is a centralized repository of transition-related information, with a mandate to support clinical programs in developing tailored transition pathways and resources. A number of tools have been developed to promote patient participation and improved communication between and among healthcare providers, the youth and families (www.sickkids.ca/good2go). The team also has a focus on research and evaluation, and supports hospital programs in designing and implementing research projects and quality improvement initiatives. Also due to the fact that more children with very complex medical needs and compromised developmental or cognitive states are surviving into their adult years, the Good 2 Go program is additionally collaborating with parents, SickKids teams and adult systems, to promote effective transition planning for youth whose condition will prevent their independence in adulthood.

8.3.3 Collaboration through Family Engagement & Representation

Family engagement and representation are key components to providing quality, ethical and safe C&FCC. Leading organizations and authors suggest that in order to improve patient safety, patient- and family-centred care principles need to be implemented, positioning children and families as collaborators in this process (The Institute of Medicine, Committee on Quality of Health Care in America 2001; O'Malley et al. 2008; Muething et al. 2007). Therefore, collaborating with children and families at SickKids has been a key priority for the organization as it strives for excellence in quality care, service delivery and improved health outcomes for all children.

The Children's Council at SickKids consists of 10 youth, aged 9–18 years, who are patients of the hospital. These children and adolescents collaborate with Child Life staff and serve as advisors to the hospital. The Children's Council is a resource to staff regarding projects and planning. The Council serves as an advocate promoting children's health, an innovator in proposing changes to improve the experience for children and families at the hospital, and a leader in offering guidance for C&FCC initiatives both in the hospital and the community at large.

The Family Centred Care Advisory Council (FCCAC) is a group of about 20 volunteers, including former patients (now adults), parents and adult family members of current or former patients, and SickKids staff. Like the members of the Children's Council, the FCCAC members advocate, lead, advise and support innovation in the area of C&FCC. They also collaborate as faculty in working with staff to educate healthcare providers around issues of communication and patient and family care. The members of the Council serve on multiple committees and short-term projects in the hospital (e.g., patient safety, quality improvement, patient-care, facilities planning). Some of these Council members also serve on committees in specific clinical programs, such as Haematology/Oncology, Neonatal Intensive Care and Critical Care.

While the FCCAC members are very active in their involvement throughout the hospital, the cultural, health literacy and socio-economic diversity that exists in the patient populations at SickKids is not yet accurately reflected by the Council itself. Strategies to reduce this disparity have been initiated. Initiatives are in place to encourage and support representation from more parents, children and additional family members who would like to participate in various roles related to education and committee endeavours across the Hospital. This work requires careful orchestration and support of patient and family representatives, with attention paid to inclusiveness and enabling diverse perspectives. Therefore, a recruitment campaign is in process to establish a network of child and family volunteers to support the requests and needs of the organization in seeking family involvement and enhancing family collaboration. The development of a formal application and screening process along with a C&FCC volunteer information package is also well underway. The package will include information about the role of child and family representatives at SickKids, along with relevant hospital policies (e.g., confidentiality, infection control) and other information required to be successful in various child and family roles (e.g., how to be reimbursed for out-of-pocket expenses). Also, in progress is the refinement of a 'family as faculty' program that will enable family members to participate in the education of clinicians. The program will include education for children and families in how to use their personal experience and real stories in promoting the uptake of knowledge and understanding of a C&FCC approach among health care providers in-training.

8.4 Challenges & Tensions in Providing C&FCC

Despite a long history of excellence and a supportive culture at SickKids, some challenges remain in fully implementing aspects of C&FCC. In the section that follows, case studies are used to help illustrate current tensions and challenges experienced in consistently incorporating the four components of our C&FCC model of care.

Case 1: *Dignity & Respect* in Considering Caregiver Burden

Reba is a 3-year-old girl with a rare congenital condition associated with multiple health problems, including sensory disturbances, airway and breathing problems, cardiac defects, feeding and nutrition challenges, and developmental delay. Reba is one of four children in her family. The family has lived in Canada for less than 5 years and resides over an hour away from the hospital. Reba is required to be followed long-term by a minimum of nine different clinical services at SickKids. Reba's father is the sole driver in the family. He tells the SickKids social worker that he is under tremendous stress, carrying the burden of providing for the family, while worried that he is at risk of losing his job due to the number of days required to take off to bring Reba to the hospital.

Dignity and respect for children and families can be endorsed theoretically and institutionally, yet the application of these commitments in practice as well as in operational governance, is inherently complex. In keeping with aims to respect the burdens experienced by families requiring services at SickKids, several highly specialized clinics have been developed to accommodate the needs of children with medical complexity. Examples include the Cystic Fibrosis (CF) Clinic that brings together internationally renowned specialists and interprofessional team members across multiple specialities to meet the care needs of children with this life-limiting illness (http://www.sickkids.ca/RespiratoryMedicine/What-we-do/Cystic-fibrosis-clinic/index.html). The interprofessional team in the CF Clinic includes physicians from numerous clinical services (e.g., Genetics, Respiratory Medicine, Gastroenterology, Endocrinology), nurses, nurse practitioners, physiotherapists, psychologists, social workers and others engaged in the care of the child with CF and their family. The team meets both before and after the CF Clinics, to coordinate care and follow-up for the child and family.

Integrated clinics in family-centred facilities demonstrate respect for the child and family by easing scheduling challenges (e.g., minimizing lost time from school or work) and enabling strong communication. However, the demand for integrated clinics among children and their families still often outweighs the health care system's capacity to address the entirety of family need. Despite admirable intentions, coordinating the busy schedules and priorities of these multiple specialists has proven difficult to orchestrate.

While it is clear that by requiring fewer clinic or hospital visits, an integrated clinic approach demonstrates respect for the time, responsibility and financial burden on patients and families, the inherent logistical challenges in delivery, especially for clinical conditions that are very rare and complex, often impose barriers to such coordination of care. The opportunity to explore the use of technologies that would enable increased access through scheduling flexibility and the availability of evening clinic appointments are included among the ideas that may help ease this burden. Such innovation complements the aim of the ongoing development of 'communities of practice' at SickKids, which will enable the establishment of more integrated family-centred clinical services.

Case 2: *Information-Sharing & Communication* in the Emergency Department

Eleven-year-old Tenzin presents to the Emergency Department (ED) with a severe headache and rash. She is accompanied by her parents and two younger siblings. Doctors and nurses in the ED are concerned that the cause of Tenzin's symptoms may be serious. They need to clarify aspects of her health history, including immunization history and exposure to infectious contacts. The team knows that if they are to effectively treat Tenzin, rapid testing and intervention are required, yet the parents speak little to no English. The team is limited as to what can be quickly explained to Tenzin and her parents in a way that they can easily understand. While preparing Tenzin for an invasive procedure, the ED team notices that Tenzin's 6-year-old brother is interpreting for his parents.

Despite the recognition that information-sharing and communication with children and families in a way that is clear and personalized are important, barriers to this C&FCC practice arise. For instance, if the language of the child or family is different from that of the health care team, information may not be clearly understood. In such instances, where the child and family do not speak English or the language of their direct care provider, SickKids staff relies on interpreter services to assist them in promoting understanding of relevant information both spoken and provided by way of written materials. Moreover, beyond the language of the family, the culture of the hospital may impose confusion and misunderstanding based on jargon or medical language. Professionals must ensure that the language used is easily understandable, and procedures should be thoroughly explained.

The Interpreter Services Department at SickKids is an on-site, service-oriented resource which strives to provide the most appropriate and highest quality language service to patients and families who receive care at SickKids and whose English is limited or who are hearing impaired. For risk and quality reasons, the Interpreter Services Department supports only the use of trained interpreters and does not rely on other hospital staff or families at the hospital (http://www.sickkids.ca/ProgramsandServices/Interpreters-Services/index.html). Although recognized as a highly regarded service, timely access to on-site interpreters for all clinical encounters at SickKids is not always possible. In situations when visits to the hospital are unplanned, staff may use a commercially available resource that enables real-time telephone interpretation support for urgent and other care needs. While an excellent resource that is available 24 hours a day, 7 days a week for over 100 languages, many staff and families report that the use of the telephone as a communication tool in a time of stress is often unsatisfactory. With this in mind, SickKids, as a tertiary care centre, is collaborating with regional health equity authorities in seeking innovative and effective ways to enhance the exchange of health-related information between practitioners, young patients and their families who may not share a language.

Case 3: *Participation* in Family-Centred Rounds

Charlie is a 4-year-old boy with a severe form of an inherited gastrointestinal condition which, among other complications, has resulted in his failure to thrive. He has been hospitalized repeatedly since his diagnosis as an infant and has undergone three major abdominal surgeries over the last year and a half. Charlie's mother has the same condition and chooses for health reasons to sleep at home, making it difficult for her to be at his bedside in the early morning when the surgical team does its rounds. The surgeon who cares for Charlie is committed to providing C&FCC. Despite substantial efforts to do so, the surgeon has not met with Charlie's mother in almost a week. Charlie's mother tells the nurse that she does not feel that her opinions are being sought about what is best for Charlie, and that important decisions about him are being made without her.

On a paediatric unit, patient-care rounds are often used to review the child's status and establish a plan of care. At SickKids, most inpatient units strive to organize interprofessional rounds that enable the input of many team members, including children and families, in a coordinated way. Some units have made the integration of children's and families' voices in rounds a priority, through the implementation of well-designed, quality improvement projects, grounded in the principle that child and family participation in rounds is a core component of C&FCC. On these units, staff actively partner with families about what level of participation they prefer in daily rounding and decision-making regarding their children's care. This practice of including children and family in rounds has been described by researchers as an important component of C&FCC for the hospitalized child, and one that promotes care coordination and enhances clinical teaching (Muething et al. 2007).

The availability of family-centred rounding, however, is uneven across the hospital. On many units, rounds still occur in the early morning hours when many parents of long-stay children in the hospital may be asleep owing to their chronic sleep disturbances, or may not be present at the bedside at all. One paediatric specialty unit, in recognition of this, has shifted their rounds from 8:00 a.m. to 10:00 a.m. and as a result has experienced both enhanced satisfaction among the health care providers and improved participation among families in rounds (Coffey, personal communication, Dec. 6, 2012). Other commonly held beliefs are that the inclusion of children and families will result in longer rounds and potentially impaired clinical teaching. However, there is growing experiential evidence at SickKids to the contrary (Nanji, personal communication, Aug. 12, 2012; Coffey, personal communication, Dec. 6, 2012), and recently published paediatric data suggests that family participation may shorten inpatient rounds and enhance family and staff satisfaction (Rappaport et al. 2012). In this observational study, the influence of family presence on the perceived autonomy of senior trainees was reported to be in the negative direction (i.e., they felt less in charge) and thus the authors concluded that this relationship requires further investigation. Family-centred rounding has also been studied by a former SickKids Clinical Nurse Specialist, who conducted a pilot feasibility trial of family-centred rounds on an inpatient unit at SickKids. The intervention included the placement of chairs in the room for all team members and the provision of a 'legend' of commonly used terms and medical jargon. When encountering a study participant, teams would sit with the family at the bedside and review the child's care plan together in an informed way. Study findings suggest that this family-centred intervention, which aimed to support family participation and improve their satisfaction with the decision-making process, was both feasible and well accepted among families and health care providers alike (LeGrow 2011).

Case 4: *Collaboration* in Child-Health Research

Mai and Danny have been bringing their son Joshua to SickKids for over 6 years. Joshua has Cerebral Palsy as a consequence of extreme prematurity and has been followed by multiple subspecialists over the years. Mai and

Danny have consented to Joshua participating in a number of research studies at SickKids. While in the clinic waiting room for a follow-up visit, Joshua's parents are approached by their nurse coordinator, who asks if they would be interested in hearing about another study being conducted by Joshua's care team. Mai and Danny are unsure what to do as they have appreciated the opportunity to 'give back' to SickKids through research, but they think Joshua is bothered by all the tests and procedures. They suggest he likely does not understand why it is important. They also wonder what the results are of the last study Joshua participated in and, overall, if their participation in research is making a difference.

In an examination of the Dana-Farber Cancer Institute in Boston, Reid Ponte and Peterson (2008, p. 451–452) stated that, "staff realized that tapping into the diverse perspectives, knowledge, and passions of patients and families not only bolstered the organization's efforts to achieve excellence in care and clinical trials but also added significantly to the quality, safety, operational effectiveness, and long-term strategic direction of organizational structures, processes, and outcomes". This invites further exploration into how children and families can collaborate to identify the (i) constitutive elements for child and family involvement in determining priorities for research and education in children's health, and (ii) best ways to translate knowledge in a manner that is discernible and engaging for children and families. Recent dialogue at SickKids has addressed how such a role for 'child and family-centred research' could be fostered. Increasing attention has been placed on effective ways to generate knowledge about research findings among children and families considering age-appropriate and digestible ways to talk about research, as well as engage children and families in research conversations with the community of science. Given the overall aims of integrated leadership in care, as well as education and research at SickKids, this application of the principles of child and family-centredness where children and families collaborate in research processes invites further development.

In an example of their commitment to knowledge translation that includes the children and families who participate in research studies, researchers at SickKids (Stinson et al. 2006) have developed an electronic diary that captures data about chronic pain from children with arthritis. A qualitative usability testing approach has been used to record data about the pain experienced by adolescents and also gather data that address ease of use of the device. In this example, adolescents were involved in the research process such that they generated ideas on how the pain diary could be improved. These findings have been broadly shared in ways that adolescents can easily understand (e.g., developmentally appropriate project summaries and interactive webcasts), and this work demonstrates that children and families can effectively collaborate with scientists in the conduct of high quality research (http://www.eouchpaindiary.ca/index.php).

Considerable translational research at SickKids has been directed toward the effective communication of research results to children and families. However, continued work seems needed in considering ways to more broadly incorporate the perspectives of children and families in research planning and priority setting.

8.5 Advancing C&FCC at SickKids

Building on previous success and establishing a baseline from which to ground innovation and advance the field, the Centre for Innovation and Excellence in Child & Family Centred Care at SickKids recently implemented a comprehensive organizational assessment and environmental scan examining current and promising practices for C&FCC. This review has built upon a sustained investment at SickKids in advancing C&FCC and sought to answer: (1) What is the state of C&FCC at SickKids?; (2) What are the (internal & external) leading practices of C&FCC?; and (3) What practices, if extended, initiated or evaluated, will continue to advance C&FCC at SickKids?

A mixed method design consisted of an extensive literature review that examined peer reviewed and online literature in C&FCC, external consultation with leading centres across North America examining best practices in C&FCC, and an internally-focused organizational assessment. Guided by core concepts of C&FCC derived from the Institute for Patient- and Family-Centered Care (2010), the organizational assessment utilized several methods for data collection including a comprehensive C&FCC survey administered to program managers across all clinical areas. Focus groups were conducted with groups of children, families and other specialty clinical or practice groups (e.g., pain team, palliative care team). Individual interviews were conducted with clinical managers, physicians and other key organizational leaders.

Preliminary findings from this organizational assessment suggest strong organizational support for C&FCC at all levels, with examples of many practices that are considered to be 'well established' and/or 'works in progress.' Recommendations were generated from a series of ranking exercises among key stakeholders including children and families, who endorsed the need for enhanced sibling support as a priority practice for the advancement of C&FCC at SickKids (Keilty et al. 2012). It was notable that sibling support was nominated as a priority practice by many informants but was especially highly ranked among the children participating in the organizational assessment. Children's voices at SickKids came through strongly as they communicated, through this organizational assessment, their concern for their siblings and suggested that SickKids can do more to meet siblings' unique needs.

Another of the priority practices identified to advance C&FCC at SickKids was the presence of children and families at nursing shift 'handover.' This practice has been suggested by O'Malley et al. (2008) who described that including families in nursing shift reports promoted patient safety and should be a goal of C&FCC. They further emphasized that involving patients and families during a hospital stay can improve patient safety in both the short and long term. These findings are in keeping with an investigation of three US acute care hospitals that have found that nurses who felt that their units had a patient-centred focus reported less frequent medical errors, as well as overall increased willingness to report errors (Rathert and May 2007). As illustrated by this literature, a linkage between C&FCC and outcomes is increasingly emerging, as demonstrated by family presence at nursing handover and patient safety. Outcomes such as patient safety hold obvious importance when considering quality care and ethical practice. Accordingly, further analysis that examines C&FCC relative to these important priorities and health care imperatives seems critically important in moving forward.

8.6 Conclusions

In this chapter, we have addressed contextual elements of C&FCC in a large urban tertiary and quaternary care facility. We conclude that C&FCC at SickKids presents both similarities and differences relative to other sectors of paediatric health care delivery. It can be acknowledged that there are situations where tensions, challenges and even conflict emerge in implementing C&FCC. Yet, overall, we believe that the necessary foundational supports and structures are in place at SickKids (e.g., child and family representation, expert paediatric providers, clinical ethicists, and executive level support for C&FCC) to promote optimal care that is sensitive to the needs of children and their families.

Despite the lack of high quality evidence for C&FCC as described more fully in Chap. 1, the growing literature examining models of care and practice in C&FCC suggests potential benefits at individual, family, facility and community levels (Trajkovski et al. 2012). At SickKids, C&FCC has invited critical discussion about ways in which core concepts of dignity and respect, information sharing, communication, participation and collaboration are instilled in systems and support. Clearly, C&FCC enhances daily care and organizational processes, and serves as a gauge to examine and build upon our aim and foundation of excellence in care to children and families.

References

Adams, S., E. Cohen, and S. Mahant, et al. 2010. The usefulness of care plans for children and youth with special health care needs. *Paediatrics and Child Health* 15:71.

Alsop-Shields, L., and H. Mohay. 2001. John Bowlby and James Robertson: Theorists, scientists and crusaders for improvements in the care of children in hospital. *Journal of Advanced Nursing* 35 (1): 50–58.

Betancourt, J. R., A. R. Green, and J. E. Carrillo, et al. 2003. Defining cultural competence: A practical framework for addressing racial/ethnic disparities in health and health care. *Public Health Reports* 118 (4): 293–302.

Betz, C., and K. Smith. 2011. Measuring health care transition planning outcomes: Challenges and issues. *International Journal of Child Adolescent Health* 3 (4): 463–472.

Canadian Paediatric Society. 2007. Transition to adult care for youth with special health care needs. *Paediatrics and Child Health* 12 (9): 785–793

Citizenship and Immigration Canada. 2011. http:www.cic.gc.ca. Accessed 10 Oct 2012.

Davidson, J. E., K. Powers, and M. Hedayat, et al. 2007. Clinical practice guidelines for support of the family in the patient-centered intensive care unit: American College of Critical Care Medicine Task Force 2004–2005. *Critical Care Medicine* 35 (2): 605–622.

Gany, F. M., C. J. Gonzalez, and G. Basu, et al. 2010. Reducing clinical errors in cancer education: Interpreter training. *Journal of Cancer Education* 25 (4): 560–564. doi:10.1007/s13187-010-0107-9.

Guerrero, A. D., J. Chen, and M. Inkelas, et al. 2010. Racial and ethnic disparities in pediatric experiences of family-centered care. *Med Care* 48 (4): 388–393. doi:10.1097/MLR.0b013e3181ca3ef7.

Hospital for Sick Children. 2010/2011. SickKids annual report. Hospital for sick children.

Hospital for Sick Children. 2011/2012. SickKids annual report. Hospital for sick children.

Institute for Patient- and Family-Centered Care. 2010. Frequently asked questions: What are the core concepts of patient- and family-centered care? http://www.ipfcc.org/faq.html. Accessed 29 Dec 2010.

Institute of Medicine, Committee on Quality of Health Care in America. 2001. *Crossing the quality chasm: A new health system for the 21st century*. Washington: National Academy Press.

Jolley, J., and L. Shields. 2009. The evolution of family-centered care. *Journal of Pediatrics and Nursing* 24 (2): 164–170. doi:10.1016/j.pedn.2008.03.010.

Karmali, K., L. Grobovsky, and J. Levy, et al. 2011. Cultural competence for improving access to quality care. *Healthcare Quaterly* 14:52–57.

Keilty, K., J. Blumberg, and K. Karmali, et al. 2012. Organizational review of child and family-centred care practices at SickKids. Poster presented at the Canadian Association for Paediatric Health Centres, Vancouver, November 2012.

LeGrow, K. S. 2011. A feasibility study of a Bourdieu-informed parent briefing intervention to improve parents' satisfaction with decision making for hospitalized children with complex health care needs. Dissertation, University of Toronto.

Muething, S. E., U. R. Kotagal, and P. J. Schoettker, et al. 2007. Family-centered bedside rounds: A new approach to patient care and teaching. *Pediatrics* 119 (4): 829–832.

National Consensus Project for Quality Palliative Care. 2009. Clinical practice guidelines for quality palliative care. 2nd ed. http://www.nationalconsensusproject.org. Accessed 15 Sept 2012.

O'Malley, P. J., K. Brown, and S. E. Krug. 2008. Patient and family-centered care of children in the emergency department. *Canadian Journal of Emergency Medicine* 10 (1): 38–43.

Pai, A. L., and L. A. Schwartz. 2011. Introduction to the special section: Health care transitions of adolescents and young adults with pediatric chronic conditions. *Journal of Pediatric Psychology* 36 (2): 129–133.

Rappaport, D. I., T. A. Ketterer, and V. Nilforoshan, et al. 2012. Family-centered rounds: Views of families, nurses, trainees, and attending physicians. *Clinical Pediatrics* 51 (3): 260–266.

Rathert, C., and D. R. May. 2007. Health care work environments, employee satisfaction, and patient safety: Care provider perspectives. *Health Care Management Review* 32 (1): 2–11.

Reid Ponte, P., and K. Peterson. 2008. A patient- and family-centered care model paves the way for a culture of quality and safety. *Critical Care Nursing Clinics of North America* 20 (4): 451–464. doi:10.1016/j.ccell.2008.08.001.

Shaw, T. M., and D. E. DeLaet. 2010. Transition of adolescents to young adulthood for vulnerable populations. *Pediatrics in Review* 31 (12): 497–505. doi:10.1542/pir.31-12-497.

Stinson, J. N., G. C. Petroz, and G. Tait, et al. 2006. e-Ouch: Usability testing of an electronic chronic pain diary for adolescents with arthritis. *The Clinical Journal of Pain* 22 (3): 295–305.

Trajkovski, S., V. Schmied, and M. Vickers, et al. 2012. Neonatal nurses' perspectives of family-centred care: A qualitative study. *Journal of Clinical Nursing* 21 (17–18): 2477–2487. doi:10.1111/j.1365-2702.2012.04138.x.

Young, J. 1992. Changing attitudes towards families of hospitalized children from 1935 to 1975: A case study. *Journal of Advanced Nursing* 17 (12): 1422–1429.

Chapter 9
The Effect of Patient and Family-Centred Care (PFCC) Delivery Model on Paediatrics at the Thunder Bay Regional Health Sciences Centre

Bonnie Nicholas

9.1 TBRHSC and PFCC Overview

The practice of delivering healthcare services according to the tenets of Patient and Family-Centred Care (PFCC) tends to provide a better experience for patients and families. In early 2009, Thunder Bay Regional Health Sciences Centre (TBRHSC) adopted the PFCC model in all aspects of its operation, from direct patient care to changes to the hospital environment.

Patient Family Advisors (PFAs) are at the heart of the PFCC model of care. Over 90 PFAs ranging in age from 8 to 85 have been integrated into everything we do, including staff hiring, board quality committees, education, program/service councils, and strategic planning. The responsibility of each department to partner with its patients and families in the development of annual action plans based on patient and family feedback ensures their needs and values are considered in the delivery of care and services. Within 18 months of starting this journey, TBRHSC has improved inpatient satisfaction scores by 6.8–21.6 % in all eight dimensions of care as reported by National Research Corporation (NRC) Picker™.

This chapter will explore the PFCC model in a smaller urban/rural paediatric centre and examine practical and ethical tensions and synergies when applying the PFCC model to a younger patient population.

Demographics

The TBRHSC is a 375-bed academic health sciences centre and a regional acute care facility located on the north shore of Lake Superior in Thunder Bay, Ontario, Canada. It provides services to Northwestern Ontario which has a population of 250,000 residents scattered over a geographical area the size

---

B. Nicholas (✉)
Thunder Bay Regional Health Sciences Centre, Thunder Bay, Canada
e-mail: nicholab@tbh.net

R. Zlotnik Shaul (ed.), *Paediatric Patient and Family-Centred Care: Ethical and Legal Issues,* International Library of Ethics, Law, and the New Medicine 57,
DOI 10.1007/978-1-4939-0323-8_9,

of France. There are 122,000 people living in the city of Thunder Bay with the remainder living in small communities as far as 850 km north of Thunder Bay, including 70 First Nations communities, 24 of which are accessible only by air or winter ice roads. This translates into 60 % of Ontario's landmass with only 2.3 % of the population.

Process of a Journey

The way we approach providing healthcare is changing. Up until recently, most healthcare professionals tended to treat the disease rather than care for the patient. However, the net result of that approach was that, in many cases, patients felt like they were left out of the process. At worst, they felt like they were part of a medical assembly line. This in itself is a good enough reason for change.

What is even more important is that we now know that outcomes tend to be much better when patients are engaged and involved in their own healthcare. To bring about that change, TBRHSC embarked on a quest for quality patient care. Three years of research, a review of leading practices, and site visits to organizations across the nation revealed that there was an approach that would meet this goal—Patient and Family-Centred Care. In 2009, the TBRHSC adapted the definition of patient-centred care (Committee on Quality of Health Care in America, Institute of Medicine 2001) to read, "Patient and Family-Centred Care is the provision of care that is respectful of, and responsive to, individual patient/family preferences, needs and values, and ensures that these values guide all clinical decisions." Recognizing the importance of family and care partners in the care process, we have developed an approach to care that is supported through collaborative partnerships with patients, families, and the care team. It is critically important that we recognize and value these partnerships in any new approach to healthcare.

There are many challenges to integrating any system-wide change, given that it must represent change to the whole model and approach to healthcare. Our first priority was to consult and engage our patients, families, staff, physicians, volunteers, and community partners in a discussion about our care. Coming together at a "Visioning Retreat" to hear stories from the experts who experienced the care—our patients and families—we created plans to innovate and integrate PFCC into our plan of care. Our objectives included: improving patient, staff and physician satisfaction, improving health outcomes, enhancing teamwork, improving quality and safety, aiming for fewer medical errors, and improving use of resources. We developed a shared understanding of Patient and Family-Centred Care and what it would look like in policy, program, teaching, research, and practice. Together, we created the core foundation for our regional TBRHSC to move forward toward the ideal of optimal care for every patient, every time.

This led to a new set of Mission, Vision, and Values statements that would reflect our commitment to PFCC. Resources and corporate strategies were developed to support PFCC practices that would engage, challenge, and support our new culture.

One of our initial corporate strategies required every unit to develop two action plans with their patients and families based on their needs and priorities. What we found initially was that there were some great ideas coming from the staff about how they felt they could improve their patients' experiences, but that did not meet the criteria for an action plan. What did meet the criteria were ideas for improvement identified by the patients and/or families that have had care experiences in their units.

9.2 PFCC in Action at the TBRHSC, Using Paediatrics as an Example

Patient Family Advisors (PFAs) are at the heart of the PFCC model of care. Creating a robust Patient Family Advisory council and integrating PFAs as partners takes careful planning and structure. A PFA can be anyone who has experienced care at the TBRHSC within the past 2 years, either as a patient or as a care partner. After the screening and application process, PFAs are orientated in the same manner as all volunteers, with additional training and support specific to the PFA role.

PFAs provide invaluable insights, and give us firsthand knowledge of how our services impact patients and families. PFAs have direct input and influence on the policies, programs, and practices that affect care and services at the TBRHSC. They have been actively engaged in over 300 working groups and committees including Board Quality, CEO selection, senior-level hires, councils, policy and program development, education, strategic planning, etc. Over 90 PFAs, ranging in age from 8 to 85, draw on their own experience to improve the experience of others. Their stories and perspectives have led us to initiate several improvement activities throughout our organization.

Our paediatric and adult PFAs have partnered with us in many ways to improve our services for our paediatric population and their parents through the development of action plans. Action plans are based on the needs and priorities of our patients and families. Some examples include paediatric transportation wagons, Family Tour Night, Bedside Communication Whiteboards, facility design, and Tele-Visitation.

Paediatric Transportation Wagons

The paediatric transportation wagon (Fig. 9.1) is an initiative brought to fruition by one of our youngest PFAs, Connor Ferguson. Surgery can be a worrisome experience for anyone, but for children it can be especially frightening. They don't always understand what's happening or why they need to go. It can be a long trip from the children's ward in 1B all the way up to the surgical unit on the third floor.

One idea to help make surgery less scary for children was to allow them to be taken through the hallways in a brand new red Radio Flyer wagon. Initially, the team

Fig. 9.1 Ethan and Emily are riding to the OR in the paediatric transportation wagon, another initiative brought to fruition by one of our youngest PFAs

came up with other ideas like a peddle car or battery-operated vehicle, but Connor didn't feel those would be safe enough in the TBRHSC hallways.

Today, the Radio Flyer wagons have totally transformed every child's journey to the surgical suite, changing a long, terrifying walk into a fun ride.

Family Tour Night

TBRHSC has 1,100 paediatric patients who require surgery every year. In the Operating Room (OR), PFAs have partnered with us to help us bridge the gap between what we thought our paediatric patients and families needed and what they actually needed. "One of the worst feelings in the world is handing your child to a stranger, listening to them cry, as they are taken away and the OR doors close," said one of our PFAs. We cannot change the fact that children need surgery, but we can change this process.

The Family Tour Night is one example of how we can improve the children's and parents' experience—and so far it has been overwhelmingly successful. It is one of the few processes that has been developed and implemented in the operating room that has shown immediate and significant improvements in our patients' and families' experience of care. PFA volunteers were instrumental in this success. They experienced the journey firsthand and were able to identify how this experience could be made better.

The tour takes the child and family on the entire surgical journey, from the admitting process to returning to their room after surgery. Nurses from the OR meet the families in the main lobby to start the tour to the paediatric unit and visit the playroom and where they will wait with their parents until they take a ride to the OR in the little red wagon. Once in the OR holding room, everyone changes into OR attire and the child is given a pair of their very own OR scrubs to take home. The children can walk or ride on a stretcher into the operating theatre, where they and their parents will be entertained by a puppet show explaining all of the machines and noises and what will happen when they come in for their surgery (Fig. 9.2). This

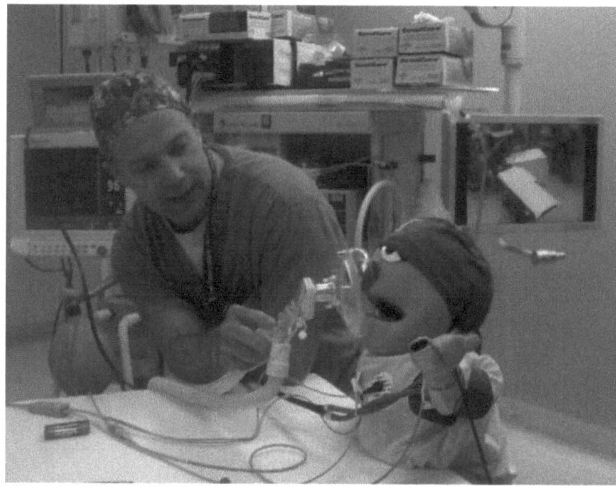

Fig. 9.2 A puppet show shows children what will happen when they come in for their surgery

also gives them an opportunity to play with flavoured oxygen masks, finger probes, "sticky pads," and other equipment. When the show is done, they take a ride to the recovery room to see where mom and dad will be waiting for them. The tour ends with snacks and playtime with the other children in the staff lounge.

When asked what they liked best, respondents mentioned the puppet show, ride, and pictures taken of the whole family "dressed up," which the parents enjoy as well. Knowing what was behind the closed doors helped dramatically to reduce the anxiety for both the child and parent and has improved the patient and family experience. Some comments we have received after the tour:

- "I thought the entire tour was excellent"
- "I have been telling everyone I talk to how great an experience it was and that it should continue for all families"
- "Made it easier to speak with our child at home about the surgery that would be taking place"

Bedside Communication Whiteboard

Paediatrics at TBRHSC has 12 beds that see an average of 370 admissions per year. Partnering with patients and their care partners is essential to provide safe, quality care and is a core concept of Patient and Family-Centred Care 2009, 2011; inpatients often feel that they are not well informed. In 2010, patients, families and healthcare providers partnered together to look at ways they could improve the communication between each other. The idea was to create an informational interactive tool that would assist all members of the healthcare team to be familiar with the collaborative plan of care between them and the patient and family. It would also provide a venue for the patient and family to communicate concerns, questions and input. What they came up with was the Bedside Communication Whiteboard. This would include:

- date and day;
- nurse, physician and other healthcare team members;

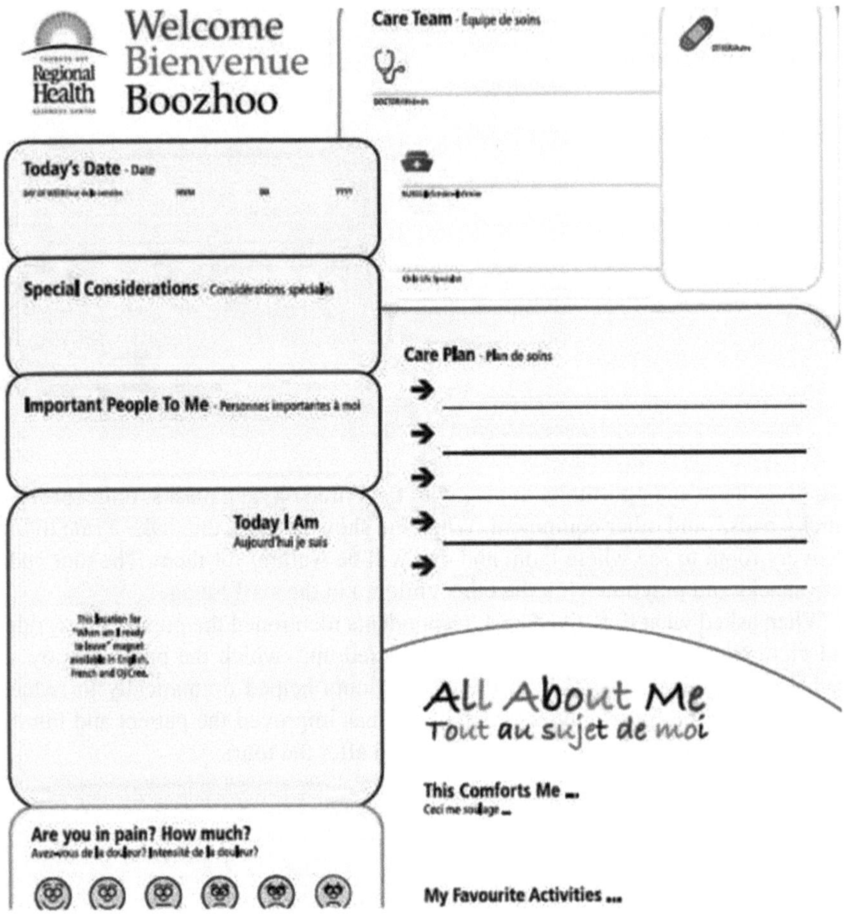

Fig. 9.3 The Bedside Communication Whiteboard assists members of the health-care team throughout the TBRHSC to be familiar with the collaborative plan of care; whiteboards in the Paediatric Unit include free space for drawing

- care plan developed in partnership with the patient and family including the patient's goal for the day;
- special considerations and assistive devices;
- colour magnet cue showing readiness to transition to home;
- free space to write notes, requests and pertinent numbers to relay to another member of the team.

When our young PFAs and families looked at bringing this concept to Paediatrics, what they considered important was a little different. The basic information about the date, care team, care plan and discharge cues remained, but the whole board was larger, more colourful, and identified "Important People to Me" along with a separate section labelled, "All about Me." There was room to draw pictures and to write about their favourite activity or what comforts them (Fig. 9.3).

Fig. 9.4 Bedside communication blackboard painted on the wall in the Child and Adolescent Mental Health Unit

Christina Purdon, Manager, Neonatal ICU and Paediatric Unit states, "Whiteboards have assisted the staff and families in having a meaningful face-to-face conversation about their child's care. It facilitates the discussion: what does your child need, what comforts your child, what does your child like to be called (favourite nickname), who is important to the child, and most importantly what does the child need to get better? What is the next step in the plan of care in treating their sick loved one? The whiteboards have helped the staff identify what is important to the child and/or family surrounding the child. If we address family priorities, we are viewed as attentive, caring, and providing good care."

A similar board (Fig. 9.4) was implemented in the Child and Adolescent Mental Health Unit (CAMHU) at TBRHSC, an eight-bed unit designed to meet the needs of youth with severe mental health issues who require hospitalization. The unit offers an assessment, stabilization and brief treatment while providing a safe and therapeutic environment for medically stable youth. Inpatient services are delivered by a multidisciplinary team that includes physicians, nurses, child and youth workers, as well as psychology and social work professionals.

In the past year, 220 youths ranging in age from 6 to 17 have been admitted, both in crisis and as elective admissions. Of these admissions, 68 % live in the city of

Thunder Bay and 12 % of the youth are from remote fly-in communities. Approximately 65 % of all admissions self-identified themselves as being First Nations. (Data derived from daily census and "How are we doing survey" distributed to both parents and youths by CAMHU staff.)

Because providing a safe, therapeutic environment is a priority in the CAMHU, there were other points to consider when bringing in the bedside communication whiteboard concept. It could not be a communication board that could be ripped off the wall and used to hurt themselves or anyone else. Their priority was that it also be cheerful and improve the appearance of the room. Patients, parents/guardians and staff worked together to create a tool to meet the needs of CAMHU's patients. The solution was a bedside communication blackboard painted on the wall. Angela Hill, Coordinator of the Child and Adolescent Mental Health Unit explained the process and priorities.

"They wanted to have the basic information, such as the care providers' names, the day and date, as well as their appointments, tests and goals. What made it a better communication tool for them was including an opportunity to express how they were feeling, listing their personal goals, and the coping skills that they were practicing during their hospitalization; for example, using distraction techniques to reduce suicidal feelings, or practicing relaxation techniques to manage their anger more appropriately. A decision not to include a discharge date on the blackboard was made, as many patients are involuntary and could be upset by a constant reminder that they did not know when they were going home. Passes off the unit with family were included as this was very important to the youth. We ask all youth and families to complete a 'How are we doing survey' at the time of discharge. This information is shared with all staff, on a monthly basis at the staff meeting. We are particularly interested in the challenges and concerns of the patient and family, as that is precisely where we need to reflect on current practice and determine if we need to look at things differently."

"Parents told us we needed to update our patient handbook and provide more accurate information about what they could expect during their child's hospitalization."

Facility Design (CAMHU)

Patients and their families complained that the environment in the CAMHU was too institutional looking. They told us it needed to be more welcoming and infused with more colour. In response to that suggestion, the patient rooms were painted a variety of colours rather than one single colour scheme throughout the unit. Colourful quilts were donated for the beds. Funding from the Thunder Bay Regional Health Sciences Foundation supported an artist to paint three large murals in the unit (Figs. 9.5, and 9.6). The main mural focuses on positive statements which were selected by patients, such as "ask 4 help" and "change your attitude change your life."

As many of the patients are First Nations, we wanted to create a culturally welcoming environment and include traditional teachings. Specific teachings may vary from

Fig. 9.5 A welcome mural in the CAMHU incorporates First Nations cultural symbols

Fig. 9.6 The main mural focuses on positive statements which were selected by patients, such as "ask 4 help"

Aboriginal community to Aboriginal community, but there are many commonalities that exist across Northwestern Ontario. One example of these teachings and skills is the "Seven Grandfather Teachings" (Kor Strategies & Associates, 2011), a set of teachings on human conduct towards others that includes: wisdom, love, respect, bravery, honesty, humility and truth (Seven Grandfather Teachings 2011; Fig. 9.7).

Fig. 9.7 A mural in CAMHU illustrating the Seven Grandfather Teachings

Tele-Visitation—Story of a Paediatric Patient in ICU

This is the story of a youth in the Intensive Care Unit (ICU) at the TBRHSC. It is emotionally very difficult for healthcare providers to care for a child in the ICU, especially when survival is hanging on a thread. You try your best to support the parents or guardians, help them understand what is happening with their child, answer their questions as best as possible, and provide them with the opportunities to just be with their child as much as they can.

In this case offering that support was a real challenge because one of the parents was unable to come to the TBRHSC—they were hundreds of miles away, unable to travel to their child, who was unresponsive and dying. However, we were able to bring the family together through a new TBRHSC Telemedicine program called Tele-Visitation.

Tele-Visitation is a videoconferencing service that "virtually transports" the patient's family to the bedside via the Ontario Telemedicine Network (OTN). Families and patients can see and talk to each other, interacting almost as naturally as if they were in the same room. First identified as a need about 10 years ago, it is an innovation that embraces the concepts of PFCC. In fact, Tele-Visitation was seen to be so much of an advance in patient care that it was identified as a Leading Practice by Accreditation Canada in 2011.

For the last 3 days of this child's life, we were able to arrange Tele-Visitation sessions so that the parent could share these precious moments with their partner and child. This allowed the parents to be virtually together in the room at a time when it mattered the most. It is impossible to measure how much this meant to this family but it is clear that they were thankful for the precious moments given to them with their child.

9.3 Evaluation, Results and Impact of PFCC on the TBRHSC

These have been some examples of how our partnership with patients and their families have resulted in profound improvements in the experience of care with our paediatric population at the TBRHSC. There are PFAs of all ages making improvements that will affect the experience of others throughout the whole organization, in both clinical and non-clinical areas.

Since the official start of PFCC in March of 2009, within the first year, patient satisfaction as reported by NRC Picker showed an overall improvement of 12% in combined scores. Improvements are evident in all Eight Dimensions of Care[1] (Picker Institute 1987) with increases ranging from 6.8 to 21.6%. All PFCC dimensions scored significantly higher than the Ontario average.

Our staff and physician satisfaction scores also showed improvement. In 2007, NRC Picker survey results revealed improvements when asked if the organization was great to work for, the percentage of positive responses among staff was 33.4% compared to our most recent results in November 2011, when the score was more than doubled, at 71%. When staff and physicians were asked if they were proud to say they were part of the organization, 64.8% of staff responded positively in 2011, compared to only 46% 2007. Physicians showed improvements from 29.3 to 62.8%. When asked if the organization inspired the best in you, staff satisfaction showed an increase of 10% and physician satisfaction doubled.

9.4 Conclusion

Developing a blueprint with key performance indicators to support the strategic integration of PFCC and the heart of the PFCC model, the PFAs, will provide the foundation for accountability, sustainability, and ongoing improvement initiatives. The intentional corporate strategic approach of our PFCC model with a robust PFA program has made profound improvements in our organization, culture, and

[1] NRC Picker™ defined Eight Dimensions of Care: Access, Respect for Patient Preferences, Information and Education, Physical Comfort, Emotional Support, Inclusion of Family and Friends, Continuity and Transition, Coordination of Care.

community. In only a couple of years, we have reached a level of integration of our PFAs that other leading organizations have not reached in a decade on their journey. Where once we questioned what we could and could not share with our patients and families, now we will not move forward without their input. Any decision that affects patient services is not approved without PFA participation. This includes policies, programs, services, education, facility design, hiring, and virtually everything we do. There is no age limitation for joining our PFA program, just a willingness to work together to make things better. Connor, one of our youngest PFAs reflects the sentiment of all of our PFAs.

"I hope to keep up my work as a PFA for women and children in the future. I hope to see more PFAs get involved, because the more people we have the better we can serve our hospital community, because at the end of the day like me, I feel very rewarded for helping patients and staff of the TBRHSC." The PFAs at TBRHSC are truly part of the fabric of our organization and for this we received the first and only Patient and Family-Centred Care Leading Practice designation by Accreditation Canada in 2011.

Patient and Family-Centred Care are not words to write on a wall; they are words we live by.

References

Committee on Quality of Health Care in America, Institute of Medicine. 2001. Crossing the quality chasm: A new health system for the 21st century. Institute of Medicine (National Academy Press). Available via IOM website. http://iom.edu/~/media/Files/Report%20Files/2001/Crossing-the-Quality-Chasm/Quality%20Chasm%202001%20%20report%20brief.pdf. Accessed 2 Dec 2012.

Institute for Patient- and Family-Centered Care. 2011. Advancing the practice of patient- and family-centered care in hospitals: How to get started… Also available via IPFCC.org website. http://www.ipfcc.org/pdf/getting_started.pdf. Accessed 2 Dec 2012.

Kor Strategies & Associates. 2011. Seven Grandfather teachings. Available at http://www.7grandfatherteachings.ca/grandteachings.html. Accessed 2 Dec 2012.

Picker Institute. 1987. Eight Picker principles of patient-centered care. http://pickerinstitute.org/about/picker-principles/. Accessed 2 Dec 2012.

Chapter 10
Creating Partnerships for Life: One Family's Story of Paediatric Patient and Family-Centred Paediatric Care

Janis Purdy

10.1 Introduction (Fig. 10.1)

Jonny is our second son and I'm not lying when I say he is special. He was born on November 14th, 2006 at Mt. Sinai Hospital in Toronto. He was sent immediately to SickKids (The Toronto Hospital for Sick Children) and spent five-and-a-half months in their Level III NICU (Neonatal Intensive Care Unit), three weeks in the CICU (Cardiac Intensive Care Unit), 3 months on the Cardiac Ward 4, and had more visits than we care to recall at emergency and on the surgical and general paediatric wards. He has had five major surgeries to date: two open-heart surgeries; an abdominal closure surgery; an emergency surgery to repair a bowel perforation; and a hernia repair. Nowadays, he is still followed by general surgery and cardiology but we are grateful that our visits are mostly planned and infrequent.

Jonny's omphalocele was first identified at his 18-week ultrasound. In normal baby development, the internal organs develop first in the umbilical cord and then begin to move into the abdomen around the four-week mark of pregnancy. Sometimes, the organs remain outside the body growing in the umbilical cord in a protective sac and the abdomen does not close. This is what is generally called an omphalocele. The size of the omphalocele depends on how many organs (e.g., liver, intestine, stomach) are outside and/or how large it is. For instance, a "Giant Omphalocele," like Jonny's, generally implies a defect that is 5 cm or more in diameter, with the liver in a central position.

Many more appointments and genetics testing confirmed that Jonny also had TOF, a serious but relatively common cyanotic heart defect. There were thankfully no associated genetic anomalies or syndromes; at least none that were (in the words of our OB/Gyn) "incompatible with life." He was born by cesarean section at Mount Sinai Hospital through their High Risk Pregnancy Unit on November 14, 2006 and admitted to SickKids NICU within hours of his birth.

J. Purdy (✉)
Toronto, Canada
e-mail: janis.purdy@provincialadvocate.on.ca

R. Zlotnik Shaul (ed.), *Paediatric Patient and Family-Centred Care: Ethical and Legal Issues,* International Library of Ethics, Law, and the New Medicine 57,
DOI 10.1007/978-1-4939-0323-8_10,

Fig. 10.1 Jonny

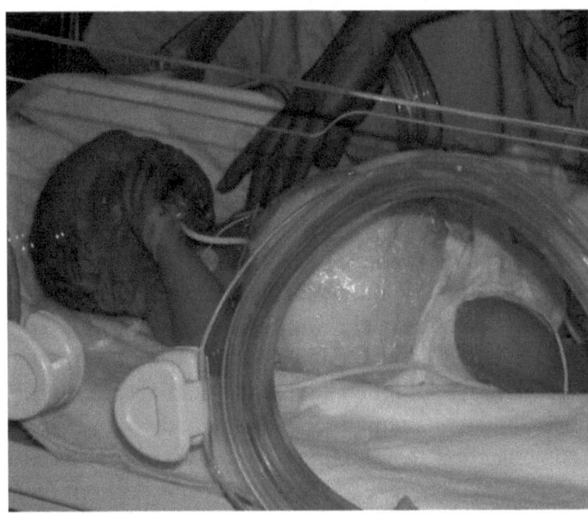

Fig. 10.2 Jonny, the day he was born

After a day or two, I was wheeled over in a chair to see him. As you can see in the photo above (Fig. 10.2), taken on the day he was born, he was a tiny and vulnerable little creature all wrapped in gauze and plastic with wires and tubes. Initially, I could only hold his hand and will him to survive.

I met a lot of people in that first week—a lot of medical professionals with titles I did not necessarily understand. I noticed that some of them seemed to look through me and some talked over my head. This was disconcerting for an adult professional like me. And being post-partum, I was particularly sensitive to tone and energy. Others, however, looked directly at me and gave me the feeling that they were really talking *with* me. Before I knew the term "family-centred care," the look in a professional's eye was the indicator I used to predict whether we were

Fig. 10.3 Janis, Jonny, Eli and Diane

going to be welcomed and included in planning or merely accepted and informed of plans after the fact.

Six years on, I can look back retrospectively and say with confidence that most of the doctors, nurses, occupational therapists and other healthcare professionals we met at SickKids were the best practitioners of family-centred care that anyone could hope for. Anywhere. And I am very grateful to them. As a result of their skills and style of practice, they were able to engage Jonathan's dazed, tired and overwhelmed parents. They integrated us as a part of his care team. And as a result of that integration we were able to be there for our son—caring for him, bonding with him, fighting alongside him, and forming important parental attachments. That has made all the difference.

However, the theme of this book is *tensions* in Patient and Family Centred Care. So, I will tell you some success stories and some stories of difficulty. And I hope in the end I will have succeeded in bringing some of the theory and practice to life.

10.2 Family Diversity

This is a photo of our little immediate family in 2007. Myself, Jonathan, Eli and Diane (Fig. 10.3). We were featured in the annual report that year and this was one of the photos taken by the SickKids photographer. You can see the atrium elevators in the background. It may seem a bit weird to be smiling and posing for family photos when your son is pale and sick, hooked up to an IV pole, with a picc line hanging from his arm. But we really were keen because the lovely photographer agreed to send us prints to keep. We had very few pictures from this time and we appreciated the chance to be documented as a family, even if things were so disjointed and often disconnected in our actual family life.

Fig. 10.4 Jonny with older brother, Eli, on a hospital sleepover

Family-centred care theory is founded on the principles of respecting families and "including patient family perspectives and values." In so doing, the whole family counts. In our case, we were a family with two urban moms (one Jewish) and a sweet and bewildered big brother (aged 4). There were two sons, two different bio-dads, and a loving—if colourful—assortment of extended family and friends. We were afraid at the beginning there might be issues. But other than a receptionist accidentally calling Diane "Mr. Purdy" a few times, we were pleasantly surprised. Only one doctor (peripheral to his care) asked us who the "real" mother was. We really appreciated that the SickKids medical professionals treated us with respect, kindness, and really did make efforts to understand and incorporate our whole, unique family.

This is Jonny with his older brother, Eli, on a hospital sleepover (Fig. 10.4). This picture was taken on the cardiac ward when Jonny was about 9 months old. You can see that his long hospitalization had affected his development. He had very "low tone" at this age. He could not sit up and was unable to control his head movements. But there is Eli, excited to be with his little brother, and so happy to have our whole family together for one night. Allowing "sleepovers" reduced our family stress and provided some "normalcy," especially for Eli. During the years that Jonny was in and out of hospital, we tried to protect his older brother as much as possible, but he is a sensitive child. He has always loved Jonny very fiercely. He didn't understand what was happening to him, exactly, but he knew his brother was sick, his parents were stressed, and his world was upside down. To this day he is very loving and protective of Jonny. Given that he only saw Jonny once a week during his time in the intensive care units, times together, especially overnights, were very special. I'll never forget the nurse on 4D who arranged these for us. Denied at other times for various reasons related to space and infection control, these overnights, to allow us family time all together, were literally the best in family-centred care.

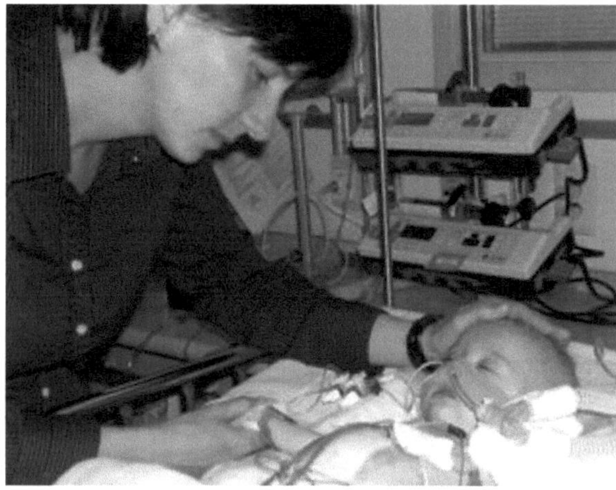

Fig. 10.5 May 23rd, 2007, after Jonny's first cardiac surgery

10.3 Family Trauma and Grief

The photo above was taken on May 23rd in 2007 when Jonny was not doing well after his first cardiac surgery (Fig. 10.5). Tetralogy of Fallot is a relatively common congenital heart defect that has four key features: a Ventricular Septal Defect (VSD); levels of obstruction from the right ventricle to the lungs (Pulmonary Stenosis); the aorta (the major artery from the heart to the body) lies directly over the VSD; and the right ventricle develops thickened muscle. The heart is also titled at a different angle. In Jonathan's case, he also has a small Atrial Septal Defect (ASD) discovered after his birth.

On the day this photo was taken, we were very worried. Jonny had not been urinating since the surgery. Despite diuretics and transfusions, he was not getting rid of fluids. As a result, he was getting very puffy. More importantly, the doctors were questioning whether there was sufficient blood flow to the kidneys. A subsequent echocardiogram found two other concerning things: (1) a small blood clot behind his right atrium and (2) that his pulmonary valve was narrower than they thought. Instead of recovering, Jonny got increasingly sicker. Four days post surgery he began to really deteriorate. On day five he crashed severely and was rushed back into surgery. Ultimately, his pulmonary valve was cut open permanently and patched. This second surgery went according to plan—which unfortunately meant that Jonny would no longer have a pulmonary valve that worked as a valve anymore—but it released the obstruction, which was causing him to deteriorate.

After the second surgery Jonny's recovery was slow. He had taken a real hit by having two surgeries within one week. His chest was a horrible wound and his heart was permanently compromised. In addition to everything, we discovered that during the first heart surgery his microscopic lymphatic duct system was torn or cut. The lymph system processes fats and because these nodes were now not working, the fats were draining into his chest cavity. His chest tubes, which are generally in

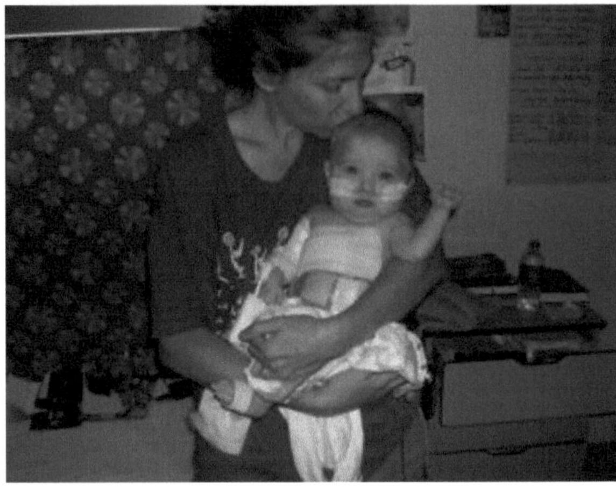

Fig. 10.6 Diane and Jonny, August 2007

for only a day or two after surgery to drain off collected blood, began showing signs of a milky substance: fats. This meant he had the rare but serious condition called Chylothorax, a type of pleural effusion, wherein excess fluid gathers in the space that surrounds the lungs. Left untreated, Chylothorax is lethal. Treatment, we were told, was uncertain. Jonny did eventually heal but it took 3 months.

In the photo above is Diane holding him in August of 2007, the day after he finally got his chest tubes pulled (Fig. 10.6). They had been in for 3 months. You can see his NG tubes, his PICC line, his IV pole with meds and TPN, his omphalocele and his chest bandages. I find this photo hard to look at because of the pain you can see in his eyes. We had no idea then that his lung had collapsed in the process of removing the tubes.

I included these photos because in hindsight I can see now that by this stage of Jonny's journey we were all in various stages of trauma. I can see it on our faces. But at the time I did not appreciate what we were going through.

Everyone reacts differently to tragic events, but grief and trauma responses include anger, fear, shock, confusion, distorted thoughts, and a whole host of other not-so-pleasant states of mind. I, for one, had ugly meltdowns and murderous impulses. I was not nice. I cried on the subway. On any given day, possibly hundreds of people in paediatric hospitals are lost in a cloud of trauma. And the ways in which they are acting/reacting are in response to this severe mental or emotional stress. I often wondered how medical health professionals working in acute environments with critically ill patients understand what their patients and families are going through. What is it like to work with that level of trauma every day? Do they get the training they need? Do they get the support they need?

When we were in the hospital we saw some parents crack. We had a friend who broke his toe kicking a door out of frustration. Security was called to escort him out. When we were in the CCU a father was walking up and down another part of the unit crying out "Why? Why? Why?" at the top of his lungs. None of us in the waiting area could see him but we could hear him. It was haunting and disturbing. I often wondered what happened with him and how the medical professionals handled it. Another time, my friend and I saw a father carrying a child about 3 or 4 years-old. They were outside the front doors of the hospital. It appeared they had been on a break to get some fresh air. But now it was time to go back. The little girl was fighting against her father, pounding with her little fists, crying "Don't take me back in there. I don't want to go back in there!" Dad had a fixed jaw and tired expression. He was trying to hold his daughter firmly and get her back into the hospital safely, a place I'm sure neither of them wanted to go. His eyes were grief stricken but firm. What strength he had! I cried that day for the little girl and her parents.

I also witnessed parents become hostile and threatening. That was tough for the medical staff to work around. One father in the NICU manifested his trauma in mistrust and fear. And perhaps there were some other mental health issues at play. He did not believe the doctors had his child's best interests at heart. He disagreed with every plan of care and demanded new doctors and different nurses. He stonewalled everything and tied them up for hours in consultations and negotiations. His energy was manic. It took over 24 h for the team to realize he was not a partner in care. I felt really sympathetic to the exhaustion of everyone on the team that day.

Some doctors, however, took the defensive too quickly. A couple we were friendly with were upset to learn that their baby might need a G-tube before discharge. They weren't convinced of this necessity and wanted a second opinion. The attending physician informed them bluntly that if they didn't consent in a timely manner he would call the Children's Aid Society and have them investigated for failing to provide the necessities of life. I came upon the mother weeping afterwards. It was unnecessarily heavy-handed. Though they ultimately found a way to move forward, I felt he had added additional trauma to their journey.

Ironically, for me, it seemed to be small things that played on my mind. An inconsiderate or thoughtless comment could have me stewing for days. A rough touch or needle poke on Jonny's arm had me poised to pounce like a cougar. My resilience was low. Small injustices—the high price of parking at the hospital, for example—really got under my skin.

Some days I was nice and grateful and calm. Some days I was weak and angry and agitated. I know I seemed dark at times and I probably treated some people brusquely without even knowing it. Some days I didn't want to talk to anyone. On other days I couldn't stop myself from talking to anyone who would listen—playing and replaying events until I had lost the interest of even the most patient listener. I understand these manifestations of trauma now when I see it in parents I work with.

No matter what day it was, however, I was always sensitive to anyone insinuating I should not be fully engaged in decisions related to my son's care. And so was my partner. The inclusive and engaging experience we'd had in the NICU taught us that we needed to be involved and consulted no matter what else was going on. The healthcare professionals had to practice around this reality, which is why I think trauma and grief are important aspects to any discussion of family-centred care. The truth is, the success of family-centred care is not measured when everyone is at their best. Not everyone has their child in acute life-and-death situations, but every loving parent is stressed when their child is sick. That is why training, support and

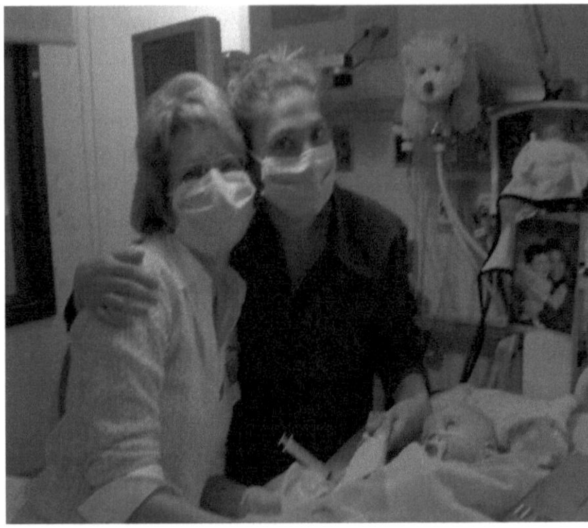

Fig. 10.7 Jennifer Young, Diane and Jonny

experience are so important for success. Effective FCC really can depend on the capacity of professionals to communicate well when their patients and families are not able to.

10.4 Collaboration

When patients and professionals share information, trust one another and work together, a most amazing kind of collaboration can develop. The photo above is of NP Jennifer Young and Diane working together in the NICU to try to clear Jonny's blocked picc line (Fig. 10.7). You can see the flush syringe there in Jennifer's hands. You can see that Diane is holding Jonny's hand and he is looking up at her. Jennifer and Diane worked very hard—for over 2 Hours—trying this and that, in order to avoid having to send Jonny down to IGT for a painful re-insertion procedure. I can't remember now if they were successful or not—I actually don't think they were. But what has stayed with me was how hard Jennifer worked and how fully Diane participated in the process. It was very genuine. They were in it together. And, as you can see from the photo, they were a great team and they really bonded that day.

10.5 The Context of the Medical Environment

The Institute for Family-Centered Care (IFCC) identifies communication—fair, unbiased and complete communication—as a core principle of family-centred care. Effective communication can be challenging in the most mundane of environments.

What if the environment in which the communication is happening is intense, frightening, urgent or traumatic? For instance, what if it's happening in a high-paced level III NICU or PICU? How do you possibly absorb or impart information when your child is critically ill? How do you have patience when you are tired, homesick and experiencing professional stress? How to communicate effectively when you do not have a lot of time? This is a challenge to family-centred care in the context of paediatric health.

Healthcare professionals at SickKids, and most other paediatric healthcare centres, are working and learning in an environment with very high expectations. They are under pressure to produce, to decide, to publish, to excel, to process vast amounts of information as quickly as they can. They are often working 12-h shifts and seeing as many patients as possible. They might be tired, especially in those fast-paced intensive care units. Some of them—residents and fellows, for example—are competing for their professional future, sometimes thousands of miles from home.

In addition, there are not always areas for compromise and consultation. There are hundreds of decisions to be made on any given day in any given paediatric hospital. Either the surgery is going to happen or it will not; the PICC line is staying in or it's coming out; we're trying one more time to get a vein or we're calling the IV team; medication is staying the same or changing, etc., etc. Someone must ultimately take responsibility for the decisions made. And when the stakes are higher, the tension in family-centred care will also rise. No one argues over an aspirin for a headache. The more uncertainty, the more frightening the intervention, the more pain, risk and possible complications, the more at stake.

I'll never forget disagreeing with a new resident in front of his team about whether Jonny had a certain infection. The resident struck me as arrogant and, I'll admit, I was not in a good mood that day. I knew the facts of Jonathan's condition very well. I was tired of the "learning" part of a learning hospital and, after so many months, I didn't want anyone "experimenting" on my son. In that moment, I saw myself as a mother protecting my vulnerable baby from unnecessary intrusion. He saw me as overstepping. In his mind family-centred care included sharing information, answering questions and maintaining respect. "Questions are one thing", he said later, "but it is not your role to second guess the doctors." For me, this was about the patient, not the doctors. My feeling was that if Jonny could talk, he would not want a needle, much less a lumbar puncture, if it wasn't necessary. Since there was plenty of room to doubt the wisdom of his assessment, I don't regret speaking up. Even specialists need to earn the respect and trust of their patients. And in the end my doubt was correct. But in hindsight, it would have gone better if I had been able to find a more discrete or diplomatic way to express my disagreement. I wasn't sensitive to the hierarchical medical model that residents operate in. Unfortunately, that was his reality. My reality at the time meant I was not at all concerned with helping him to "save face." But disputing him so directly in front of his peers and superiors created bad blood between us and affected our working relationship for the future. It was exhausting but that was the reality of the environment sometimes. On days like that, it was physicians and nurses with high level FCC skills, experience and competencies who were most needed and appreciated.

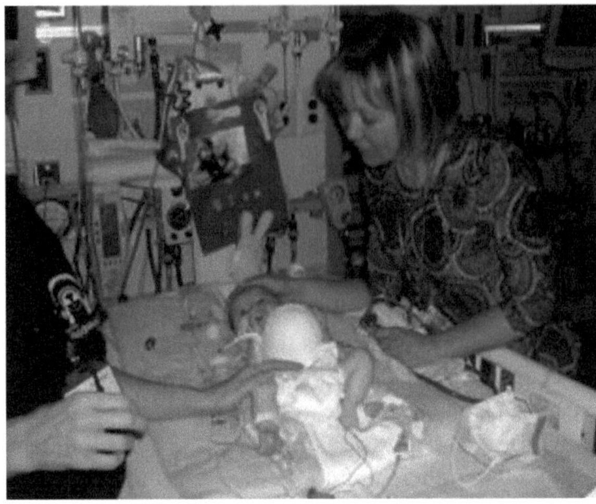

Fig. 10.8 Diane, Jonny and Kim Dionne

10.6 Effective Listening and Communication

The photo above includes Jonathan, my partner Diane and our nurse practitioner Kim Dionne (Fig. 10.8). This photo was taken right after a "wound" dressing change (note the omphalocele in the middle of his body all newly wrapped and netted). Kim is patting his head to soothe him as he was crying. Jonny had to have his omphalocele dressings changed twice a day in the early months and it was very uncomfortable for him. He kicked his legs and cried throughout the 40 min. When we spoke to one nurse about this she replied, "It doesn't hurt him. Babies often cry out of fear." It's hard to explain how I could tell this wasn't a fear cry. Another nurse said to me, compassionately, "I know, mom, it's hard to watch. Take solace in the fact that he won't remember." Kim Dionne, however, saw things differently.

Kim began assisting with the dressing changes in order to learn more. She observed Jonny carefully, getting to know him and his responses. Already a neonate expert, she consulted with a colleague who was a pain specialist, and she engaged us, his parents, to try to learn more. She really put in the effort to figure out how to make him more comfortable during dressing changes.

It was a trial-and-error process, but a combination of sucrose, a well-timed bolus of morphine and distraction with the use of baby DVDs did the trick. It made his life more comfortable from that point on. And we were much more comfortable working on the dressing changes not feeling like we were causing him such excruciating pain. Diane, especially, became more confident and competent.

You can see Diane in the photo above on the left (she is in the middle) with our nurse practitioner Carol McNair and wound care nurse Theresa Allen (Fig. 10.9). With this new confidence and the encouragement of the team, Diane eventually moved from assisting with his dressing changes to taking the lead, which, since the dressing protocol was new to each new shift of nurses, became medically necessary.

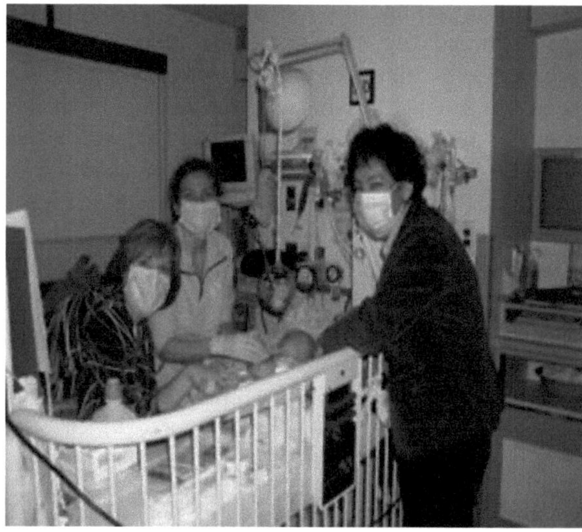

Fig. 10.9 Diane, Carol McNair and Theresa Allen

Fig. 10.10 The day Jonny left the NICU

It was a wonderful example of family-centred care that led to real family-integrated care.

The photo above was taken the day Jonny left the NICU (Fig. 10.10). From left to right: Kelly Roddy, Jonny's fierce and passionate core nurse; Dr. Jonathan Hellmann, his primary neonatologist; Diane, Jonny's Mama with tears in her eyes; and Carole McNair and Jennifer Young, his wonderful nurse practitioners. Jonny is in the stroller, unaware that he is about the leave the hospital for the first time since his birth. These people were some of the many who became very special to us. I wanted to share this photo because the notion of sharing "complete, accurate and unbiased information" reminded me of these special human beings who all practice superb family-centred care. And in particular, a story about Dr. Hellmann.

Late one evening after Jonny had been in the NICU about 3 months, Dr. Hellmann was leaving the unit and I was the only mother at bedside in our room. It was quiet and dark. As Dr. Hellmann passed by, with his coat over his arm, on his way home, he noticed me. Instead of waving and smiling—which would have been nice enough—he came over to say hello and see how I was doing. I told him, honestly, I was distressed. Jonny had been getting sicker and no one seemed to know definitively what was going on. The suspected culprit, necrotizing enterocolitis (NEC), was both fearsome and perplexing. NEC is a horrible condition in which the lining of the intestinal wall dies and the tissue falls off. The cause for this disorder is unknown. But it is most commonly seen in premature babies, especially newborns. Jonny was not premature nor was he a newborn. I knew Jonny was very sick but NEC just didn't make sense to me. Despite my doubts he was under an intense course of treatment.

Dr. Hellmann first listened while I shared my concerns. He gave me his full attention, as if he had nowhere else to be that night. When he did speak, he told me a story about an infection that had moved through an NICU that he had previously worked at. Many babies had become infected and some had died as a result. He admitted that the doctors had been unable to successfully predict or completely understand why it was that some babies became infected while others did not. And of those who became infected, why some were able to fight it off and others succumbed. "You see," he said, "we know a lot, we doctors, but there is an aspect to this work, that we must admit we don't know, that may not be knowable. We do the best we can."

With that he began a line of thinking that was as moving as it was fascinating: the mystery of the human spirit and the sometimes indiscriminate nature of life and death. I was captivated and amazed. I'd never heard a neonatologist talk this way—with such honesty and candor. And, eventually, I understood the deeper meaning of his story. I let go of my suspicion that the medical professionals were keeping information from me or were somehow not doing enough. I understood then, that they were genuinely uncertain about Jonathan's condition but were doing the best they could. It was my introduction to the concept of "prognostic uncertainty." Having seen a baby die, under these conditions, I knew, one would take every precaution to ensure that it did not happen again, even if those precautions were invasive and painful in the short-term. It was hard learning for me but it was important, intellectually and therapeutically. I will always be grateful to Dr. Hellmann for taking a detour that evening on his way home to spend half an hour with a mom and her sick baby.

10.7 Patient-Centered Care: WWJD (What Would Jonny Do?)

Babies, infants and toddlers are not able to effectively comprehend or communicate their wishes. So parents, as their legal alternate decision-makers (ADM), must try to do it for them. Doctors, however, are legally and ethically responsible for the best

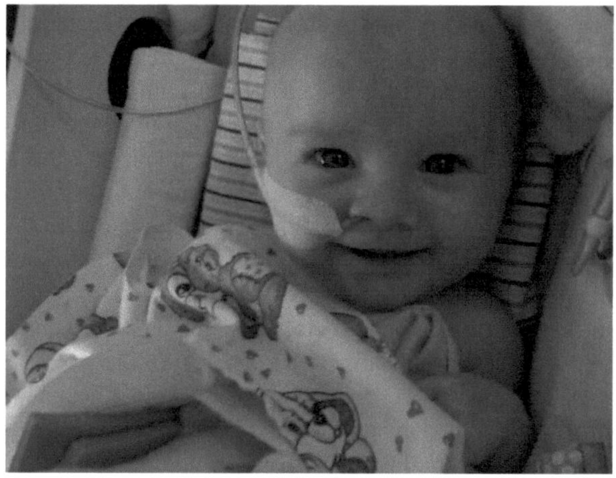

Fig. 10.11 Jonny

interests of their patients. What do you do when there is profound disagreement? Parents are susceptible to emotional confusion. Doctors are susceptible to clinical distance. Can either really know the better course of action when the stakes are high? Are either capable of understanding the wishes of an infant, the right course of action, without working with the other?

Throughout his early life Jonny was always tiny and innocent. I often felt helpless to protect him from all the daily pain necessary to keep him alive. I sometimes wondered if what we were doing to him was right. I would look at him as if from a distance sometimes. "What do you want us to do?" I would ask silently. "If you could speak," I did on occasion think, "would you tell me to let you go?" And then I had to ask myself, if he did ask me to let him go, would I be strong enough to honour his wishes?

Fortunately, I was saved the torment of having to actually answer those questions. At SickKids, and since, I have met parents who had to survive the passing of a child. No pain in life compares. I can't imagine, and would never pretend to know what that feels like. Jonny did come close a few times. More than once he "circled the drain," as Kelly Roddy put it. But each time, his strength of spirit—his life spirit—spoke louder than words. He hung on through unbelievable circumstances. And that is how we ultimately understood what Jonny wanted. Jonny taught us all that if we collaborate, work together and listen closely, babies and young children can communicate in a way that is beyond words (Fig. 10.11).

10.8 Family-Centred Care and Decision-Making

As I mentioned above, when Jonny was 3 months old he was in the NICU and he began to get sick. Sicker than normal. Not tolerating feeds; vomiting; losing weight; green bile coming up his NG tube. Eventually the attending physician told us they suspected Necrotizing Enterocolitis (NEC), a fearsome medical condition wherein the bowel disintegrates. The treatment prescribed included lumbar punctures, lots of needle pokes for daily testing, and going off NG tube breast milk and back to TPN or IV nutrition. I felt that robbing him of the life-giving nutrition in mother's milk would reduce his resilience, kill his ability to ever feed orally, and that the TPN would very likely cause liver damage if not failure. I wanted a diagnosis that made more sense. As I wrote above, the tensions in family-centred care rise when the stakes are high and there is no one, clear answer. But this was an emergency situation. The decision to provide treatment for "suspected NEC" was approved and moved on. I've already told you how I came around to accept that decision with help from Dr. Hellmann. Here is the rest of the story.

A few days later, on February 1st, 2007, Diane was holding Jonathan in her arms. He still had his oxygen and his low SATs but he was calm and smiling. Then he suddenly started to scream and writhe. His SATs crashed into the 50s, his heart rate went into the 200s. He turned grey. His core nurse, Kelly, kept trying to up the oxygen levels, but quickly realized the situation was dire and mobilized others: Respiratory Therapy, Surgery, Nursing, Neonatology, everyone. First, they had to intubate Jonny so he could breathe. It took a number of tries during which his heart rate dropped to around 50. He had to be "bagged," which is a horrible thing to see. Kim Dionne finally got the tube in, and by the time I arrived Jonny was on a ventilator but in great distress and very unstable. He was rushed to get x-rayed and Jonny's surgeon, the inimitable Dr. Annie Fecteau, came back to report that they could see that his bowel had perforated! Just as they had feared! He would have to go into surgery immediately. Dr. Fecteau quickly outlined three difficult possibilities: he might have very little healthy bowel left and they would close him up to live until he died; he might not be able to survive surgery; they might be able to remove the damaged bowel and he would survive, but with a stoma for the rest of his life. Sign here, we'll see you after surgery. And with that he was whisked away.

I was in shock. I thought to myself, this is it. A heart defect, a giant omphalocele and now NEC. How could he possibly handle it all? In addition, now that the perf had happened I felt guilty for having doubted the diagnosis.

Diane and I waited anxiously in the surgical waiting area for 4 h, fearing the worst and hoping for a miracle. And here's the surprising twist: we got one.

I was floating in my own body when Dr. Fecteau came back to report the results of surgery. Jonny came out of surgery. He survived. He did have a bowel perforation and it was quite devastating. But the surgeons saw no signs of NEC. What? He didn't have NEC but he got the thing that NEC causes? She said it was a spontaneous perforation and they were not able to define a clear and definitive cause at that point. Huh? And, unbelievably, the incredible surgeons had managed to find to point of perforation and repair it without having to take out any more of his bowel. So, there was no need for a stoma. What?

Part of the good fortune was that the treatment for NEC had contributed to saving his life, because when the perforation happened he didn't have any contents in his bowel, so there was no toxic leak into his abdomen. So, the diagnosis was technically not correct but the treatment for that diagnosis saved his life. Had he not had

Fig. 10.12 Jonny

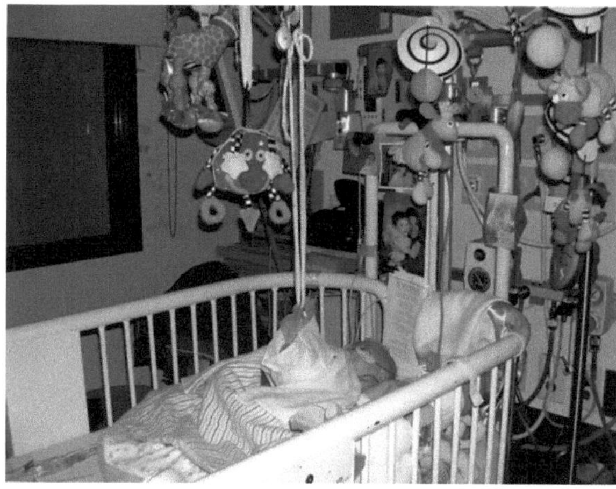

the treatment for the NEC diagnosis, he would have died from what he did have. Had the doctors held off the treatment until a more definitive diagnosis, he would have likely died. And I would have felt guilty forever.

Every day at SickKids, medical professionals wrestle with the weight of uncertainty and unbearable choice situations. Sometimes the way forward is not clear. Sometimes, and I know this now, luck and fortune and the mystery of the individual life spirit can be unpredictable deciding factors. Thankfully, the doctors and surgeons in the NICU worked with us as partners and included us in planning and decision-making. But in the end—and this is the most challenging balance—we were spared bearing the full weight of decisions that might have crushed us emotionally.

10.9 The Last Story—Collaboration Leading to Innovation!

After Jonny recovered and came back from surgery there was another challenge, where the collaborative aspect of family-centred care may have saved Jonny's developing brain and our family's sense of hope and optimism.

In the course of bowel perforation surgery, all the new skin that had grown around his omphalocele and the organs inside it—3 months of growth—was lost. There was nothing to protect his abdominal organs that were still outside his body. He came out of surgery with two silos stitched into the fascia of his abdomen, tied together, and held aloft by suspension You can see the photos above (Figs. 10.12 and 10.13). When the dressing was off you could see his liver through the plastic. It was almost incomprehensible. He was heavily medicated. He was artificially paralyzed. The only thing he could move was his little arms.

Once we got over the shock and horror of the perforation and the surgery, and then the euphoria of having not lost him, we spent a lot of time staring at this bizarre contraption thinking, "How do we get around this?" I couldn't see a way forward.

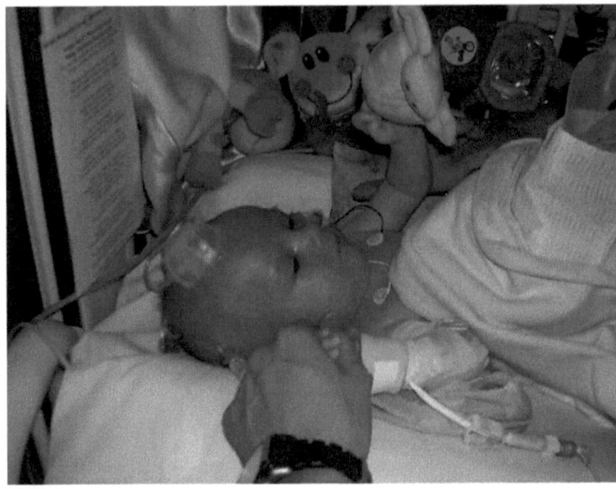

Fig. 10.13 Jonny

As the days passed into weeks, I began to really worry about losing Jonny in a different way. He seemed disconnected. Not so present anymore. He spent every moment—hour after hour—lying on his back staring up at the ceiling tiles. He didn't smile.

I began to worry about his immobility and its effect on his neurodevelopment. He was three-and-a-half-months old. I knew that this was a vulnerable time for a baby's developing brain. The lack of stimulation and lack of diverse experiences was concerning. Were there going to be long-term effects, I wondered. Might we be losing him mentally? I would speak about it to anyone who would listen. I was encouraged to have patience. Yet Diane and I could not help worry. Was there any way he could be more mobile at all, I wondered? If he could see more sights and sounds, even get to a window with sunshine, perhaps his little brain could be engaged, I thought. Many people listened sympathetically but it was a hard situation.

One day Kim Dionne came back to the bedside with a little sketch on a napkin. "Is this the kind of thing you're thinking of?" she asked. I looked at her drawing and smiled. "Yes! Yes!". The sketch was a kind of portable small bed, so he could be immobilized in his body, but move around within the bed; see sunshine, get new sights and sounds, be with his family. We showed her sketch to Jonny's Dad, James Anderson, who is an Architect. He made professional level drawings that we in turn showed to our carpenter friend Gerry Rochon, who has a daughter who works at SickKids. He built a little box based on the drawings and measurements, which we in turn then fitted into a wagon.

Now, most of the NPs, nurses and some doctors in the NICU knew what was going on. Whatever they thought privately, they were very encouraging with us and almost conspiratorial. I am so grateful for that. We were careful, on their advice, not to leak it to the surgeons until the last moment, lest the whole project be quashed.

I'm sure many of them thought we were crazy, but at least Diane and I had a positive focus for our restless, anxious energy. Finally the day came when we wheeled

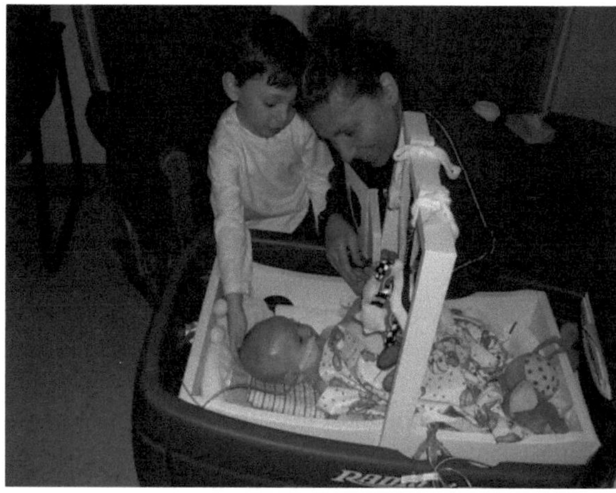

Fig. 10.14 The portable Jonathan Purdy-Flacks carrying device!

Fig. 10.15 The portable Jonathan Purdy-Flacks carrying device!

in the wagon and box. The surgeon was called. A group stood around the bed. What did they think? Would it work? Could we try? Dr. Fecteau agreed that we could try. Carefully!

A team of seven people—four holding him flat, two supporting the silo, and one managing the lines—slowly, gently, moved him from the bed to the wagon. It was a collaborative effort from start to finish. And below are two photos (Figs. 10.14 and 10.15) of what we created: The Portable Jonathan Purdy-Flacks Carrying Device! From that day forward Jonny was able to be safely moved around the unit and even out into the hallway. In the photo below on the left, you can see that we were able to visit with family in the care-by-parent room. And in the photo on the right with James and Jonny, you can see we were out in the hallway with windows and natural light. It was incredible! What a happy development for all of us in a terribly difficult time.

10.10 Conclusion

We were at SickKids for a long time and had a lot of experiences with Jonny in and out of hospital. We don't underestimate how difficult it can be to be truly family-centered in a paediatric context, but I hope the experiences we've shared can illustrate how crucial it can be in the life of the patient and the family. For our family, sharing knowledge led to collaboration. Collaboration led to innovation. And innovation led to healing. Through this experience, and others like it, I came to understand family-centred care in a very visceral way. We understand that these are theories and principles that are sometimes difficult to translate into practice. In different situations FCC practice can look quite different. It needs to be adaptable like that. And it can be toughest in critical life-and-death situations or uncertain/unclear situations. But when it is the most challenging it can also reap the greatest rewards.

Chapter 11
Supporting the Developing Autonomy of Youth Patients: Reflections on the Role of Healthcare Providers

Christy Simpson

11.1 Introduction

From the perspective of youth patients and decision-making about healthcare, several questions related to the value(s) of patient and family-centered care arise. It is relatively well-recognized that changes to, and transitions in, how care is provided for youth, corresponding to the development of their abilities to make their own health decisions, are required. Yet, determining the balance between protecting youth, i.e., protecting those who do not yet have (legal) capacity to make their own decisions, and demonstrating respect for youth decisions, when this capacity is present, can be difficult. Connected to this, there is discussion about the role of the family, the parents[1] who have been making decisions for their children, and the need, over time, for parents to move from being the primary decision-makers into more of a secondary role, where they instead support their youth's decision-making. It is expected that the healthcare team will identify, direct and assist youth patients and parents with these changes and transitions.

However, there is limited discussion in the literature on patient and family-centred care, adolescent health, and/or clinical ethics that is available to help support the healthcare team in enacting their role in providing assistance and direction to youth patients and parents with respect to decision-making, especially in terms of negotiating some of the challenging ethical questions and tensions that may arise. Specifically I am interested in the role of the healthcare team and individual healthcare providers as *actors* and *participants* in these transitions and changes for youth patients and their parents. For example, while patient and family-centred care directs much of the attention onto the desires and needs of youth patients and parents, there is much less discussion about what healthcare providers should reflect on and

[1] I will be using *parents* in this chapter to denote the decision-makers for youth who do not yet have the capacity to make their own healthcare decisions. This includes any caregiver or person with authority to act in this role.

C. Simpson (✉)
Department of Bioethics, Dalhousie University, Halifax, Canada
e-mail: christy.simpson@dal.ca

R. Zlotnik Shaul (ed.), *Paediatric Patient and Family-Centred Care: Ethical and Legal Issues,* International Library of Ethics, Law, and the New Medicine 57,
DOI 10.1007/978-1-4939-0323-8_11,

anticipate with respect to their own professional responsibilities and obligations, especially if they have been caring for a patient and family over the span of several months or years. With growing numbers of children with chronic conditions who are living longer, such as with spina bifida and cystic fibrosis, the time period for being followed by a healthcare team has also been substantially lengthened. I will argue that this context of care requires us to consider in more depth the relationships that healthcare providers develop with youth patients and with parents, and that this requires additional ethics analysis and reflection.

I am particularly interested in exploring these aspects of providing care for youth patients and their parents for several reasons. First, as part of my role in providing ethics support at a paediatric health center, I have participated in clinical ethics consultations related to the care of youth patients. Both in these consultations and in more informal conversations with healthcare providers about some of challenges they face in providing care for chronically ill patients and their families, several issues related to the nature of patient and family-centred care have arisen. These include questions related to respect for developing decision-making abilities and autonomy, and how this should be balanced with the duty to care and protect those who are not capable of making their own decisions. Further, these issues are interwoven with considerations about the appropriate role of the healthcare team, often linked closely with personal reflections on what it means to care for a patient over much of their lifespan, witnessing the changes in this patient and her family over time, all with the goal of providing the best care possible. Second, in discussing these types of issues with ethicists at other paediatric health centres, I have come to appreciate that these questions are relevant for many involved in providing care for chronically ill youth. Third, at a broader societal level, I am interested in, and variously concerned by, some of the ways in which youth are typically characterized in popular media, i.e., the "just you wait until they are teenagers—then watch out!" phenomena. How and whether these characterizations influence care, including how the healthcare team may engage with youth patients in some instances, is beyond the scope of this chapter, yet it is something that I believe deserves further examination.

In the following, I offer reflections related to the healthcare team's role in the care of chronically ill youth, focusing specifically on the time frame when these patients are older adolescents, and the time for transition to adult-oriented health services is approaching shortly or within a few years. It is further assumed that the healthcare teams have been caring for the youth patients, at a minimum, for several years, many from early childhood. The relationships between the healthcare team, the parents, and the youth patient have had many years to take shape. The *looking ahead* to the time of transition is both a recognition of the developing decision-making abilities of the youth[2] and a time of more obvious change within these relationships about who makes decisions about care and treatment—unsettling, if you will, patterns of care and interactions that have built up over time.

[2] Recognizing the complexity in this reflection on the healthcare team's role, I am further going to assume that the developing decision-making abilities of youth with chronic illness are still occurring along the same rough trajectory as healthy adolescents, even if this may occur at a slightly different pace due to illness.

Finally, as a last introductory comment, the following analysis is preliminary and represents an attempt to capture and describe some reflections about providing care for chronically ill youth; I invite the reader to respond and take these reflections further.

11.2 Setting the Context—Sample Cases

Case #1 Jonas has been coming to the hospital for several years for ongoing checkups and care related to his cystic fibrosis (CF). He is 16 years old and has a good relationship with his parents. Either one or both of his parents have come to all of his appointments, and they feel very comfortable with the physicians and healthcare team. Likewise, the healthcare providers look forward to seeing Jonas and his parents, as they seem to be able to find a bright side or joke in anything that may be going on. As Jonas has been getting older, he is participating more in the discussions about test results, what is happening, and any treatment decisions related to managing infections and the like. Jonas recently called the main nurse for the CF team and requested a separate appointment, one that his parents would not know about. While the nurse set up this appointment, she was a little concerned that Jonas did not want his parents involved. At the appointment, Jonas shares that he has been feeling ill and that something is *off*. He does not want to worry his parents, so this is why he came in alone; he feels that it is time that he did more of this on his own. Follow-up tests show that Jonas has a lung infection and will need to start a course of antibiotics. At Jonas' next appointment, his mother is there too and it is inadvertently revealed that Jonas had been in previously by himself. Jonas' mother expresses her disappointment and frustration with the healthcare team by not keeping her and her husband in the loop. "He is our child, and we need to know what's happening to him," she says. While the healthcare team believes they did the right thing in respecting Jonas' wishes, believing he does have capacity to make his own decisions, several team members express the feeling that they missed something and feel guilty about *deceiving* his parents.

Case #2 Darlene was diagnosed with diabetes when she was 10 years old. While it has been a challenge at times for Darlene and her family to manage her diabetes through diet and insulin, they have been managing fairly well for several years. Darlene is now 17 years old and she has been asked to come and meet with a few members of the healthcare team to talk about how things are going with her self-care; Darlene, her parents, and the healthcare team have been working towards this over the last year. For the most part, she has been doing quite well at managing her diabetes. From the test results over the last month, Darlene's blood sugars have varied much more than normal and for several days at a time. The team members want to explore with Darlene whether there is something that is going on that they can help with. They are concerned about these recent variances. Darlene sits with her arms crossed and says, "you are just like my parents. You keep harping on my blood sugars without giving me a break. Can you just back off a little bit and let me sort this out? How am I going to ever learn to do this on my own if everyone keeps interfering?"

Case #3 Ryan is 15 years old and has spina bifida. He has been attending meetings between his parents and the healthcare team for a couple of years, and has recently started asking more questions about his care. In today's meeting, Ryan tries to ask several questions, but keeps getting cut off by either his parents or a few members of the health team, who both seem to want to minimize the discussion about the future and what might happen down the road. The focus is on the more immediate issue about whether his shunt is blocked or not. Ryan is clearly frustrated by this and following the meeting indicates to his social worker that he is not sure whether he should go to these meetings, as it seems no one wants to listen to him.

These cases are fictional and are included for the purposes of setting the context for the following discussion about patient and family-centred care for chronically ill youth. While the intent is not to conduct a full analysis of these cases, and the reader has likely already identified possible approaches to these cases, I present these cases as examples of some of the everyday care situations that may occur and to help illustrate ways in which the healthcare team may be involved.

11.3 Paediatric Patient and Family-Centred Care, and Developing Autonomy Considerations

We can understand patient and family-centred care to be an approach to providing healthcare that recognizes the importance of the family in the patient's life and the difference that family involvement and consideration can make for providing the best care possible for patients (Arango 2011; Guion et al. 2010; Harrison et al. 1997; Kuo et al. 2012). This approach places an emphasis on building strong working relationships between patients, families and healthcare providers. With a specifically *paediatric* patient and family-centred care approach, there is clear recognition of the role that family plays throughout the life of the infant, child and youth along the pathway to adulthood. As described in the American Academy of Pediatrics (AAP) (2012) statement on patient and family-centred care (p. 395):

> Practitioners of patient- and family-centered care are keenly aware that positive health care experiences in provider/family partnerships can enhance parents' confidence in their roles and, over time, increase the competence of children and young adults to take responsibility for their own health care, particularly in anticipation of the transition to adult service systems.

Recommendations for paediatricians arising in this AAP statement refer to the need to respect the privacy and decision-making abilities of youth with capacity, and the need to work closely with patients, families and other members of the healthcare team to successfully transition youth and young adults to adult healthcare services (see Recommendations #6 and #10, specifically, in AAP (2012)). This recent statement is helpful in that it acknowledges the changing decision-making abilities of youth patients and makes clear reference to the role of the paediatrician and other healthcare providers in the transition process to adult services. Yet, little direction or acknowledgement of potential difficulties in applying and living up to the expectations outlined in this statement is provided. In the following, I discuss several interrelated aspects of caring well for chronically ill youth during the transition period.

One of the inherent challenges in paediatric patient and family-centred care is recognizing that while the patient is the focus of care, decisions about care for children and some youth patients are made by their parents, and that this decision-making responsibility will transition to the youth patient as her capacity and decision-making abilities develop. This transition to the youth patient taking a primary role in decisions about her health is often stated as simply as I have just done, which elides the variable and changing nature of developing decision-making abilities. For many, if not all, youth, this ability to make one's own healthcare decisions is something that will—during the development phase—likely move between being able to make one's own decisions on some days (and having the recognized authority to do so), and not being able to do this on other days, thereby still requiring parental consent. Youth patients' involvement in, and the ability to make, decisions can be further affected by their illness, exacerbations in side effects, changes in medication, and the like. If, as it is often argued from an ethical perspective, capacity is not age-related nor necessarily something that one either has for all or no decisions, but is understood as being decision-specific, the shifting balance between who makes what decisions when for youth patients needs to be closely attended to by physicians and other healthcare providers.

Given that the ability to make one's own decisions as a youth does not necessarily proceed in an orderly fashion from 1 day to the next or from one decision to the next, this variability creates an additional layer of complexity in the relationships healthcare providers have with youth patients and parents, especially with respect to preparing all involved, healthcare providers included, for the shifts in who makes the final decision(s) about care, what personal health information will be shared and with whom, etc. The ability of healthcare providers to negotiate space for decision-making that supports both youth patients and their parents, while being clear about their obligations to ensure confidentiality, for example, as in Case #1, requires time and attention. Further, given that healthcare providers are used to sharing all health information with the parents, the concern about deceiving parents or leaving parents out if requested by the youth patient can create real pressures on healthcare providers. For example, it can be difficult to not be able to share information in the way they had been able to previously with parents, both as a result of the relationship that has been established with the parents and the need to consciously shift patterns of communication that have been formed. Parents may also be quite confounded by this change, especially if this possibility/eventuality has not been discussed earlier.

These types of shifts in relationships between healthcare providers, youth patients and parents are often marked, as described, by the recognition that the youth patient has capacity to make their own decisions about their healthcare and that it is now the youth's decision about what personal health information will or will not be shared. Yet, it is also the case that assessing the capacity of youth patients can be quite difficult. As discussed by, for example, Dahl (2004) and Diekema (2011), this assessment is complicated, in part, due to relatively recent evidence that the prefrontal cortex of older adolescents and young adults continues to develop and mature into their mid-twenties. The prefrontal cortex contributes to the ability to engage in high-level reasoning, planning, control impulses and assess risks and rewards. As such, there is a legitimate concern that youth patients may not fully appreciate or account for the longer term risks and outcomes of their decisions. This needs to be taken seriously and factored into capacity assessments for youth. However, I also want to probe into other factors that could potentially influence capacity assessments for youth in ways that may not always be fully acknowledged or scrutinized. For example, this includes considering whether healthcare providers may be overly restrictive or less inclined to identify youth patients with capacity, based on a stance that is more paternalistic or protectionist. Exploring this aspect also acknowledges the power of the healthcare provider in making these assessments and what this means relative to the ability of youth patients to participate more fully in, and make decisions about, their healthcare.

Accordingly, it is important to consider the approach to care in many paediatric health organizations and the types of relationships that will likely be created early on with patients with chronic illness, which develop while children are young(er). As Rosen (1995) comments, "...pediatricians frequently support and nurture patients rather than challenge them. They are more likely to be prescriptive than collaborative and less likely to involve the young person in determining management plans" (p. 14). In these contexts, there may be a greater possibility that the relationships between healthcare providers and the patient has elements of a more parental-like or family-like relationship where the focus is understandably, initially, on the relative vulnerability of the child and the limited ability to participate in decisions about their care. Could this history influence whether the capacity of youth patients is appropriately acknowledged? For example, if a healthcare provider has a long-standing relationship with a patient and her family, say from the age of 6 to 16, how might this influence the determination of capacity? Depending on the length and depth of the relationship with the youth patient and her family, it is possible that the healthcare provider may not as readily recognize that capacity is present as the relationship to date has been predicated on the youth patient not having the ability to make her own healthcare decisions. In other words, the default assumption up to this point in time is that the youth patient does not have capacity (contrasted with adults where the opposite is true) and the burden, if you will, is on the patient to prove that she does have capacity or the abilities to make this particular health decision.

My (perhaps over-) emphasis of these points helps draw attention to the fact that healthcare providers must also make transitions in their relationships with youth patients and their parents—transitions that may not always be well-identified or discussed in relation to what is expected of healthcare providers as well as youth patients and their respective roles in healthcare decision-making. As stated earlier, the relationship with the youth patient is changing in terms of the youth taking on the primary decision-making role, while the relationship with the parents will become more secondary, over time. Whether this is acknowledged, supported or, conversely, potentially undermined as in Case #3, this sends signals to youth patients about whether they are welcomed and encouraged to participate in their own care. And, as the focus turns to transitioning the youth patient to adult-care, there is a need to acknowledge that, as Conway (1998) states, "[j]ust as parents must allow their child to move on, so the paediatric caring team must loosen its ties with the patient…" (p. 209). However, this *loosening of ties* may be a challenge for the healthcare team, especially if there are concerns about self-care as illustrated in Case #2.

Taking this a step further, with respect to Case #2 and similar types of care management situations, Rosen (1995) offers the following insight for consideration, "[e]xcess dependency may be inadvertently encouraged and can become self-fulfilling when providers are skeptical of adolescent self-care…" (p. 14). Allen and Gregory (2009) also suggest that, "…whilst diabetes management carries high levels of responsibility in terms of self-management, ultimate power and control are maintained by health professionals through surveillance and this appears to be informed by a culture of distrust" (pp. 164–165). Thus, an important aspect of providing care for youth patients includes the need for healthcare providers to reflect on how their attitudes and perceptions of these youth patients, including the health-related behaviours of these patients, may create a situation where it becomes very difficult for youth patients to *prove* or demonstrate their ability to be in charge and to be able to make their own decisions. This type of reflection could further help ensure that these expectations of *proof* do not become excessive or inappropriate, i.e., require an unduly high demonstration of the ability to make one's own decisions.

Interestingly, a recent study suggests that during puberty there may be an increased insulin resistance due to changing hormonal levels which may contribute to the difficulty in achieving optimal management of blood sugars (Murphy et al. 2006). In other words, even if a youth patient is doing everything right with respect to managing their blood sugars, this insulin resistance may be a confounding factor which makes it appear that they are not actually doing what they should. With this evidence in mind, how we think through what it means to be able to demonstrate that one is able to make and follow through with one's own health decisions and self-care becomes essential. While behaviours following a decision may be indicators of a lack of understanding or potentially even a lack of capacity, putting a heavy emphasis on "follow through" with one's decisions seems to potentially set the bar very high for youth patients. Some may argue that this is justified, given the attempt to find a balance between protecting those who are not capable and supporting those who are capable—and for those who are in the process of developing decision-making abilities, we may very well want to err on the side of protection. Yet, we should not ignore the question of how information, such as provided in the above research study, could impact or influence one's view of whether a youth patient has capacity. This further encourages deeper consideration of our assumptions about what and how data, such as blood sugar levels and health-related behaviours, is utilized and potentially influenced by both positive and negative assumptions about the relative decision-making abilities of youth (patients) and the developmental stages through which they are passing. Even if the above-cited evidence is disproved or changes over time, the point is that there needs to be a certain level of openness to other factors that could be influencing the youth patient's ability to demonstrate her capacity and commitment to self-management.

In other words, as I have indicated above, there is a possibility that the healthcare team is taking on more of a parental or protectionist role than may be explicitly acknowledged, as represented by the setting of high or overly restrictive parameters with respect to what youth need to be able to do and demonstrate in order to be seen as having capacity. The response to this point to not to become overly lax about what we discuss with and expect of youth patients; it is more about how we can better balance what we know about the development of capacity and decision-making abilities of youth with appropriate attention to the need to protect or support these patients. This includes a critical assessment of how our own biases and assumptions about youth may have the potential to negatively affect this process of assessment and support.

Along these lines, Allen and Gregory (2009) discuss how transitioning youth to the adult system is framed as a problem, and invite reconsideration of whether this framing is appropriate for this period of care for older adolescents and, even, whether it is this framing itself that contributes to some of the challenges that healthcare teams may experience when working with youth patients. Specifically, Allen and Gregory ask whether the focus on transitioning youth to the adult system, in terms of supporting their developing decision-making abilities, tends to overlook and not understand the needs that youth patients currently have. Are we too future-focused, i.e., focused on what youth *will* need to be able to do and structure their care along these lines to such an extent that this both places additional pressures on youth patients and has the potential to ignore or overlook their *current* health or decision-making needs? This is an interesting challenge for patient and family-centred care; i.e., what is the appropriate balance between the needs of the patient now and the anticipated needs of the patient in the future?

The challenge of finding this balance is especially relevant for youth patients, if we appreciate that what they do or do not do in terms of self-care can have significant impacts on their health years later. Given that one of the motivating reasons for the growing field of adolescent medicine was to better identify, understand and employ approaches to care that work well with and for youth patients and young adults, especially in terms of better facilitating the actual transition to adult-care, Allen and Gregory's position is to call for a second look at the emphasis that is currently placed on the *problem* of transitioning. They do not dispute that the original and continued focus on transitioning youth patients successfully to adult-oriented care settings is fundamentally about providing better patient and family-centred care, but question whether we have gone too far in this approach such that we are not attending sufficiently to what youth patients need in the present as part of their care.

Adding to these considerations is the need to appreciate the potentially mixed messages youth may receive about what it means to grow up and be an adult. For example, while there is a need to establish independence and a separateness that is part of growing up, we also recognize the value of interdependence—the ability to rely on, and seek support and guidance, from others. Indeed, many adults make decisions with the important people in their lives or, at a minimum, with a sense of their needs in mind. As Pinto (2004) points out in her exploration of gender and age in healthcare, "[t]o be a woman is to be connected; to be an adult is to be separate… The question many girls seem to ask is: How can I distinguish myself and be in relationship at the same time?" (p. 80) We need to be attentive to how expectations related to independence and what it means to be able to make decisions on one's own may be shaped by, among other things, gender, culture, and, for healthcare providers, one's work environment. In this sense, while patient and family-centred care focuses on the need for care to be responsive to these aspects of patient and family life, little is specifically said about healthcare teams being aware of their own cultural biases or understandings of what it means to be a youth and to mature, develop independence, etc. Greater appreciation of these aspects of providing care for youth patients and their parents could be useful, especially in situations of conflict.

A second example of what I am trying to identify for consideration comes from the seeming expectation of rebellion as a part of adolescence. If a youth gets along with her parents, this is seen as a cause for surprise rather than seeing this as part of the spectrum of what is possible during the teenage years. As such, this raises questions for me about whether some of the programs for youth patients or the high degree of focus on transitioning may be implicitly based on an expectation that there will be these types of problems, i.e., rebellion—possibly contributing to a sort of self-fulfilling prophecy in some senses. Especially if the healthcare team has taken on more of a parental role as discussed earlier, could this be part of what is being played out between youth patients and healthcare teams (see also Allen and Gregory 2009), as Case #2 suggests, and not be just a symptom of a youth developmental stage?

11.4 Concluding Remarks

In the above, I have explored several different aspects of the role of healthcare providers with respect to the developing decision-making abilities of youth patients. In specifically considering youth patients who are approaching the time of transition to adult-oriented health organizations and who are chronically ill, this exploration enabled the opportunity to further flesh out the relational aspects of providing care for youth patients and their parents. As the role of the parents is moving into the background, and the role of youth patients is taking up more of the foreground as part of recognizing the growing decision-making abilities of these youth, there are pressures and tensions within these relationships that require additional ethical reflection and analysis. In particular, I have highlighted questions and concerns related to the nature of capacity assessments for youth patients, the possible (negative) role of assumptions about teenagers, and the need to further assist youth patients in navigating some of the possible mixed messages about what it means to become an adult. Whether the increased focus on the transition period for youth patients helps or hinders assessment of the degree to which healthcare providers are engaged appropriately in their relationships with youth patients and parents has also been considered. As we continue to explore what providing patient and family-centred care means generally, and in the context of caring for youth with chronic illness specifically, it is my hope that these reflections contribute to continuing to improve care and support for patients, parents and healthcare providers.

Acknowledgments Many thanks to Lori D'agincourt-Canning for several helpful discussions as I worked on this chapter and for directing me to the articles by Diekema (2011) and Dahl (2004), among others, as part of a previous presentation we did together. Lori's insights helped shape some of the above reflections related to transitions in relationships with and between patients, families, and healthcare providers. Thanks to Tamatha Campbell as well for her assistance with appropriate formatting and referencing for this chapter. Finally, I want to acknowledge the Ethics Collaboration between the Department of Bioethics, Dalhousie University and the IWK Health Centre for enabling opportunities to engage with youth, parents and healthcare providers with respect to care issues.

References

Allen, D., and J. Gregory. 2009. The transition from children's to adult diabetes services: Understanding the 'problem'. *Diabetic Medicine* 26:162–166. doi:10.1111/j.1464-5491.2008.02647.x.

American Academy of Pediatics, Committee on Hospital Care and Institute for Patient- and Family-Centered Care. 2012. Policy statement: Patient- and family-centered care and the pediatrician's role. *Pediatrics* 129 (2): 394–404. doi:10.1542/peds.2011–3084

Arango, Polly. 2011. Family-centered care. *Academic Pediatrics* 11:97–99.

Conway, Steven P. 1998. Transition from paediatric to adult-orientated care for adolescents with cystic fibrosis. *Disability and Rehabilitation* 20 (6/7): 209–216.

Dahl, Ronald E. 2004. Adolescent brain development: A period of vulnerabilities and opportunities. *Annals New York Academy of Sciences* 1021:1–22. doi:10.1196/annals.1308.001.

Diekema, Douglas S. 2011. Adolescent refusal of lifesaving treatment: Are we asking the right questions? *Adolescent Medicine: State of the Art Reviews* 22 (2): 213–228, viii.

Guion, W. Kent, Shelley Mishoe, Gregory G. Passmore, and Pamela Witter. 2010. Development of a concept map to convey understanding of patient and family-centered care. *Journal for Healthcare Quality* 32 (6): 27–32.

Harrison, Christine, Nuala P. Kenny, Mona Sidarous, and Mary Rowell. 1997. Involving children in medical decisions. *Canadian Medical Association Journal* 156:825–828.

Kuo, Dennis Z., Amy J. Houtrow, Polly Arango, Karen A. Kuhlthau, Jeffrey M. Simmons, and John M. Neff. 2012. Family-centered care: Current applications and future directions in pediatric healthcare. *Matern Child Health Journal* 16:297–305. doi:10.1007/s10995-011-0751-7.

Murphy, H. R., G. Rayman, and T. C. Skinner. 2006. Psycho-educational interventions for children and young people with Type 1 diabetes. *Diabetic Medicine* 23:935–943.

Pinto, Kristina C. 2004. Intersections of gender and age in healthcare: Adapting autonomy and confidentiality for the adolescent girl. *Qualitative Health Research* 14 (1): 78–99. doi:10.1177/1049732303259674.

Rosen, David. 1995. Between two worlds: Bridging the cultures of child health and adult medicine. *Journal of Adolescent Health* 17:10–16.

Part IV
In Context

Chapter 12
Participation in Paediatric Research Studies

Nancy Walton

12.1 Introduction

Approaching care from a child and family-centered care approach or a patient-centered care approach has implications for involving children in research. Adopting a clear patient-centered approach implies that researchers pay special attention to the autonomy of individual participants. Individual autonomy is, of course, one of the most important considerations in research involving competent adults. However, when research involves children, there are a number of approaches that may be used, which may each incorporate the family to varying degrees.

What does it mean to include the family in research involving children? Typically, by inclusion, we are talking about inclusion in the decisions about whether to take part in research and the process of informed consent: two phases in the research process that involve the most complex and important interactions between researcher and participant. Involving others in these processes is often important—not just for children. Participants may require assistance from those in their family or social circle to help them make a decision about participation. Other participants may require assistance from their family members in understanding the information provided in the informed consent process. In research involving adults, family involvement may be seen, for example, in cases of participants who have cognitive impairments, or older adults with some degree of dementia. Typically families will tend to be more involved when patients require either informal support with consent processes or more formal involvement of another person in this kind of a process, through the use of a proxy decision-maker.

The issue of protection of participants and their best interests, rooted in the respect for dignity of persons, is also relevant to the discussion of how families are included in research involving children. A common default is to put in place more extreme protections for research involving children, in the form of gatekeepers,

N. Walton (✉)
Daphne Cockwell School of Nursing,
Ryerson University, 350 Victoria Street,
Toronto, ON M5B 2K3, Canada
e-mail: nwalton@ryerson.ca

R. Zlotnik Shaul (ed.), *Paediatric Patient and Family-Centred Care: Ethical and Legal Issues,* International Library of Ethics, Law, and the New Medicine 57,
DOI 10.1007/978-1-4939-0323-8_12,

complex consent-assent processes and strategies to ensure voluntariness, and a favourable risk-balance ratio. But this protection has served, in many cases, to exclude children from research that is considered to involve a higher risk or harm or no direct benefit to individual participants. The exclusion of children, to fulfill the purpose of protecting them, has also worked to historically deny children, as a group, access to the possibility of important benefits that could be gleaned through research. Protection of children (and all research participants) is something that researchers and research ethics boards must pay close attention to, but it must also be tempered by considerations of the just distribution of both the burdens and benefits of research, across groups in society.

Determining best interests is often not as straightforward as it may seem when the family is at the centre of care. In a patient-centered care (PCC) model, the important components of respect for autonomy, treating the patient as a unique person, and the mutual participation and shared power between healthcare professional and patient implies that best interests can be determined and articulated by the patient him or herself, and that due consideration is paid to the expressed wishes of the patient (Sine and Sharpe 2011; Mead and Bower 2000).

In the case of children and adolescents, however, determining best interests of the child is necessarily more complex, especially from a perspective that views the child within the context of the family system. In a family-centered care approach, the family's interests, values and context are considered to be important in the provision of care, and families work together with healthcare professional in the best interests of children. As children grow and develop, their role as partner in care is enhanced and their inclusion in decision-making processes is facilitated through an approach that respects the child as an individual at a particular developmental stage within a family system (King et al. 1998; Kuo et al. 2012; Maternal and Child Health Bureau 2005). But there are many stages in which parents may disagree with children about best interests or the prioritization of interests. School age children and, in particular, adolescents may have significantly different views on what constitutes their best interests, and this may be a source of conflict within a family. When it comes to decisions about participation in research, these difference in the processes of evaluating what constitutes the best interests of the child and determining who has a say in this may become apparent through negotiation of the consent/assent processes and deciding who provides informed consent for the child to participate.

In this chapter, the topic of involving children in research will be explored, from both a family-centered care perspective and a patient- or child-centered care perspective. What it means to include parents and families in research involving children will be discussed, with consideration of the principles and directives of both a family-centered and a patient-centered approach. Values and value conflicts within families will be discussed, as this often provides a source of tension for decisions about how families should be included in research involving their children. The topics of consent and protection will be explored within the context of research involving children. Finally, a few important key topics in the area of research involving children will be identified that require further deliberation and discussion, including: the reporting of incidental findings, privacy and confidentiality, and the role of the parent in the consent process.

With research involving children, the presence of a family, parent, or guardian is almost inevitable. How these persons are involved in the processes of research may, however, vary noticeably among different institutions and approaches. The differences in care approaches between institutions that clearly purport either family-centered care or patient-centered care may not be as obvious in the context of research. In some cases, the approach to research involving children may be quite similar regardless of the institutional perspective on the approach to care. In other institutions, there may be a clear encouragement or discouragement of families being involved in particular parts of the research process. Furthermore, there are extensive ethical guidelines and legislation governing the conduct of research involving children, and the care perspective may have far less of an effect upon how families are included in the research process, when compared to ethical guidelines and laws.

12.2 Assumptions

Let us first examine some of the assumptions that we are acknowledging in this chapter. Each of these assumptions may have an effect upon how involved the family may be involved in research on children and adolescents.

First, we are not only discussing clinical or behavioural research in this chapter, although that will be a significant focus. To not also consider the kinds of issues that arise in social or educational research involving children and adolescents would be to fail to acknowledge a significant proportion of research on this population. Social research on children and childhood may be carried out by researchers from a variety of disciplines including psychology, sociology, social anthropology, education and other social sciences (Morrow and Richards 1996). In this chapter, we will be discussing research across a spectrum of potential risks for harm, from very low risk studies to higher risk studies that may involve therapy, or medical interventions. Involvement of families in research involving children and adolescents is not only relevant to the context of higher-risk clinical research, but must also be considered in other kinds of research with different inherent levels of potential risk.

Second, we must also note that children may be approached to take part in research in a variety of environments and contexts. While children who are ill or in hospital are an often-studied population, children living with chronic illnesses or disabilities in community or outpatient settings may also be participants in research, as well as healthy children. Children in daycare or school settings may be approached to take part in educational or social science research. Community-based organizations or agencies may well recruit children for social research or function as conduits through which researchers from other environments may access children. When we limit our discussions of children's involvement in research to a clinical context or to clinical environments, we overlook a significant and growing body of social research involving children, in which children are approached to make decisions about research from a variety of sources and settings.

Third, it is important to note that the family unit itself or relationships within the family may be the focus of researchers' interest. Researchers may well be interested in involving the entire family or specific members of a family in the same study. Research on parenting or parent-child attachment focuses on both the parent and child. Genetic research often necessarily involves collecting data from more than one family member in a single research protocol. Topics of questioning in many kinds of research may include asking about the nature and quality of relationships in the family. Research that directly involves more than one member of a family and includes a child understandably adds complexity to an already often-complex interaction between the researcher, child and parent or guardian.

Fourth, it is important to note that many of the guidelines, principles and discussions of family-centered care and patient-centered care address the *care of patients*, rather than the *treatment of research participants*. Much of the discussion about family-centered care and patient-centered care has not been adequately developed to also directly address research participants, rather than patients receiving treatment and care. This is, of course, much the same as what we see in the literature dealing with the consent of children: there is much discussion of models and guidelines for seeking the consent of children in their healthcare, but, for the most part, these guidelines and models are then taken from the context of informed consent for treatment and care, and subsequently applied to the discussion of informed consent for participation in research. Even though there are overt similarities between these two contexts when discussing clinical research, there are also significant differences, and this can create a tension in applying these guidelines in a research context.

Finally, we should assume that by "children" we are using a widely accepted definition of child, that refers to a person under the age of 18 (United Nations 1989). This, of course, encompasses a broad spectrum of persons: from toddlers to school age children to independent adolescents, all at varying levels of competence and cognitive development. The implication of this is, of course, that there is no one set of guidelines or principles that can easily apply to *all* children who may be involved in research. Extensive work examining children's decision-making capacities, cognitive and developmental stages, from Piaget's work to modern-day explorations of the sociology and psychology of childhood, demonstrate that children, as a group, are quite obviously not uniform or standard (Kanner et al. 2004). While we acknowledge the gradual growth of capacity in terms of decision-making and cognitive abilities over time, even children of the same age may have striking differences in these abilities, as a result of their life experiences, health state, education, and other determining contextual factors quite separate from chronological age.

12.2.1 Family-Centered Care and Research Involving Children

> Family centered care is a way of caring for children and their families within health services which ensures that care is planned around the whole family, not just the individual child/person, and in which all the family members are recognized as care recipients. (Shields et al. 2006, p. 1318).

Family-centered care (FCC) is an approach to providing care for children within the context of the family. As part of a family, children are directly affected by changes, strengths and tensions within the family system. Neglecting to acknowledge this notable influence, from a FCC perspective, means that there is a failure to consider the context of the child and the situations, relationships and systems in which the child exists. The periods of time that children spend in hospital or in the healthcare system are often analogized to be mere snapshots of their entire lives, which, to further the analogy, would be seen as a feature length film. These mere snapshots mean that, from a FCC perspective, it is difficult to impossible for healthcare teams to really know the child, and parents and families, by virtue of the key roles they play in the life of the child, are the experts. Parents are the continuous presence and constant in children's lives while healthcare professionals' involvement may be transient and will likely not be as prolonged, consistent or sustained in terms of relationships.

A FCC approach is a contrast to the more traditional paternalistic model of care, in which healthcare professionals are felt to be the experts, and parents are either irrelevant or lacking in special knowledge. This traditional model assumes that parents feel that they are ill-equipped or not motivated to be highly involved in their children's care, that they wish or need to be told what to do, and that they will do whatever is recommended without question or hesitation. Instead, a FCC approach acknowledges the unique expertise and special knowledge that parents and families have about their children. By virtue of this special knowledge, a FCC approach acknowledges that families should be partners and collaborators in care and that their unique needs and attributes, strengths, tensions and coping skills must be considered in all facets of care delivery (King et al. 1998).

The use of a FCC approach to care has implications for other aspects that might be involved in care or exist alongside care, such as research. Certainly, much clinical research is embedded within care delivery, and may involve or have an effect upon care, procedures, treatment or medications that are delivered to the child. Thus, the FCC approach must be considered, and adapted to processes that are inherent to the conduct of research. From a FCC approach, parents and families should have involvement in decisions about their child taking part in research. How that is actualized depends more upon the ethical guidelines and norms that are in place, as well as legislation regarding decision-making and children. These guidelines, norms and laws are typically enforced through the process of ethics review by an institutional Research Ethics Board (REB). REBs are mandated to consider such elements of research protocols before they begin, in which researchers are required to articulate and provide rationale for the research design and methods, processes of recruitment and selection of potential participants, the informed consent process, the balance between potential risks of harm and potential individual and societal benefits, how voluntariness is ensured and how the privacy and confidentiality of potential participants are being protected.

While REBs are guided by federal guidelines for the ethical conduct of research (Tri-Council Policy Statement 2, 2010), they also interpret these guidelines at a local level, keeping in mind the norms of the community that they serve and the interests and special needs of the populations within these communities. REBs in children's hospitals, clinics or agencies are often seen as "gatekeepers" for access to children for the purposes of research and may have quite different expectations and stringent requirements for researchers who wish to study children. An REB in a children's hospital or agency that practices FCC would need to attend to the kinds of legislation and ethical guidelines regarding the inclusion of children in research and respect for the dignity, welfare and integrity of the individual child. In addition, the REB would also need to consider the families who are at (and may well have purposefully sought care at) an institution that fosters and promotes active family inclusion in decision-making and the view of the child within a family context. This, of course, may create tensions for an REB and researchers who may find that the obligation to adhere to one set of guidelines while also considering local norms create serious challenges in recruitment, consent and other decision-making processes within the conduct of research involving children.

12.2.2 Patient-Centered Care and Research Involving Children

Patient-centered care (PCC), according to the Institute of Medicine, is defined as care that incorporates respect for and is targeted to the unique individual who comes to the healthcare system with unique contextual and diverse needs. PCC is care that is provided within a context of compassion, and respect for dignity (Greene et al. 2012; Institute of Medicine 2001; McClimans et al. 2011). In the U.K., Lord Darzi (a world-leading surgeon at Imperial College in London and most recently, the Parliamentary Under Secretary of State in the House of Lords at the Department of Health, appointed by the British Prime Minister to lead a review of and report on the future of the National Health System in the UK) notes that patient-centered care can empower patients and give them more control over their healthcare (and in some cases, health spending) in order to allow them to be involved in decision-making that incorporates their specific needs and contexts of their lives (Lord Darzi 2008). The personalized care and attention to more than simply the biophysical needs of patients that are both important components of PCC imply that care is less paternalistic, but more responsive to the individual, voiced needs of patients (Bridson et al. 2003; McClimans et al. 2011; Mead and Bower 2000). While the foundations of PCC are typically broad moral values (respect, dignity, compassion), the justifications given for adopting a PCC approach by institutions are more often related to policies and outcomes, including efficiency, choice, quality, safety and effectiveness (Institute of Medicine 2001; Lord Darzi 2008; McClimans et al. 2011).

From a PCC perspective, clinicians and researchers are obligated to consider patients within the specific context of their lives—and to consider how different contextual factors will affect decisions about care, treatments, or involvement in research. While a FCC approach and a PCC approach may differ in terms of *who* is considered in processes of power and decision-making, both approaches aim to turn the more traditional paternalistic model of healthcare on its head by sharing decision-making power between healthcare professionals and patients. In the context of research involving children, most guidelines for the ethical conduct of research support an approach that shares the fundamental elements of PCC, prioritizing respect for dignity, autonomy, integrity of person and welfare above other competing priorities.

Some might argue that a PCC approach is more aligned with current thinking about how children should be actively involved in consent and assent processes to take part in research—i.e., that their individual welfare and dignity should be the foremost consideration in all aspects of the research process. However, the requirements for including children in decision-making about taking part in research is much less related to the overarching approach to care (a patient-centred or a family-centred approach), but rather to legislation, best practices and ethical guidelines.

12.3 How We View Children

Children may be viewed in many sorts of ways, but we can group these views into three main perspectives: the view of children as the same as adults, the view of children as different from adults, and the view of children as *similar* to adults but not the same (Punch 2002). While many may support the notion that treating children the same way as adults recognizes them in a respectful way, treating them as we would competent and grown adults fails to consider the social reality of children (Punch 2002; Morrow 1999). Relationships between adults and children involve inevitable power differences, and these also exist in the adult researcher-child participant relationship. Failure to acknowledge these power differentials when considering the important social context in which children live is inconsistent with both a PCC and FCC approach. From a PCC approach, we must consider the social reality of patients and their unique situations and trajectories. From a FCC perspective, we must consider the family, relationships and hierarchies that exist within the patient's family. While these approaches may differ in a number of ways, both require practitioners to consider the important social context in which their patients exist.

Most children are, arguably, used to having the power of adults exerted over them and, by virtue of their position in an adult-constructed and adult-run society, are not accustomed to being afforded the same privileges as adults (Alderson and Goodey 1996; Punch 2002). They are also habituated to being considered as particularly vulnerable, and being considered as being in need of protection from adults (Alderson and Goodey 1996; Beauchamp and Childress 2008; Hall et al. 2001; Punch 2002) Adult researchers must confront these realities in any approach to involving children in research. How each researcher views childhood will necessarily have an impact upon how children are treated as participants, how and from whom consent is obtained, and what kinds of methods are used in research. The growing body of literature supporting the fact that children have both "social and moral competence" (Alderson and Goodey 1996, p. 114) challenges a purely protective approach to children being involved in research, by acknowledging that children are much more competent than previously thought, and that assumptions about children's competence based solely on age is an outdated and perhaps unhelpful constructs.

Exposing children to risk of harm must, of course, be tempered with the expected benefits, to individuals or groups. Certainly children have the potential to experience, for longer periods of time, the effects of either benefits or harms (Royal College of Paediatrics and Child Health [RCPCH] 2000). It is commonly understood that research involving children should be based upon a number of clear guiding principles and that researchers who wish to involve children in research must be attentive to the kinds of legal and ethical obligations to which they will be held (RCPCH 2000).

12.4 Guiding Principles for Research Involving Children and Adolescents

As recently as the twentieth century, many researchers simply did not involve children in clinical research. In many cases, participation in clinical research was, for the most part, limited to competent adults, and it was felt that children were particularly vulnerable when exposed to risk of harm. It is also the case that, in the recent past, it was felt to be unethical to expose children to potential risk or harm without direct benefit to individual participants. However, that perspective has notably evolved today. With the evolution of the idea that children are "active citizens" (Lambert and Glacken 2011, p. 781) in their lives, it is also acknowledged that they should be able to be included in research, and be allowed to exercise self-determination about what to participate in, and how to do so. This evolved notion of children acknowledges that they are unique individuals, with their own values, interests and preferences, rather than simply "small adults" (RCPCH 2000, p. 177) whose preferences, values and beliefs are not yet fully formed and therefore do not merit consideration—a view which implies that they are merely incomplete versions of the adults they will one day become (Twycross 2009). This attitudinal shift (the recognition of children as unique persons, rather than simply small not yet fully-formed adults) also implies a reconceptualization of what how we think about the distribution of benefits and burdens of research across the population. Up until recently, many medications, treatment and modalities for diseases of childhood had been tested exclusively on adults, and children had been felt to be too vulnerable to include in clinical research that might not provide direct benefit to them (although this is not the case in research involving adults). However, excluding children from research on diseases and illnesses of childhood means that while they are protected from the burdens of research participation, they are denied potential benefits (Hall et al. 2001; RCPCH 2000; TCPS 2010). It may be considered to be an injustice to deny benefits to entire groups in an attempt to protect them from potential risks or harm. The lack of specific knowledge about particular drugs or modalities of treatment and their effects on children means that, in some cases, children were in fact being regarded as "small adults" (RCPCH 2000, p. 177) when being treated with medications that had undergone testing only in an adult population. Clinical practice guidelines for paediatric research should ideally be based upon research that is carried out on children in order to best serve the interests of and improve the healthcare for children. Furthermore, children are especially positioned to experience the longest-lasting benefits from research and, as a result of this, should be supported to take part in ethically sound research (RCPCH 2000).

It is generally assumed that children should only be involved in research if the research is necessary in order to improve the health of children and the same research cannot effectively use adults, animals or laboratory models (Gill et al. 2003; RCPCH 2000). Before a child can be included in research, care must be taken to ensure that children are recruited and engaged in research in ways that are appropriate to their developmental stage, and with processes that are neither coercive nor involve undue influence. Research protocols involving children should be evaluated with careful attention to the balance of potential risk of harm and benefit and that the best interests of the individual child are considered. Further, the values and best interests of the individual child who is a potential participant should be paramount in terms of considerations, by ensuring that the child clearly can exert his or her wishes in terms of decision-making to take part in the research (Cameron et al. 2011; Gill et al. 2003; Kanner et al. 2004; Neill 2005; Punch 2002; Showalter Salas et al. 2008). Most current guidelines note the role of the parents in decision-making as supportive in terms of facilitating the child's decision-making process with respect to taking part in research (Gill et al. 2003; Neill 2005). The role of the parents as articulated in the literature clearly situates the child as the centre of care, with the family playing a more supportive and facilitative role. From a FCC perspective, this may necessarily mean that the interests, wishes and values of parents are considered secondary to those of children, but this is not clear from the literature.

Ethical guidelines and discussions tend to focus on two main foci: *informed consent* and *protection* of child participants (Morrow and Richards 1996). The Ethics Working Group of the Confederation of European Specialists in Paediatrics notes fundamental guiding principles for the ethical inclusion of children in biomedical research (Gill et al. 2003, p. 456). These guidelines are similar to those intended to protect adult participants, but are considered to be stronger obligations for researchers when children are involved in research. They are as follows: First, the dignity of the child-participants must be respected. Second, the best interests of the child must be protected. Researchers must have processes and checks in place to ensure that the child is taking part in the research to further their interests or because participation aligns with the child's values, rather than for reasons such as to please others, or out of feelings of obligation to those in positions of power, such as parents or healthcare professionals (TCPS 2 2010). Third, the child must not be exposed to harm as a result of taking part in the research. While reasonable risks of harm may be inherent in participation, clear steps to mitigate potential harms and reduce children's exposure to harm must be taken. Fourth, the privacy and confidentiality of the child must be respected. Privacy may include protecting physical privacy during clinical research procedures and also ensuring social and psychological privacy by not exposing children's views or opinions without their permission or in a way that might cause them to be harmed, embarrassed or disenfranchised. The confidentiality of the child must be ensured through the safe and secure storage of all documents and records related to the research. In addition, children's confidentiality must be maintained in any dissemination or publication of research findings (de Lourdes Levey et al. 2003; Gill et al. 2003; Groene 2011; RCPCH 2000).

The question of whether or not children should be permitted to take part in research that does not offer any direct benefit to individual child participants has been debated at length in the literature. Some argue that it is unethical for children to take part in research that does not offer them potential direct benefit. Others state that research that does not offer a possibility of direct benefit to children is not inherently unethical, and that children should still be offered the opportunity to take part in such research, if the informed consent process is managed appropriately and potential harms minimized (RCPCH 2000). If we deny children the opportunity to participate in ethically sound research that may not offer the individual participants any potential direct benefits, it follows that we may be denying children as a group the potential benefits that might be a result from such research. Longitudinal research of cohorts of children, observational research of children and research involving evaluation of diagnostic assessments offer no possibility of benefit to individual participants, but may offer significant potential benefits to children as a social group (RCPCH 2000).

Discussions of risk-benefit balances, however, must be tempered to acknowledge that there may be differences between risk exposure for ill children and well children (Beigay 2007). The degree of potential risk of harm that a healthy child may be exposed to may be different than the acceptable risk of harm for an ill child, depending upon the nature of the research and the potential benefits (Harrison and the Canadian Pediatric Society Bioethics Committee 2004; Beigay 2007).

12.4.1 The Important Role of Values

Children develop values over time and with the influence of many factors in their lives. Typically, parents and guardians play the most important role in the development of fundamental values, as very young children generally adopt or follow the rules and norms of a family. Healthcare providers consider parents to be the experts about very young children's preferences and wishes, which are more often than not situated within the values of the family (Hallström and Elander 2003). Other kinds of influences upon value development may include: peer groups, culture, socioeconomic status, neighborhood, school, health state and religious beliefs or practices. The dominant discourses that children are exposed to may also have an effect upon their values and beliefs. Additionally, the kinds of mentors, both formal and informal, that children have or seek out, can have a tremendous impact upon the development of values. Teachers, coaches and family friends are examples of the kinds of mentors children may have outside of their parents, with whom they may seek to align their values. This may be especially true of older children and adolescents who, as part of typical adolescent rebellion, are seeking to explore values outside of those of their families and to challenge established norms.

As children seek to develop their own worldview and establish their own moral autonomy, they may quite naturally wish to explore new experiences and 'try on' new values. While very young children live with the moral authority of others over them, they gradually move into a state of moral exploration as they move towards moral autonomy (Yeo et al. 2010). Moral exploration may involve some degree of perceived rebellion or secrecy from parents as adolescents try to establish themselves as morally autonomous and thereby distinct from their parents. If, in an ideal world, participants take part in research because it aligns with their values, this presents an inherent dilemma in cases where children's values may differ from those of their parents. Children may value participation or an opportunity to take part in research while parents might have misgivings. In this case, whose values trump? The literature on FCC provides little guidance for this kind of dilemma, however, researchers, especially those working with adolescents, will inevitably encounter values conflicts within families.

One of the foundational concepts that underpins a FCC approach includes the acknowledgement that parents know their children best and should be actively involved in their care (King et al. 1998; Kuo et al. 2012). While this may be the case for very young children, adolescents may not consider their parent to have expert knowledge of their child's interests and wishes.

Some also argue that the challenge which the evolving values of children presents is that children have not yet had an opportunity to establish clear values and beliefs about what is most important to them. Therefore, they are felt to be unable to take part fully in decision-making, as they lack the ability to evaluate choices in light of authentic and personal values that have been sustained over an appreciable amount of time (Alderson 2007). Theories of moral autonomy note that we achieve a state of moral autonomy through moral exploration and subsequent adoption of what we decide to be our own values, rather than of those who have moral authority over us (Yeo et al. 2010). It is important to remember that one does not simply go from being under moral authority to being morally autonomous overnight. It is an iterative process that progresses in different ways for different persons. The concept of the *mature minor* suggests that children, at a variety of ages, may have decision-making capacity arising from their evolving moral autonomy. These are children who, by virtue of their experiences and cognitive stage, may have a more well developed set of authentic values, such that they can express and seek opportunities that align with those values.

Respect for values is an important consideration from both a PCC perspective and a FCC perspective. The Institute of Medicine's definition of PCC focuses on respect for values as a key factor in providing respectful and responsive care by noting that PCC is "care that is respectful of and responsive to individual patient preferences, needs and values" (Institute of Medicine 2001). The earliest attempts to articulate the guiding principles of PCC note "respect for patients' preferences and values" (Luxford et al. 2011, p. 1) as one of seven core components (Gerteis et al. 1993; Luxford et al. 2011).

According to The Institute for Patient- and Family-Centered Care (IPFCC) the two main guiding principles of FCC are dignity and respect (IPFCC 2010; Kuo et al. 2012). Operationalizing these two guiding principles implies honouring family choices, cultures, perspectives, knowledge, beliefs and values (IPFCC 2010; Kuo et al. 2012). Discussions of FCC note the importance of respecting values of families but say little about how to approach differing values within a family system.

12.4.2 *Protection*

Over the past two decades, significant attention has been directed towards the protection of human participants in research. According to the Tri-Council Policy Statement for the Ethical Conduct for Research Involving Humans (TCPS 2) (2010) researchers and REBs who review and approve research involving human participants must balance the societal need for research participation with the protection of those same persons. The apparent tension created by this balancing act is even more pronounced in research that involves those persons who are considered to be more vulnerable (Hall et al. 2001; Lambert and Glacken 2011; TCPS 2 2010)

As a result of their perceived level of vulnerability, their often-restricted social power and their unique social position, children are felt to be in particular need of the protection of responsible adults (de Lourdes Levy et al. 2003; Hall et al. 2001; Hallström and Elander 2004; Lambert and Glacken 2011). In the case of research, responsible adults may include their parents, healthcare professionals involved in the circle of care, and researchers who are accessing children. As Gill notes (2004, p. 54),

> Children involved in research are of special concern because of their age specific peculiarities, the protection of their developmental potentials, the respect of their increased vulnerability and fears, and the biological differences between children and adults as mentioned above, that can be summarised as "best possible protection and promotion of the best interests of the individual child".

Those children who are youngest and at an earlier stage of development are felt to need the highest level of protection, particularly from their parents who are felt to have significant insight into their best interests and the protection of those interests (de Lourdes Levy et al. 2003).

12.4.3 *Informed Consent*

Informed consent is the most important principle in relation to participation in research. It is complex and dynamic, and typically reflects an agreement between the researcher and the participant that is ongoing and based on mutual trust and respect for dignity. Free and informed consent, as a process, is considered to have two elements: volitional ("free") and cognitive ("informed") elements. The volitional element of informed consent requires that persons be able to voluntarily agree to take part in research, free from undue influence, pressure, and feelings of obligation or coercion. The cognitive aspect of informed consent involves an understanding of the information provided, and a clear ability to both demonstrate that understanding and apply the information to one's own situation. Finally, consent can be provided and withdrawn at any time, and the dynamic and ongoing nature of consent must be respected (Roberson 2007; TCPS 2010).

> To ensure that the best decisions are made for children and adolescents, these decisions should be made jointly by members of the healthcare team, the child or adolescent's parents, and sometimes the child or adolescent. Children and adolescents should be involved in decision-making to an increasing degree as they develop, until they are capable of making their own decisions about treatment. (Harrison and the Canadian Pediatric Society Bioethics Committee 2004, p. 99)

From a legal and an ethical perspective, in Canada, decision-making capacity and the ability to provide informed consent, particularly in decisions regarding health is not necessarily tied to age (Harrison and the Canadian Pediatric Society Bioethics Committee 2004; Jackman and McRae 2012). This is reflected in the doctrine of the *mature minor*, in which it is acknowledged that a child's ability to understand the nature and consequences of decisions regarding healthcare options is not solely dependent upon age, but upon other complex and contextual factors (Harrison and the Canadian Pediatric Society Bioethics Committee 2004; Jackman and McRae 2012). When a child can demonstrate understanding of the treatment and the nature of consequences, they may be found to be a *mature minor* and parental consent is not typically required (Doig and Burgess 2000). While there are provincial and territorial variations on the rights of children to provide consent for treatment decisions, the mature minor doctrine is adhered to across most Canadian jurisdictions (Gilmour 2002; Harrison and the Canadian Pediatric Society Bioethics Committee, 2004; Jackman and McRae 2012).

Children are felt to be in the process of developing their decision-making capacities. These developing skills should not be ignored, by excluding children from decision-making about their care or treatment. Ideally, all children should be included in decision-making, to the extent that they are able, with consideration for the kind of decision being made. Depending on the nature and gravity of the decision, what children are being asked to do, the child's developmental stage and cognitive capacity, and the level of risk and possible consequences of the procedure or treatment, children's involvement in decision-making can be adjusted, on a kind of sliding scale (Doig and Burgess 2000; Gaylin 1982; Morrow and Richards 1996). Even very young children can indicate either willingness or, on the other hand, objection to treatment or interventions, and their assent to or dissent from treatment must be considered and respected. Providing an opportunity for children to be heard is a legal obligation, according to Article 12 of the United Nations Convention of the Rights of Children (Lambert and Glacken 2011; United Nations 1989).

In terms of consent for participation in research, there still exists a certain amount of ambiguity or uncertainty about children's involvement in the informed consent process (Committee on Bioethics 1995; Lambert and Glacken 2011; Simpson 2003). Many parallels are made between the ability of children to provide consent for medical treatment with their ability to provide informed consent to take part in research. While there may be similarities in terms of requirements for providing informed consent in the two contexts, there are obvious differences in what may be at stake, in most cases. Furthermore in many cases, guidelines for children's consent to medical treatment is not entirely applicable in many types of social research on children, outside of the medical or clinical context.

The Tri-Council Policy Statement (TCPS 2) notes, in Article 3.10, that children have varying and evolving decision-making capacities of judgment and self-determination, and may require parental consent if they do not fulfill the requirements for full capacity. However, the TCPS 2 also clearly notes that children without full capacity for decision-making should have an opportunity to provide assent or dissent. While their assent may not be enough to allow them to make an independent decision to participate without the consent of a parent or guardian, their dissent does supersede parental consent for participation in research. Researchers must be able to assess children for signs of objections to participating. The American Academy of Pediatrics notes four key responsibilities that paediatricians have in ensuring children can provide assent (American Academy of Pediatrics 1995). While these guidelines are somewhat dated and are aimed towards those seeking informed consent for clinical procedures, they are still relevant today and they align nicely with expectations for researchers seeking the informed consent of children to participate in research. First, researchers must facilitate understanding of the research at a level appropriate to the child's developmental stage. Second, they should be transparent about informing the child what he or she might experience as a result of procedures involved in participation. Third, the researcher must have relevant experience in working with children in order to assess both the child's understanding of the situation, as well as factors that may have an influence upon the ability of the child to make a voluntary decision, such as pressure, influence or coercion. Finally, researchers must seek an active affirmation of participation from the child. It is not enough to accept silence, "passive resignation" or a lack of objections as affirmation (Hoehn and Nelson 2004, p. 781).

Some guidelines indicate an approximate age for assent (Foreman 1999; Gill 2004) but best practices would have researchers consider both the child's age, their developmental stage and other factors that might contribute to their ability to have an appreciation of the procedure (Lambert and Glacken 2011). An example of another factor would be a child who lives with chronic illness and who has tremendous insight into the disease process as well as his or her individual experience of living with the illness. This might, in fact, render a younger child able to provide full consent due to their level of understanding and knowledge as long as the child can demonstrate sufficient levels of understanding (appropriate to the risk of harm involved in the research), and has the cognitive capacities to understand the nature and consequences of the proposed research (Lambert and Glacken 2011). As such examples help to demonstrate, the development of moral autonomy can progress at different rates among different children and is not simply dependent upon chronological age.

Parent and Guardian Involvement in Consent The literature shows that parents have preferences when it comes to being part of the process of involving their child in research. It also shows that there are a number of factors that can have an impact upon parents' ability to provide consent or their willingness to allow their children to provide consent for self (Rothmier et al. 2003). When involved in the decision of whether or not to enroll their child in a research study, parents prefer clarity and thoroughness in the provision of information, time for decision-making and support from the healthcare team during the process of decision-making (Cameron et al. 2011). Other studies show that parents wish to hear more about potential risks of harm and benefit, but less about either the actual procedures involved in the research study or information about how voluntariness is ensured (Cameron et al. 2011; Eder et al. 2007; Tait et al. 2002). In one study examining mothers providing consent for children who had been newly diagnosed with cancer, researchers reported that mothers felt the risks of enrollment were not adequately explained to them even though they did enroll their child in the study (Pletsch and Stevens 2001; Ruccione et al. 1991).

The literature also supports the fact that many parents may enroll their children in clinical research studies out of altruism, not out of a desire to access possible incentives or compensation (Langley et al. 1998; Rothmier et al. 2003). Other literature suggests that learning more about the disease or illness may be a motivating factor for parents and children when making a decision about whether or not to enroll their child in a clinical research study (Rothmier et al. 2003).

It is clear that whether a PCC perspective or a FCC perspective is taken, parents may well be involved in the decision about whether or not to engage their child in research. From a FCC perspective, parents are one of the units of care, and their values, preferences and decisions are an important consideration in determining what is in the best interests of the child. From a PCC perspective, children may require or wish the involvement of their parents in decision-making and these wishes must be respected but the key decision-maker is always the child.

For children who may be able to provide assent, but not full consent, to participate, the parental role may be that of decision-maker, with the noted caveat that a lack of active assent on behalf of the child will very likely render the parental consent invalid, most especially in a case in which there is felt to be no potential direct benefit to the individual child by involvement in the study. In cases of assent, parents are asked to consider the best interests, welfare, inherent dignity and integrity of the child when decision-making and should ensure that they can take time to provide fully informed consent on behalf of their child. Parents can also assist with providing information to their child, at a level at which the child can understand, in order to provide an active assent. Parents can also advocate for children by informing researchers about their child's preferences or dislikes and can alert researchers to sign that a child is no longer happy to continue an activity.

Parents whose children can provide fully informed consent may still play a role of advocate and facilitator, but this is may be less likely in social or educational research, where children are often taking part in research in environments where their parents are not typically present, e.g., at school. Children who are in hospitals, by virtue of being acutely ill or having a chronic illness, may have more day-to-day involvement with their parents who are at their bedside and researchers are approaching children in an environment in which their parents typically also spend a great deal of time. In cases of clinical research where children can provide full informed consent, parents' wishes and preferences may be put aside in order to ensure that the morally autonomous child can exercise his or her will, and make an independent decision. From a FCC perspective, this may understandably create tension or friction between all players: children, healthcare providers and parents. While a FCC perspective considers the entire family unit as the unit of care and family participation as a key element, in situations in which children can provide fully informed consent, their autonomy must be respected. The parent whose child is in an environment in which FCC is practiced and in which their child is able to provide full informed consent for him or herself to take part in research may feel an understandable disconnect between the overall approach to care and the practices of obtaining informed consent.

12.4.4 Protecting the Privacy and Confidentiality of Children

Best practices in protecting privacy and maintaining the confidentiality of children in research are not that different than the kinds of recommendations in place for adults, but there are unique problems related to privacy and confidentiality that may be accentuated in the inevitable tension between respecting families and respecting the autonomy of individual children involved in research (Sheahan et al. 2012).

Privacy and confidentiality are often used interchangeably, but are different concepts. Privacy is described, in the TCPS 2, as the right of persons not to be subject to intrusion or interference from others, and this right extends to one's body, the places and spaces that one occupies, as well as opinions, thoughts and communications with others (TCPS 2 2010). As noted previously, in a research context, privacy can relate to bodily privacy during research or clinical procedures, psychological or social privacy (Gill et al. 2003; Groene 2011). Confidentiality, in a research context, is the obligation of others to keep trusted information safe and to ensure that records and documents are secure, and protected from either exposure, or unauthorized use or access throughout the life of the research project (de Lourdes Levy et al. 2003; TCPS 2 2010). Researchers must have clear plans for secure data storage, and legitimate use, especially of data that are identifiable, for the entire life of the data, until it is (also securely) destroyed in order to keep promises of confidentiality made to participants. Participants should receive full information about how their privacy will be protected and how confidentiality will be maintained, at a level that they can understand (de Lourdes Levy et al. 2003; Gill et al. 2003; Lambert and Glacken 2011).

In research involving children, data may be kept for long periods of time in some disciplines and if not managed properly and securely, the risk of exposure of trusted information may have a negative impact upon an adolescent or young adult who previously took part in research, whose information may be exposed without their permission through a breach of confidentiality. While this is a concern for clinical research, it is also a concern for social and educational research. Children may offer their opinions or views within the researcher-child relationship of trust and be informed that confidentiality will be maintained and that others, outside of the researcher, would not have knowledge of the specific views of the child. In qualitative research involving sensitive topics, exposing a child's view in a way that does not protect their privacy, has the potential to cause potential harm, embarrassment or to damage relationships, and this can have devastating effects. While this is a concern also for adults taking part in research, the risks related to confidentiality may not be clear to children and must be explained at a level that they can understand.

Just as adults, children involved in research in which genetic material is stored over long periods of time should be informed about the potential risk of storage and it should be explained to them the implications of storing data either without a link to their identity or with a link back to them. The presence of such a link implies particular kinds of obligations for researchers who enroll children as participants in studies that involve the procurement and storage of genetic materials for testing. Often, children who are only able to provide initial assent for storage and subsequent testing of their identifiable genetic material are provided with an opportunity to reaffirm or withdraw their agreement through an active consent process once they reach an age and developmental stage to offer full consent for this.

Trust that researchers will protect the protection of privacy of children involved in research is paramount for both children and families. From a FCC perspective, the storage and use of genetic materials has implications for the family beyond just the individual child. Parents of young and school-age children who are supporting their children in an assent process should ensure that the implications of the storage of genetic materials is well understood and that this is communicated to the child a second time, once he or she reaches an age at which full informed consent can be provided independently.

12.4.5 Reporting of Incidental, Unexpected or Critical Findings

From either a FCC or a PCC perspective, for research involving children, a two-way flow of information between researcher and child is implied. Participants may voluntarily provide their opinions, views, time, biological material, their bodies for diagnostic testing or clinical research procedures, and even their genetic or family information. Researchers must provide information about the research, the risk and benefits, the alternatives to participation, the voluntary nature of participation, the ways in which privacy and confidentiality will be maintained, and information about how the data will be disseminated used and stored. They also must answer any questions or concerns that participants may have. In FCC, this flow of information may, at some points, likely involve the family. For younger children, parents will want to be involved in the provision of information from the researcher and may well provide assistance and support to their children while they provide information, answer questions or undergo procedures involved in research. With involvement of parents and agreement of the children, the information flow is not typically restricted.

However, in cases of unexpected or incidental findings that might have significant implications for the health or welfare of a child, how the information should be provided to children and parents is less straightforward. Researchers have an ethical obligation to inform participants of anything that is unexpectedly noted or discovered that might have an implication for their security or interests (TCPS 2 2010).

Consider the example of the 12-year-old female adolescent who undergoes testing as part of a research project and is found to have a sexually transmitted infection or is pregnant. Principles of respect for dignity would recommend that an incidental finding like this be disclosed to the child or adolescent first, without involvement of the parents. It is not unexpected to find that children request privacy from their parents, and this is can become an ethical issue if not managed sensitively (Delaney et al. 2010). From a FCC perspective, this might be problematic and, understandably, many parents may be uneasy with the process in a case like this. An important and proactive strategy is for researchers to anticipate of the kinds of potential incidental findings that might be encountered in a study. In some research, this is quite straightforward. For example, research involving computerized tomography of the brain, or extensive blood work could possibly make known other pathologies through examination of images or testing of the blood. In social research that may involve asking children specific questions about their home life and parents, a child may disclose potential abuse, neglect or situations of concern. There may be legal obligations for mandatory reporting involved in some incidental finding, and the process of anticipating a wide spectrum of possible findings allows the researcher to have a clear plan in place describing how these kinds of findings may be handled. This plan must be detailed and thorough and must be described in full to the child and the family as part of an informed consent process.

12.5 Conclusion

The involvement of children in research is a topic that will certainly continue to be a topic of discussion and deliberation. With the evolution of ideas about the moral autonomy and developing capacities of children alongside the increased interest of researchers in children, the sociology of childhood and the experience of children across a variety of contexts and environments, it is an inevitability that children will continue to be sought after to take part in many different types of social and educational research. Further, with the recognition that children are not simply small adults, their individual experience and unique perspectives will be sought out even more than before. Their involvement in health research will be driven by the motivation to cure childhood diseases, better treat chronic illnesses earlier in life and, ensure that medications, treatments and procedures are tailored to the unique needs of children. Healthcare professionals, researchers from a wide spectrum of disciplines, Research Ethics Boards, parents and children will continue to grapple with the issues raised by the involvement of children in research, such as how and from whom to obtain consent, how to ensure voluntariness, and whether the protection of children should override both the need to provide benefits through research and the involvement of parents when children can provide full consent for self. While FCC and PCC approaches provide clear guidance for *the provision of care* in particular healthcare environments and institutions, they may also provide some insight into issues that research involving children brings to our attention.

References

Alderson, P. 2007. Competent children? Minors' consent to health care treatment and research. *Social Science and Medicine* 65:2272–2283.

Alderson, P., and C. Goodey. 1996. Research with disabled children: How useful is child-centered ethics? *Children and Society* 10:106–116.

American Academy of Pediatrics Committee on Bioethics. 1995. Informed consent, parental permission, and assent in pediatric practice. *Pediatrics* 95 (2): 314–317.

Beauchamp, T., and J. Childress. 2008. *Principles of biomedical ethics*. 6th ed. New York: Oxford University Press.

Beigay, T. M. 2007. Children in research: Human subject considerations for the inclusion of children as research participants. *Progress in Transplantation* 17 (1): 54–56.

Bridson, J., C. Hammond, and A. Leach. 2003. Making consent patient centered. *British Medical Journal* 327:1159–1161.

Canadian Institutes of Health Research, Natural Sciences and Engineering Research Council of Canada, and Social Sciences and Humanities Research Council of Canada, Tri-Council Policy Statement: Ethical Conduct for Research Involving Humans (TCPS 2). December 2010.

Cameron M. A., L. E. Marsillio, L. F. Cushman, and M. C. Morris (2011) Parents' perspectives on the consent approach for minimal-risk research involving children. *IRB* 33 (6): 6–12

Delaney C. , M. Spriggs, C. L. Fry, and L. Gillam. 2010. The unique nature of clinical ethics in allied health pediatrics: Implications for education. *Cambridge Quarterly of Healthcare Ethics* 19:471–480.

De Lourdes Levy, M. , V. Larcher, R. Kurz, and Members of the Ethics Working Group of the CESP. 2003. Informed consent/assent in children. Statement of the ethics working group of the Confederation of European Specialists in Paediatrics (CESP). *European Journal of Pediatrics* 162:629–633.

Doig, C., and E. Burgess. 2000. Withholding life-sustaining treatment: are adolescents competent to make these decisions? *Canadian Medical Association Journal* 162 (11): 1585–1588.

Eder, M. L., A. D. Yamakoski, P. W. Wittmann, and E. D. Kodish. 2007. Improving informed consent: Suggestions from parents of children with leukemia. *Pediatrics* 119 (4): e849–859.

Foreman, D. M. 1999. The family rule: A framework for obtaining ethical consent for medical intervention from children. *Journal of Medical Ethics* 25 (6): 491–496.

Gaylin, W. 1982. The competence of children: No longer all or none. *Hastings Center Report* 12:33–38.

Gerteis, M., S. Edgman-Levitan, J. Daley, and T. L. Delblanco, eds. 1993. *Through the patient's eyes: Understanding and promoting patient-centered care*. San Francisco: Jossey-Bass.

Gill, D., and Ethics Working Group of the Confederation of European Specialists in Paediatrics. 2003. Guidelines for informed consent in biomedical research involving paediatric populations as research participants. *European Journal of Pediatrics* 162:455–458.

Gill, D. 2004. Ethical principles and operational guidelines for good clinical practice in paediatric research. Recommendations of the ethics working group of the Confederation of European Specialists in Paediatrics (CESP). *European Journal of Pediatrics* 163:53–57.
Greene, S. M., L. Tuzzio, and D. Cherkin. 2012. A framework for making patient-centered care front and center. *The Permanente Journal* 16 (3): 49–53.
Groene, O. I. 2011. Patient centeredness and quality improvement efforts in hospitals: rationale, measurement, implementation. *International Journal for Quality in Health Care* 23 (5): 531–537.
Hall, J. M., P. E. Stevens, and P. K. Pletsch. 2001. Team research: Using qualitative methods: Investigating children's involvement in clinical practice. *Journal of Family Nursing* 7 (1): 7–31.
Hallström, I., and G. Elander. 2004. Decision-making during hospitalization: Parents' and childrens' involvement. *Journal of Clinical Nursing* 13:367–375.
Harrison, C., and Canadian Pediatrics Society Bioethics Committee. 2004 Position Statement: Treatment decisions regarding infants, children and adolescents. *Paediatric and Child Health* 9 (2): 99–103.
Hoenh K. S., and R. M. Nelson. 2004. Advising parents about children's participation in clinical research. *Pediatric Annals* 33 (11): 778–781.
Institute for Patient- and Family-Centered Care. 2010. Patient- and family-centered care core concepts. http://www.ipfcc.org/faq.html. Accessed 9 Dec 2012.
Institute of Medicine (US) Committee on Quality of Health Care in America. 2001. *Crossing the quality chasm: A new health system for the 21st century*. Washington, DC: National Academies Press.
Jackman, M., and A. McRae. 2012. Medical decision-making and mature minors. Royal College of Physicians and Surgeons of Canada. http://www.royalcollege.ca/portal/page/portal/rc/common/documents/bioethics/section1/case_1_5_2_e.html. Accessed 10 Dec 2012.
King, G., M. Law, S. King, and P. Rosenbaum. 1998. Parents' and service providers' perceptions of the family-centeredness of children's rehabilitation services. *Family-Centered Assessment and Intervention in Pediatric Rehabilitation* 18 (1): 21–40.
Kanner, S., S. Langerman, and M. Grey. 2004. Ethical considerations for a child's participation in research. *Journal for Specialists in Pediatric Nursing* 9 (1): 15–23.
Kuo, D. Z., A. J. Houtrow, P. Arango, K. A. Kuhlthau, J. M. Simmons, and J. M. Neff. 2012. Family centered-care: Current applications and future directions in pediatric health care. *Maternal and Child Health Journal* 16:297–305.
Lambert, V., and M. Glacken. 2011. Engaging with children in research: Theoretical and practical implications of negotiating informed consent/assent. *Nursing Ethics* 18 (6): 781–801.
Langley, J. M., S. A. Halperin, E. L. Mills, and B. Eastwood. 1998. Parental willingness to enter a child in a controlled vaccine trial. *Clinical and Investigative Medicine* 18:86–93.
Luxford, K., D. Gelb Safran, and T. Deblanco. 2011. Promoting patient-centered care: A qualitative study of facilitators and barriers in healthcare organizations with a reputation for improving the patient experience. *International Journal for Quality in Health Care* 23 (5): 510–515.
Lord, Darzi. 2008. *High quality care for all: NHS next stage review final report*. London: The Stationary Office.
Maternal and Child Health Bureau. 2005. Definition of family-centered care. http://www.familyvoices.org/work/family_care. Accessed 9 Dec 2012.
McClimans, L. M., M. Dunn, and A. Slowther (2011) Health policy, patient-centered care and clinical ethics. *Journal of Evaluation in Clinical Practice* 17:913–919.
Mead, N., and P. Bower. 2000. Patient-centeredness: A conceptual framework and review of the empirical literature. *Social Science and Medicine* 51:1087–1100.
Morrow, V. 1999. "It's cool, …'Cos you can't give us detentions and things, can you?": Reflections on researching children. In *Time to listen to children*, eds. B. Milner and B. Carolin, 203–215. London: Routledge.
Morrow, V., and M. Richards. 1996. The ethics of social research with children: An overview. *Children and Society* 10:90–105.

Neill, S. J. . 2005. Research with children: A critical review of the guidelines. *Journal of Child Health Care* 9 (1): 46–58.

Pletsch, P. K., and P. E. Stevens. 2001. Children in research: Informing consent and critical factors affecting mothers. *Journal of Family Nursing* 7 (1): 50–70.

Punch, S.. 2002. Research with children: The same or different from research with adults? *Childhood* 9 (3): 321–341.

Roberson, A. J. 2007. Adolescent informed consent: Ethics, law, and theory to guide policy and nursing research. *Journal of Nursing Law* 11 (4): 191–196.

Rothmier, J. D., M. V. Lasley, and G. G. Shapiro. 2003. Factors influencing parental consent in pediatric clinical research. *Pediatrics* 111 (5): 1037–1041.

Royal College of Pediatrics and Child Health. 2000. Guidelines for the ethical conduct of medical research involving children. *Archives of Disease in Childhood* 82 (2): 177–182.

Ruccione, K, R. F. Kramer, I. K. Moore, and G. Perin. 1991. Informed consent for treatment of childhood cancer: Factors affecting parents' decision making. *Journal of Pediatric Oncology Nursing* 8 (3): 112–121.

Sheahan, L., M. Da Silva, C. Czoli, and R. Zlotnik Shaul (2012) A Canadian perspective on a child's consent to research within a context of family-centered care: From incompatibility to synergy. *Journal of Clinical and Research Bioethics* 3 (1): 1–7.

Shields, L., J. Pratt, and J. Hunter. 2006. Family centered care: A review of qualitative studies. *Journal of Clinical Nursing* 15:1317–1323.

Showalter Salas, H., Z. Aziz, N. Villareale, and D. S. Diekema. 2008. The research and family liaison: Enhancing informed consent. *IRB* 30 (4): 1–8.

Simpson, C. 2003. Children and research participation: who makes what decisions. *Health Law Review* 11 (2): 20–29.

Sine, D, and V. A. Sharpe. 2011. Ethics, risk and patient-centered care: How collaboration between clinical ethicists and risk management leads to better patient care. *Journal of Healthcare Risk Management* 31 (1): 32–37.

Tait, A. R., T. Voepel-Lewis, A. Robinson, and S. Malviya. 2002. Priorities for disclosure of the elements of informed consent for research: A comparison between parents and investigators. *Paediatric Anaesthesia* 12 (4): 332–336.

Twycross, A. 2009. An interprofessional approach to the ethics of undertaking research with children. *Nurse Researcher* 16 (3): 7–20.

United Nations. 1989. UN Convention on the Rights of the Child. http://www.crin.org/resources/treaties/CRC.asp. Accessed 10 Dec 2012.

Yeo, M., A. Moorhouse, P. Khan, and P. Rodney (2010) *Cases and concepts in nursing ethics*. 3rd ed. Peterborough: Broadview Press.

Chapter 13
Clinical Genetics Practice: Integrating Paediatric Patient and Family-Centred Care

Cheryl Shuman and Riyana Babul-Hirji

The diagnosis of a genetic disorder in a child has rippling effects throughout the family. Much of the impact mirrors the psychosocial adaptation and pragmatic issues that arise any time a child falls ill; however, when there is an underlying genetic contribution to the disorder, additional issues typically emerge. Initially the genetic contributions to a disorder may not be recognized or acknowledged, nevertheless, genetic factors are estimated to account for 96 % of chronic paediatric disorders in children and prompt 71 % of paediatric hospital admissions (McCandless et al. 2004). Therefore, concerns related to a genetic diagnosis are likely to surface more often than might otherwise have been anticipated.

Key issues inherent to the diagnosis of a genetic disorder in a child include communicating and addressing potential risks to family members. By its very nature, genetic medicine and genetic counselling involve the family unit. Whereas in previous years family members might have been offered empiric information regarding their own potential risks, our increasing ability to define the etiology of genetic disorders via genetic testing now also offers them more actionable options. The increasing availability of genetic testing may prompt some families impacted by the diagnosis of a genetic disorder to reframe their notion of family. That is, families may need to consider communicating genetic information to more than just first degree relatives (e.g., nuclear family members). The intricacies involved in the provision of genetic counselling as applied to patient-centred and family centred care models will form the basis of this chapter.

The advent of new genetic/genomic testing technologies, such as microarray and whole genome/exome sequencing (described below), holds tremendous promise for resolving the diagnostic odyssey encountered by many families when a specific diagnosis and etiology cannot be determined. These technologies are also contributing to our understanding of chronic, multigenic, and multifactorial conditions, such

C. Shuman (✉) · R. Babul-Hirji
Clinical and Metabolic Genetics, The Hospital for Sick Children, Molecular Genetics, University of Toronto, Toronto, Canada
e-mail: cheryl.shuman@sickkids.ca

R. Babul-Hirji
e-mail: riyana@sickkids.ca

R. Zlotnik Shaul (ed.), *Paediatric Patient and Family-Centred Care: Ethical and Legal Issues,* International Library of Ethics, Law, and the New Medicine 57,
DOI 10.1007/978-1-4939-0323-8_13, © Springer Science+Business Media Dordrecht 2014

as autism spectrum disorder and diabetes, where the cause of these complex diseases is thought to involve an interaction of both genetic and environmental factors. Current thinking holds that genomic testing will be transformative in predicting and managing healthcare for children and their family members. In fact, many propose that genomic testing, specifically whole genome sequencing, will be offered widely to individuals including perhaps all newborn babies, to identify genetic alterations before the development of symptomatology; the results of such testing would then be used for anticipatory medical management with the intended goals of improving overall health and well-being, and reducing healthcare costs. At present, the utilization of genetic and genomic testing by non-genetics healthcare providers indicates that this testing has already transitioned into mainstream medicine.

Genomic testing, as with the implementation of many new testing modalities, presents a number of ethical and practical challenges. Some of these are not entirely new nor even unique to the field of genetics. Examples of challenges associated with genomic testing include managing the volume of data generated, interpreting multiple results of uncertain or unknown clinical significance, disclosing incidental findings (i.e., findings detected that are not related to the original indication for testing), and recontacting patients/families as re-interpretation and new information become available (Tabor et al. 2011; Sijmons et al. 2011; Ormond et al. 2010). The landscape of genetic medicine is changing dramatically and the provision of care to children and their families in this new context merits thoughtful consideration and planning involving all stakeholders.

When genetic information becomes available on a child, by way of a clinical diagnosis or through the use of genetic testing, can either of the two health care delivery models, i.e., (i) paediatric patient centred care and (ii) family centred care inform clinicians in terms of providing optimal care? How do clinicians balance what might be in the best interest of the child when the child's diagnosis and/or test results have implications for other family members? Is one health care delivery model more relevant to paediatric genetic medicine than the other? This chapter will discuss the application of these models in the context of genetic counseling and genetic testing, using whole genome sequencing as an example in a paediatric setting.

For the purposes of this chapter, patient-centred care is defined as focusing on the needs of the patient and his/her world. The clinician-patient relationship is viewed as a collaborative partnership where the patient is encouraged to be actively involved in decisions involving his/her own medical care (Epstein and Street 2011). In contrast, the family-centred perspective values the needs of the patient's family in that it respects "the family's values, environment, culture, resources, needs and strengths, as well as viewing these characteristics as assets for patient care and treatment plans" (Hernandez et al. 2006). In a paediatric genetic medicine setting, it can be difficult to disentangle the needs of the child from that of his/her family, especially given the potential implications of genetic alterations to the personal health of other family members. In order to clarify some of the concepts presented, we begin with a brief review of basic genetics and testing terminology.

13.1 Background—Genetic Concepts

Genes determine the instructions for how our bodies develop, grow and function. It is currently recognized that environmental factors can influence the expression of genes and this has become an important area of study but will not be addressed in this chapter. Each gene comprises a specific stretch of DNA (deoxyribonucleic acid) and is encoded by four bases: adenine (A), guanine (G), cytosine (C) and thymine (T). Simply put, the specific sequence of these bases is critical for the correct cellular instruction. Genetic errors such as the replacement of one base with another or having too few or too many copies of genes can result in a problem with development (birth defects, developmental delay, etc.) or the functioning of our cells and bodies (cancer, diabetes, etc.). Such errors are typically referred to as mutations. However, many changes in the sequence can represent benign variants that contribute to each of us being a unique individual. The term "genome" refers collectively to all of the genes in an individual.

It is currently estimated that humans have approximately 25,000 genes which are 'lined up' on structures called chromosomes. We typically have 23 pairs of chromosomes, or 46 chromosomes in total, and these pairs are numbered from the largest (1) to the smallest (22). The first 22 pairs are the same in males and females, while the 23rd pair is the sex chromosomes; females usually have two X chromosomes (i.e., 46, XX) and males usually have one X and one Y chromosome (i.e., 46, XY).

13.2 Background—Genetic Testing

When a child is suspected of having a genetic condition, genetic testing may be undertaken to determine or to confirm a diagnosis based on clinical presentation.Such testing would be considered diagnostic and may include targeted genetic testing-looking for mutation(s) or change(s) in the specific gene responsible for the genetic condition in question.

Diagnostic testing approaches include chromosome analysis to screen for extra or missing chromosomal material. This testing would report an individual's chromosomal make-up or karyotype; an example of a relatively common karyotype alteration is trisomy 21 Down syndrome, where instead of the usual two copies of chromosome number 21, there are three copies. A far more sensitive approach currently utilized for the detection of extra or missing chromosomal material is microarray testing, also known as array-based comparative genomic hybridization (aCGH). This testing scans the whole genome to detect changes in the dosage of genetic material at a much higher resolution than conventional karyotype analysis. Such alterations are referred to as copy number variants (CNV). Another method to detect even more subtle genomic variations (single nucleotide polymorphisms or SNPs) either in the clinical or research arena is a SNP array. Detecting such alterations (i.e., CNVs and SNPs) may assist in determining the cause or possible contributors to an individual's physical and/or intellectual disability. However, these testing modalities may also reveal incidental genetic alterations which could have reproductive or health implications not only for the child, but also for other family members. Incidental findings may not align with one's clinical presentation; that is to say, the testing result may suggest that the individual has a condition which s/he clearly does not. The interpretive expertise of the clinician is critical for contextualizing any findings detected on testing.

As the cost continues to decrease, even newer genetic testing technologies such as exome or whole genome sequencing are transitioning from the research to the clinical arena, such that they are now becoming incorporated into testing algorithms. These tests allow for the analysis of all of the gene coding regions (exome sequencing) or of the entire genome (whole genome sequencing) for an individual, rather than the evaluation of one or a limited number of genes known to be associated with a specific disorder. Such testing not only generates huge volumes of data requiring sophisticated bioinformatics analysis for interpretation, but is likely to reveal incidental or secondary findings, and with increased frequency in comparison to microarray testing. Some of these findings may have potential health implications for which current medical management in childhood may or may not exist. Additionally, whole genome or exome sequencing undertaken on a child may unintentionally reveal genetic alteration(s) associated with adult-onset disorders; this issue of predictive testing is discussed further in the case example below. When should such unanticipated findings be disclosed and to whom? Parents can consent to testing for their children, but there are inherent ethical challenges when considering disclosure of such results to the parents, including preserving the child's autonomy and privacy.

Another issue related to genetic testing involves changes that may be detected in a chromosomal region involving a gene(s) not previously reported (i.e., not published or catalogued in one of the databases curated and accessed by scientists and clinicians) and therefore classified as being of unknown or uncertain clinical significance. In such situations, the information available for dissemination is limited. Family studies (i.e., potentially obtaining DNA from biological parents, sibs, and other family members) are often recommended to ascertain if other family members, especially those not affected with the disorder under investigation, may have the same alteration. This can assist with determining if the change(s) identified is pathogenic (i.e., disease causing).

Genetic testing implications for family members also arise when a pathogenic mutation(s) is identified in a child as the clinician may recommend cascade testing for other family members. This would require that the parents inform other family members of both the diagnosis in their child, and the availability of genetic testing for the familial mutation. Such testing may be informative for diagnostic purposes since the condition may be present in other family members but with a milder presentation. This information may also provide relevant information for future reproductive risks (e.g., carrier status) depending on the inheritance pattern. Communication of this information to other relatives may pose challenges for some families, both logistically (e.g., may not have any contact) and psychosocially (e.g., stigmatization). Furthermore, consideration must be given to providing such information to unsuspecting relatives who, if given the choice to know, may have chosen otherwise.

In addition to pursuing genetic testing for the purposes of a diagnosis, additional genetic testing categories include predictive genetic testing, carrier testing and population screening. Predictive testing may be undertaken for individuals at increased risk for developing a disorder based on their family history. Ideally, testing is first initiated on the individual(s) in the family diagnosed with the disorder in question. If a genetic mutation were identified in that individual, then this would facilitate predictive testing for other interested but as yet asymptomatic family members. The absence of the familial mutation would allow the clinician to offer reassurance to the asymptomatic family member about not being at increased risk to develop the condition. Predictive genetic tests are often undertaken in adult patients for adult onset conditions (familial forms of breast and/or ovarian cancer, Huntington disease, early onset Alzheimer disease, etc.). In contrast, predictive testing in children may be considered for conditions where anticipatory medical management may be possible (e.g., cancer syndromes such as Li Fraumeni or von Hippel Lindau, cardiac disorders such as hypertrophic cardiomyopathy or long QT syndrome [arrhythmia], etc.). Generally, predictive testing for adult onset disorders is not undertaken in childhood unless there are medical management recommendations that might improve outcome. If no such medical management exists, testing is typically postponed until such time as the child has the capacity to participate in the decision-making process about the test and to provide informed consent (Clarke 1994; ASHG Social Issues Committees 1995; Arbour and Bioethics Committee of the Canadian Paediatric Society 2003; Borry et al. 2009). Along the same continuum, direct to consumer (DTC) testing could impact the provision of medical care if parents were to ask for testing of their children based on their own desire for such information.

Carrier testing is primarily relevant for reproductive planning and therefore the same principles for predictive testing for adult onset disorders would apply for carrier testing of children. However, if carrier testing results might impact medical interventions, exceptions to these guidelines would be entertained. One notable exception pertains to carrier or premutation testing for fragile X syndrome, a common cause of intellectual disability in males. Given that some premutation carriers can have learning and psychological challenges and for which early intervention may be of benefit, testing for premutation status in these cases may be undertaken under the auspices of medical management.

13.3 Ethical Issues and Care Models

When undertaking genetic testing in a child, consideration must be given to key ethical tenets such as respect for autonomy (Beauchamp and Childress 2001a). However, with the integration of whole genome/exome sequencing in the clinical arena, and the potential amplification of ethical issues in general, a guiding medical ethics framework is necessary. Challenges in developing such a framework include balancing parents' desire/rationale for testing and the child's right to autonomy and privacy. As well disclosure of results would necessitate consideration of what information specifically to provide, to whom and when; and the implications of testing results for extended family members.

The guiding principles of paediatric patient-centred care and family-centred care have been articulated previously. At first glance, one might assume that a paediatric patient-centred care model would not resonate in a specialty where the potential for familial transmission is a key concept, however, elements of this model are in fact highly applicable. In the family-centred care model, the family and healthcare provider are meant to work in partnership to make decisions that ideally should be in the best interests of the child. Difficulties with this model can arise, however, when there are competing interests of different family members, whether or not articulated, which can interfere with the child's best interests. As well, family members may align in their stated interests and decisions but these may differ from the recommendations of the clinician(s). Is it possible to draw on the strengths of both models, i.e., an integrated approach, to guide best clinical practice? We present a case, involving whole genome sequencing, to facilitate the above discussion with a focus on disclosure of incidental findings.

13.3.1 Case 1

An eight-year-old child, Lisa, was referred for genetic assessment and genetic counselling because she was born with a cleft palate (opening in the roof of her mouth) and was also recently diagnosed with a mild intellectual disability. She attends the initial assessment with both of her parents, who are reportedly healthy and have no similar learning or physical issues. Lisa's mother is 35 years of age and her father is 38 years of age. As well, Lisa has two brothers, one of whom is 12 years old the other 5 years old; neither brother attends the appointment and the parents note that they are both healthy and without similar issues. The family history is non-contributory, with both parents having healthy sibs, all of whom are younger than themselves. The maternal grandparents are alive and in relatively good health but the paternal grandparents died in a motor vehicle accident a number of years ago. After obtaining informed consent from the parents, whole genome sequencing is undertaken on Lisa and a genetic alteration is identified which would account for her clinical presentation; specifically, the report notes a submicroscopic deletion of chromosome 22. This is a well-documented finding in the literature and referred to as 22q11.2 deletion syndrome. Incidentally, a different genetic change associated with an increased risk for breast and ovarian cancer in adulthood is also found, that is a deletion at chromosome region 13q13.1. This region houses the BrCa2 gene, and mutations of this gene can result in a significantly increased risk for breast and ovarian cancer. If this deletion at chromosome region 13q13.1 were inherited from one of Lisa's parents, this could confer an increased risk for that parent to develop cancer.

13.3.2 Case Discussion

The utilization of whole genome sequencing in the above case was undertaken for the purpose of establishing a diagnosis. Testing did in fact reveal a pathogenic change (i.e., deletion of 22q11.2), and provided a diagnosis consistent with Lisa's clinical findings. Unexpectedly, the report also noted the detection of an incidental, predictive finding associated with a potential increased risk of an adult onset condition (i.e., breast and ovarian cancer) in an eight-year-old child. In the current climate, parents seeking testing for a BrCa1 or BrCa2 mutation in their young child because of a positive family history of breast and/or ovarian cancer would be engaged in a discussion about deferring this request until such time as the child is old enough to participate in the decision-making process regarding the pros and cons of testing. This approach is based upon an extensive review process with consensus being reached by a number of professional organizations and resulting in the published guidelines noted earlier in this chapter. These policy statements stipulate that when there is no medical benefit to be gained from testing for an adult-onset condition in childhood, testing of minors should be deferred. Although the deletion on chromosome 13 involving the BrCa2 gene would not lead to revised medical management in childhood, it could pose significant health implications for one of Lisa's parents and, potentially, their extended relatives. Screening and surgical options exist for adults with a BrCa1 or BrCa2 mutation and these interventions can significantly impact health outcomes.

Disclosure to Lisa's parents of the pathogenic deletion of 22q11.2 would be undertaken, but should the incidental finding of the potential increased risk for breast and ovarian cancer also be disclosed? In considering this question, one might first look to published guidelines regarding the informed consent process and the delineation of such issues, including the plan for results disclosure (Canadian Institutes of Health Research 2010; Thorogood et al. 2012; Clayton and McGuire 2012). In the case of offering whole genome sequencing in a research context, the research and clinical teams would develop a plan regarding disclosure of test results including incidental findings. The clinicians/researchers involved would then review the possible outcomes with the family, prior to initiation of testing, and ask them to consider what they would prefer to learn and the potential implications of categories or 'bins' of information (Berg et al. 2011). The informed consent process in such research protocols can serve to guide the transition of whole genome sequencing from the research realm to the clinical arena; however, of equal importance would be to look at established healthcare models, as these might inform clinical practice. What follows next is a discussion of the practical applications of the paediatric patient-centred and family-centred healthcare delivery models in trying to address the health and ethical constructs that might inform best clinical practice given the issues inherent in our case.

13.4 Healthcare Delivery Models and Genetic Care

13.4.1 Paediatric Patient-Centred Care Model

As mentioned previously in this chapter, the premise of the patient-centred care model involves the active participation of the patient in decisions involving his/her health and medical care. In the adult healthcare world, active participation of the patient is plausible but even in this setting, additional implications arise in genetic/genomic medicine given the potential relevance of pathogenic alterations to one's family members. Nonetheless, such issues would be discussed with the individual presenting for consultation and testing, and while s/he would be encouraged to discuss these implications, his/her autonomy need not be breached. Issues regarding discretion vs. healthcare providers' duty to warn at-risk relatives have been considered in the genetic/genomic context, and generally the privacy of the patient is maintained (Knoppers et al. 1998; Lacroix et al. 2008).

In the paediatric healthcare setting, we considered two potential approaches regarding disclosure of testing results pertaining to incidental findings which are medically actionable in adulthood. Firstly, one could infer that the patient-centred care model would entail communicating relevant information to children, at their level of understanding, encouraging children to participate in medical decision-making and seeking their assent. Obtaining assent vs. consent from children pertains to their having some capacity for decision-making but this can vary even between children of the same age. As such, healthcare providers would need to ascertain the particular capacity of each child. Therefore, to invoke the patient-centred care model when the patient is a young child/an infant with no decision making capacity or where the child does not have the intellectual capacity to engage in such a discussion is simply not feasible.

Another approach would be to not disclose the incidental finding(s) but rather to 'lock' them away in the child's medical record until such time as the child has the intellectual and emotional maturity to decide whether or not to receive them (i.e., capable minor or adult). In this regard, the parents might be viewed solely as a proxy for safeguarding undisclosed information on behalf of their child. This would imply that the parents would not be engaged in any decision-making regarding the incidental test result(s) for their child.

In the case scenario posed above, one would need to work out the practical issues of locking away Lisa's incidental genetic testing information and construct a plan for future contact and communication. Given the basic tenets of the paediatric patient-centred care model, locking away Lisa's test results regarding the incidental finding would maintain her privacy and autonomy but might this negatively impact the health and well-being of her parents and other family members? Moreover, one might anticipate that negative health sequelae in one of Lisa's parents could subsequently impact her own well-being.

Our case is further compounded by Lisa's young age and her intellectual disability, as she may have a very limited understanding, or none at all, of the testing process and results. Given this, perhaps the risks inherent in disclosure of incidental test results to Lisa's parents can be rationalized, i.e., breaching her privacy and autonomy. Implicit in the discussion regarding the deletion involving the BrCa2 gene is the potential increased risk of cancer in one of her parents and possibly their extended family members. Although the paediatric patient-centred care model advocates that the role of the healthcare provider is to respect the patient and his/her needs, genetic test results transcend beyond the individual patient. Key ethical tenets of beneficence and non-malificence can easily be extended to preserving the child's well-being by having both parents remain in good health (Beauchamp and Childress 2001b). Contextualizing the incidental finding for the parents with respect to their own health could have the unintended benefit of invoking anticipatory healthcare for one of them. These significant health benefits can also be extended to other family members. Potential discord arises in the genetic/genomic medicine setting when one tries to isolate the genetic test results and ignores the potential benefits to other family members.

13.4.2 Family-Centred Care Model

In contrast to the paediatric patient-centred care model where the primary focus is the child or patient, the family-centred care model views the child within the context of his/her family. The family, with its own goals, values and belief systems is seen as a source of strength for the child and integral to promoting the child's health and medical care. In other words, the family-centred care model "recognizes the family as central to the child's life (and) accepts parental expertise as an information resource for a child and an important component of health care delivery" (Brown et al. 2008).

Defining what constitutes "the family" is fundamental to extrapolating this model to genetics/genomic medicine. The definition of family can vary from a two-parent household to one where the child is cared for and raised in an extended family (aunts, uncles, grandparents, etc…). How the family is defined in this model has relevance to its application to the case scenario involving Lisa. In a two-parent family, the family-centred care model would look to Lisa's parents for their expertise as they would know their daughter and her needs best. This family-centred care model would support disclosure of the incidental findings to Lisa's parents given their involvement in decision-making for her and in facilitating interventions. What is not so clear is how this model would inform practice with regards to communicating genetic test results to other family members (outside of the two-parent household) who might also be at risk. With regards to their extended family, one can hypothesize that if the extended family is seen as the expert in Lisa's care, then communication of the incidental findings would occur in their presence, and by so doing, would accomplish the goal of sharing test results with at-risk relatives. Given that some family members have the "right not to know," disclosure to the extended family would have implications for breach of individual family member's right to autonomy and privacy. Additional complexities pertaining to communication and decision-making can arise when dyadic coalitions are formed that include any two of the following stakeholders: capable minor, parents, extended family members, healthcare providers (Gilbar 2005). These discussion points are beyond the scope of this chapter but are important issues to touch upon. Equally possible with the extended family version of the family-centred care model is differing expert views on what is in the best interest of the child and divided consensus with regards to disclosure of incidental findings. Granted this could also occur with the two-parent version of family; however, the likelihood is greater when more family members are involved.

13.5 Summary

Overall, the principles guiding the paediatric patient-centred and family-centred care models satisfy specific requisites and appear to be somewhat independent in their applicability to our case example with Lisa. Neither model adequately guides best clinical practice where genetic/genomic medicine is concerned, given that genetic disorders affect not only the child but also have implications for the extended family. Perhaps the designations of paediatric patient-centred and family-centred care are somewhat arbitrary distinctions in the discipline of genetics/genetic counselling. The American Academy of Pediatrics (AAP) recently issued a Policy Statement (Pediatrics 2012) which advocates for the involvement of the family, as well as the need to work in partnership with paediatric patients themselves in decision-making about their own medical care where developmentally appropriate. Recommendations outlined in this policy statement include "respect for families' unique insights into and understanding of their child's behavior and needs….and [to] appropriately incorporate family preferences into the care plan" (Pediatrics 2012). Additionally, paediatric healthcare providers should actively involve children, including children with physical and/or intellectual disabilities in decision-making about their own health by adapting the information to their level of understanding (Pediatrics 2012).

Although the AAP patient and family-centred care model integrates a number of highly relevant constructs of the two models (i.e., the *paediatric patient-centred care model* and the *family-centred care model*) discussed in this chapter, the issues inherent in caring for a young child with a genetic condition and his/her family continue to pose challenges not addressed by any of these healthcare delivery models. Perhaps it is unrealistic to assume that existing healthcare delivery models will address all of the underlying nuances specific to each medical specialty and, conceivably, it is not necessary to have a "one size fits all" model. However, with the genomic era of personalized medicine upon us, there is a pressing need to develop a model that will help shape best clinical practice guidelines for optimal patient and family care in genetics.

References

Arbour, L., and Bioethics Committee of the Canadian Paediatric Society. 2003. Position statement (B 2003-01): Guidelines for genetic testing of healthy children. *Paediatric Child Health* 8 (1): 42–45.

ASHG Social Issues Committees. 1995. ASHG/ACMG REPORT: Points to consider: Ethical, legal, and psychosocial implications of genetic testing in children and adolescents. *American Journal of Human Genetics* 57:1233–1241.

Beauchamp, T., and J. Childress. 2001a. *Principles of biomedical ethics*. 5th ed. 57–103. New York: Oxford University Press.

Beauchamp, T., and J. Childress. 2001b. Principles of biomedical ethics. 5th ed. 113–214. New York: Oxford University Press.

Berg, J. S., M. J. Khoury, and J. P. Evans. 2011. Deploying whole genome sequencing in clinical practice and public health: meeting the challenge one bin at a time. *Genetics in medicine: official journal of the American College of Medical Genetics* 13 (6): 499–504. doi:10.1097/GIM.0b013e318220aaba.

Borry, P., G. Evers-Kiebooms, M. C. Cornel, A. Clarke, and K. Dierickx. 2009. On behalf of the Public and Professional Policy Committee (PPPC) of the European Society of Human Genetics (ESHG). Genetic testing in asymptomatic minors Background considerations towards ESHG Recommendations. *European journal of human genetics: EJHG* 17 (6): 711–719. doi:10.1038/ejhg.2009.25.

Brown, K., S. E. Mace, A. M. Dietrich, S. Knazik, N. E. Schamban. 2008. Patient and family-centred care for pediatric patients in the emergency department. *CJEM: Canadian Journal of Emergency Medical Care* 10 (1): 38–43.

Canadian Institutes of Health Research. 2010. Tri-Council policy statement: Ethical conduct for research involving humans. Ottawa, Canada.

Clarke, A. 1994, Oct. The genetic testing of children. Working party of the clinical genetics society (UK). *Journal of medical genetics* 31 (10): 785–797.

Clayton, E. W., and A. L. McGuire. 2012. The legal risks of returning results of genomics research. *Genetics in Medicine: Official Journal of the American College of Medical Genetics* 14 (4): 473–477. doi:10.1038/gim.2012.10.

Epstein, R., and R. Street Jr. 2011. The values and value of patient-centered care. *Annals of Family Medicine* 9:100–103. doi:10.1370/afm.1239.

Gilbar, R. 2005. *The status of the family in law and bioethics: The genetic context*, 185. England: Ashgate Publishing Limited.

Hernandez, V. R., K. Selber, and M. S. Tijerina. 2006. Visioning family-centered care in genetics: What parents and providers have to say. *Journal of Genetic Counseling* 350. doi:10.1007/s10897–006-9032–9.

Knoppers, B. M., C. Strom, E. Wright Clayton, T. Murray, W. Fibison, L. Luther, P. R. Billings, L. Godmilow, and B. Handelin. 1998. Professional disclosure of familial genetic information. *American Journal of Human Genetics* 62:474–483.

Lacroix, M., G. Nycum, B. Godard, and B. M. Knoppers. 2008, Feb. 26. Should physicians warn patients' relatives of genetic risks? *CMAJ* 178(5). doi:10.1503/cmaj.070956.

McCandless, S. E., J. W. Brunger, and S. B. Cassidy. 2004. The burden of genetic disease on inpatient care in a children's hospital. *American Journal of Human Genetics* 74 (1): 121–127.

Ormond, K. E., M. T. Wheeler, L. Hudgins, T. E. Klein, A. J. Butte, R. B. Altman, E. A. Ashley, and H. T. Greely. 2010. Challenges in the clinical application of whole-genome sequencing. *Lancet*. doi:10.1016/S0140–6736(10)60599–5.

Pediatrics. 2012. Patient- and family-centered care and the pediatrician's role. *American Academy of Pediatrics* 129:394–404.

Sijmons, R. H., I. M. Van Langen, J. G. Sijmons. 2011. A clinical perspective on ethical issues in genetic testing. *Accountability in Research* 18 (3): 148–162. doi:10.1080/08989621.2011.575033.

Tabor, H. K., B. E. Berkman, S. C. Hull, and M. J. Bamshad. 2011. Genomics really gets personal: How exome and whole genome sequencing challenge the ethical framework of human genetics research. *American Journal of Medical Genetics Part A* 155:2916–2924.

Thorogood, A., B. M. Knoppers, W. J. Dondorp, and G. M. W. R. de Wert. 2012. Whole-genome sequencing and the physician. *Clinical Genetics* 81:511–513. doi:10.1111/j.1399-0004.2012.01868.x.

Chapter 14
End-of-Life Decision-Making in the Neonatal Intensive Care Unit: Serving the Best Interests of the Newborn within a Family-Centred Care Framework

Jonathan Hellmann

14.1 Introduction

In this Chapter I will explore how recognition of the stress endured by parents of infants in the neonatal intensive care unit has led to the ready adoption of a family-centred care (FCC) approach to neonatal care. Family-centred care is an attempt to both ameliorate the negative components of parents' NICU experience and improve the outcome of infants admitted to this environment. These initiatives have led to enhanced parental participation and empowerment in the care of their infants. However, it has also raised the potential for tension in decision-making with parents, particularly in end-of-life negotiations where the focus on the best interests of the infant, i.e., a patient-centred focus, may limit a strict adherence to parental decision-making authority. I will endeavour to advance a nuanced interpretation of both family-centred and a newborn-centred care philosophy in order to derive the positive synergy of these concepts when harmoniously integrated in a shared decision-making process. This, I believe, is necessary if we are to avoid conflict between parents and the healthcare team when challenging decisions for critically ill infants in the NICU are required.

14.1.1 The Contextual Reality of the NICU

The NICU is the embodiment of modern technology, with the capability to provide complex care to critically ill newborn infants. As such it holds great promise to enable infants who may have died in earlier times to survive, and to do so with every chance of a normal developmental outcome. The parents of critically ill and extremely premature infants have great expectations of the technological capabilities of intensive care, as does society at large. It is therefore incumbent on all the

J. Hellmann (✉)
The Hospital for Sick Children, 555 University Ave,
M5G 1X8, Toronto, ON, Canada
e-mail: jonathan.hellmann@sickkids.ca

R. Zlotnik Shaul (ed.), *Paediatric Patient and Family-Centred Care: Ethical and Legal Issues*, International Library of Ethics, Law, and the New Medicine 57,
DOI 10.1007/978-1-4939-0323-8_14, © Springer Science+Business Media Dordrecht 2014

healthcare professionals (HCPs) working in this environment to use this technological armamentarium in a responsible and ethically sound manner.

It is always an emergency and almost always a surprise for parents when their newborn infant is admitted to an NICU. Parents are usually young and healthy and often have had no prior experience with the healthcare system. They experience a wide spectrum of feelings, including shock, anxiety, guilt, disbelief, anger, and denial that this is really happening to them. They are faced with the shift in their healthcare delivery from the personal one-on-one relationship they may have enjoyed with their antenatal caregiver to a technological world with diffuse and often fragmented care by a variety of HCPs.

Parents enter this medical arena with its multiple caregivers, multi-tiered teams and fast-paced environment where the focus is on acute, curative care and the physician and team approach is that of intensivists, i.e., those who are highly competent in acute management of neonatal disorders, but for whom the responsibility for longer-term care may be less imperative. In addition to the culture of acute care medicine, the staff of a particular unit may have evolved their own 'culture' of care. This is usually unspoken and even the HCPs working in the environment may not be fully aware of the way this 'hidden curriculum' (Hafferty 1998) affects the way they provide care and make decisions. Thus, at first glance, the NICU is a very intimidating and frightening new world for parents. In addition, the parents and caregivers are 'moral strangers' in that they often have different traditions and beliefs, they may be from culturally diverse backgrounds, and may not share the language of communication.

Neonatal caregivers and parents need to overcome the challenges inherent in this context and not lose sight of the fact that this is occurring within the framework of the underlying prognostic uncertainty regarding the outcome of the infant. Despite these obstacles, meaningful, therapeutic and trusting relationships are necessary to facilitate ethically sensitive decision-making when that becomes the focus of interaction between the parents and the healthcare team.

14.2 Family-Centred Neonatal Intensive Care

The profound shift towards a more democratic and participatory process of decision-making in healthcare in general, and the specific recognition of the stresses parents undergo when their infant is in an NICU, have led to the ready adoption of family-centred approaches to neonatal care (Moore et al. 2003; Gooding et al. 2011). The principles of family-centred neonatal care were first described by Harrison in a seminal article in *Pediatrics* (Harrison H 1993). Family-centred care acknowledges the sick newborn infant's place within the social unit of the family and helps shape policies, programs, facility design and day-to-day activities in the NICU. FCC initiatives are directed at reducing the negative psychosocial impact of hospitalization on the parents, supporting families to participate in the care of their infant to the level they desire, and ensuring that families are viewed as integral members of the healthcare team. FCC shapes the framework wherein decisions between parents and caregivers are made, and in addition considers the effects of a decision on *all* family members, their responsibilities towards one another, and the burdens and benefits for each family member.

14.2.1 Attachment and Family-Centred Care

Attachment theory is critical to the incorporation of FCC into neonatal intensive care. Bowlby first described attachment as a "lasting psychological connectedness between human beings" and postulated that the earliest bonds formed by children with their parents or other caregivers have major developmental effects that continue throughout their lives (Bowlby 1969, 1988). Bowlby believed that infants form an "internal working model" about what to seek and expect from relationships, based largely on the extent to which their own early feelings are reciprocated (Hrdy 1999). Fundamental to Bowlby's work is the concept that mothers who are available and responsive to their infant's needs establish a lasting sense of security in their children (Daneman and Daneman 2012). Thus, countering the potential negative effects of the loss of early maternal attachment propels neonatal healthcare professionals to optimize the environment in which neonatal intensive care is practised.

Parents have preexisting attachment styles which may affect their capacity to deal with the admission of their infant to an NICU, as well as to accept help from HCPs. For the mother of a premature infant admitted to an NICU there have already been multiple losses, including the many anticipated weeks of the pregnancy; the 'hoped-for' child; the perfect birth; control over the birth process; confidence as a mother; normal parental holding, touching and bonding at the time of birth; and the possible loss of breast feeding (Woodroffe 2006). The NICU may also trigger unexpressed emotions of other previous life losses (Dyer 2005). Shah et al. have shown that unresolved grief related to preterm birth is associated with the development of insecure infant-mother attachment: mothers with resolved grief after preterm birth were 2.9 times as likely to have securely attached infants, compared with mothers with unresolved grief (Shah et al. 2011). Grief resolution is only one component of attachment and what makes dealing with this aspect of parental care so challenging is that no single, simple consistent model has evolved that can help predict which parents will encounter difficulties in emotional regulation in response to their infant's hospitalization. Nevertheless, it is important for neonatal HCPs to identify parents at risk and intervene in as attachment-appropriate a way as possible. It is equally important that HCPs not lose sight of parents who are less demanding or in obvious need, as all parents will have some degree of loss of attachment in the NICU.

The stresses that parents undergo related to the admission of their infant to an NICU have been well documented (Shaw et al. 2006). Franck categorised four domains of stress (Franck et al. 2005):

- The physical state and behaviour of the infant
- The complexity of language and of care
- The threat of loss
- The altered role of parents

Aspects of infant state and behaviour that give rise to parental stress include the infant's breathing pattern, colour, evidence of pain, bruising, cuts or incisions, their small size, the needles and tubes *in situ*, and their restless or jerky behaviour. Elements of the alteration in the parents' role include their inability to hold their baby, their feeling of helplessness in providing care, and their inability to feed or protect their child.

The needs of NICU parents encompass (Cleveland 2008):

- Accurate information specific to their baby
- Inclusion in the infant's care and decision-making
- Vigilant watching-over and protecting the infant
- Contact with the infant
- Being perceived positively by the nursery staff
- Individualized care
- Reassurance and a therapeutic relationship with the staff

14.2.2 Accurate Information and Aspects of Communication with Parents

Underlying all components of FCC is the content and manner of communication with parents. Parents require information about their infant, the diagnosis and prognosis, the available treatment options (including, where relevant, the option of no treatment), the risks and benefits of each option, and the limits of technology. How information is communicated influences parents' understanding, their ability to discuss issues openly, and their ability to participate effectively in a decision-making process. Transparency in a physician's thinking and reasoning helps parents to build an understanding of the infant's condition and tempers unrealistic parental expectations (King 1992).

Before focusing on the specific issue of communication within an end-of-life discussion, it is important to describe the general elements of effective communication and the development of a trusting relationship with parents. This may limit some of the inhibitory factors that arise in communication in an end-of-life decision-making interaction.

Elements of effective and respectful communication with parents by NICU staff include:

- Creating an environment for communication that encourages parents' participation and their becoming as fully informed as possible
- Identifying and attempting to remove barriers that limit parents' role in communication, e.g., language and physical distance
- Using plain language and being mindful of medical jargon and acronyms
- Encouraging parents to seek clarification of information at any point
- Assessing family communication preferences and attempting to communicate accordingly

- Being open, honest and truthful at all times
- Demonstrating concepts using diagrams and pictures
- Convening formal meetings with both parents when important decisions need to be made
- Providing information as accurately as possible and with as much certainty of diagnosis and prognosis as possible
- Explaining the concept of prognostic uncertainty
- Encouraging information-seeking
- Striving to keep parents informed of any special investigations that are planned in the management of their child
- Recognizing the need for processing and absorption of information and that repeated communication is vital
- Attempting to be consistent in communication in the face of staff changes and handovers
- Recognizing and respecting cultural, spiritual and religious diversity
- Listening to parental concerns
- Exploring parents' hopes and fears
- Being proactive in communication in any clinical situation in which a poor outcome is predicted
- Explaining the concept of shared decision-making
- Being mindful of the power imbalance in the relationship
- And all the while attempting to build trusting, respectful relationships

14.2.3 Spectrum of Communication

It is critical to appreciate the difference in communication when implemented as an information-providing process as opposed to when the intention is to develop a relationship with the parents. In the former mode the responsible physician/team tends to give the facts only, in the belief that facts are neutral and being as objective as possible is 'safer' territory. This is done in a way that does not involve family beliefs and generally does not offer guidance. In this less desirable form of communication, characterized as a one-way conversation, parents are simply made aware of the possible outcomes for their infant.

Providing information	Developing relationships
'just give the facts'—facts are neutral	Facts need interpretation/judgment
'objective' is safer	Enter moral territory
Do not explore family beliefs	Deliberative-interactive process
Little guidance for parents	Guidance, share views/make recommendations
Parents simply aware of possible outcomes	Elucidate parental concerns and explore/challenge parental views
One-way	Two-way conversation

In the preferred mode of communication the aim is to develop a sound parent-physician/team relationship. Here facts need interpretation and judgment, and a deliberate-interactive process (Emanuel and Emanuel 1992) is undertaken in a two-way conversation in which views are shared, and parents' concerns are elucidated, explored, and even challenged at times. Development of a sound, mutually respectful relationship between parents and the healthcare team facilitates discussion of the issues when an EOL decision is required.

There are obvious contextual factors in the NICU that challenge the achievement of a mutually satisfying parent-physician/team relationship, but there are also physician-related factors that may inhibit the development of such a relationship: physicians may be reluctant to express their personal views or may avoid prognostication and discussion about outcomes for fear of damaging the relationship with parents. Nevertheless, it is incumbent on physicians and the team to develop a 'therapeutic alliance' with parents and to engage in relationship-building with them. The responsibility for the relationship always rests with the physician and team.

14.2.4 Parents and the Healthcare Team

As neonatal intensive care is provided by numerous professionals with expertise in many disciplines, all team members need to be aware of differences in responsibility in regard to communication with parents. The content, nature and responsibility for each type of communication must be defined to minimise fragmented and inconsistent sharing of information. In order to maintain consistency and continuity (two of the most important measures of parent satisfaction with care) (Conner and Nelson 1999; Heyland et al. 2002), interprofessional meetings with all the practitioners involved with the family become extremely important. Throughout the 'roller-coaster' journey for parents in the NICU, it is important for the team to ask "Who is providing parents information, building relationships, dealing with uncertainty, and testing reality?" And "Where in the spectrum of NICU acculturation are the parents, and how ready would they be if an end-of-life decision were to become necessary?" It is important to note that parental acculturation may not be in synchrony when one parent is able to spend more time in the NICU than the other, and that this has the potential to create tension between the parents (De Rouck and Leys 2009), particularly at times of EOL decision-making.

14.2.5 Other Components of FCC

There are many other components of FCC that aim to improve parents' experience in the NICU. These will not be detailed here but include: written information in the form of books, information sheets and websites; the personalization of each infant's care space with toys and family photos; the promotion of breastfeeding and opportunities for skin-to-skin contact; parent and family support programs in the form of "buddy" parents (Preyde and Ardal 2003), and programs for siblings and/or fathers. In some units there is also an increasing trend where former NICU parents take on roles to facilitate family involvement in care as well as validate and diminish parents' feelings of alienation and perception of the uniqueness of their suffering.

The design of newer NICUs has also been influenced by FCC philosophy: single-family rooms allow families to develop strong relationships with their baby and with each other; they promote privacy for interaction with the baby and uninterrupted opportunities for skin-to-skin contact and, more fundamentally, a sense of parental control and appreciation that the child belongs to the family (White 2011).

14.2.6 Scientific Evidence to Support FCC Initiatives

There is a growing body of evidence to support FCC. Developmental care initiatives have shown that skin-to-skin practice in premature infants accelerates EEG signs of brain maturation (Scher et al. 2009); the 24/7 presence of the family decreases the length of stay and reduces the incidence of bronchopulmonary dysplasia (Ortenstrand et al. 2010); and training parents about the neurosensory needs of their babies improves white matter development (Milgrom et al. 2010).

Developmentally appropriate care is in accord with the hypothesis that early-life exposure can shape lifelong health trajectories in humans. Early adversity alters the chemistry of DNA in the brain: traumatic experiences cause methyl groups to affix to genes and disable them, preventing the brain from regulating the stress response. In animal studies Meaney found signs that methylation patterns can be reduced by parental nurturing (Meaney et al. 2007). This type of epigenetic study helps us to understand the links between physical disease and the environment and how early life exposure may affect later life events (Szyf 2009). Thus the potential for altering some of the negative effects of the NICU experience serves to strengthen the commitment to family-centred neonatal care.

14.3 End-of-Life Decision-Making in the Neonatal Intensive Care Unit

The commitment to a family-centred philosophy of care in the NICU is strongly determinative when it comes to EOL discussions, making it imperative that family interests are incorporated in any decision affecting the newborn. It is precisely at this intersection that tension may arise between an FCC approach and the pursuit of the best interests of the infant regarding decisions to limit or withdraw life-sustaining medical treatment (LSMT).

14.3.1 *Withholding and Withdrawing Life-Sustaining Medical Treatment in the NICU*

Withholding LSMT involves a decision to omit treatment that is not considered beneficial; **withdrawal** involves removing or discontinuing treatment that has not achieved its beneficial intent. These concepts are not considered different in moral terms, for if it is right (or wrong) to withhold treatment deemed to be ineffective, it is equally right (or wrong) to withdraw this same treatment when it becomes clear that the treatment is no longer effective. However, the moral complexity of actually withdrawing life-supportive interventions from a patient cannot be ignored (Bliton and Finder 2002). Not only morally but practically and psychologically it is far more challenging to withdraw treatment once momentum has been generated than to withhold treatment *a priori*.

14.3.2 *Criteria to Withhold or Withdraw LSMT*

Any one of the following considerations on their own may be determinative:

- The inevitability of death, where it is highly likely that the infant will die whether or not intensive care is continued
- When treatment is not meeting or will not meet the goals set for that treatment
- When an extremely poor quality of life is predicted for the infant: quality of life considerations encompass cognitive and neurodevelopmental outcome, the potential for motor disability or other physical handicap (e.g., vision, hearing), and longer term concerns such as behavioural and learning difficulties or school problems. It also considers the need for repeated or prolonged hospitalization, surgery or medication, and the potential for pain and suffering. Quality of life considerations may also include less concrete medical states, such as the capacity for meaningful and potentially enjoyable interaction with other people and the environment. Despite the complexity in determining the quality of a life, a poor quality of life prediction is a valid consideration whether treatment should be initiated or continued (Wall and Partridge 1997; Wilkinson 2006; Isaacs 2011; Feltman et al. 2012).

14.3.3 *The Rightful Decision-Makers*

Those participating in an EOL decision-making process are those who bear the greatest burden of care and conscience (the parents); those with special knowledge (the responsible physician and relevant sub-specialists); and those with the most continuous, committed and trusting relationship with the patient and parents (members of the healthcare team directly involved in the care of that infant) (Mitchell 1984). While extended family members or religious leaders identified by the parents and others may be present, they are usually not considered primary decision-makers. However, Ladd and Mercurio suggest that giving a voice to family members whose lives will be affected by the decision may be a fair approach, especially with young mothers who will need their family's help (Ladd and Mercurio 2003).

The Principles Underlying EOL Decision-Making Practice in the NICU at the Hospital for Sick Children, Toronto are:

1. That decisions conform to a standard of judgment that is in the best interests of the infant
2. That there is respect for parents' moral and legal authority for healthcare decisions for their young children
3. That parents' values and preferences are incorporated in a shared decision-making process
4. That consensus be derived between the rightful decision-makers before implementation of a decision
5. That medical technical expertise not dominate *value* considerations

14.4 The Best Interests of the Infant Standard of Judgment

The obligation to promote the best interests of patients who are not able to make autonomous healthcare choices is embodied in the concept of the best interests of the newborn infant. This is the moral and legal standard of judgment which establishes the primacy of duties to infants and ensures that other interests not override those of the infant. Pursuing an infant's best interests implies determining the potential for the success of the proposed treatment; the risks involved in the treatment; the degree to which the treatment will, if successful, extend life; the pain and discomfort with the proposed treatment; and the anticipated quality of life with and without the treatment (AMA 2010). Despite the acceptance of the best interests of the infant as a standard of judgment, the subjective nature of the assessment as well as the fact that the assessment is being done by surrogate decision-makers must always be acknowledged. Best interests is not only a standard of judgment but also a standard for possible intervention when it is perceived by the rightful decision-makers that the infant's best interests are not being served by a particular course of action (Kopelman 2005).

The challenge for the NICU physician/team is to make recommendations based on their assessment of the infant's best interests. Decisions should be based on the best evidence available: it is important to acknowledge that medical evidence is not static and that generalized evidence must always be interpreted, analysed and applied to the *specific* patient. Some may fear that exercising clinical judgment is perceived as paternalistic, non-scientific and inexplicable. However, via a process of clinical reasoning and both scientific and humane judgment (Downie and

MacNaughton 2000) the application of the relevant facts to that specific patient *and family* can, and should be determined to support a certain recommendation.

The 'right' degree of physician directiveness in this interaction is dependent upon the uniqueness of the patient, the parents and the physician/team. This may range from:

i) an interaction with no *a priori* assumptions and an exploration of parental views with an open-ended outcome, usually when there is a great degree of prognostic uncertainty, to
ii) a more directive but still open-to-negotiation physician recommendation, to
iii) a situation where the parents are fully informed about the medical assessment and a stronger recommendation is made based on the greater certainty of a very poor predicted outcome, with little choice offered to parents. This latter situation is exemplified by the "confession" case described by Humikowski of an infant on a ventilator with irregular spontaneous respiration, non-reactive pupils and no withdrawal response to pain (Humikowski 2012). Many physicians feel the need to ask "Do we have your permission to proceed?" (to disconnect the ventilator). Clark's commentary on this case states that "using respect for parental authority as an excuse to request permission to withdraw life-sustaining interventions rather than providing directive recommendations within the context of parental goals is ultimately more harmful than helpful" (Clark 2012).

14.4.1 Challenges to the Best Interests Standard of Judgment

The notion of best interests of the patient engenders much debate (Kopelman 2005; Wilkinson 2006; Hester 2007): How should family interests be incorporated? Is *best* misleading (Holm and Edgar 2008) or unattainable or unknowable due to the great difficulty in accurate prognostication, and how should physicians act when parents are not available as sounding boards for discussion? In addition, the concept is very individualistic and tends to ignore cultural differences.

Family Interests It is well argued that it is legitimate for both physicians and parents to consider *family* interests in the assessment of the infant's best interests (Hardwig 1990) because when the total 'good' of a patient is considered it is axiomatic that the child's interests intersect with the family's, as its own 'moral community.' There is often a reluctance to acknowledge "the ethical salience of the needs and interests of the parents and other children and family members" (Truog and Sayeed 2011). Truog and Sayeed suggest that this distorts the process of decision-making such as, for example, when parents are concerned about the impact that a severely disabled child might have on their ability to emotionally and financially care for their other children, or the potential impact upon the stability of their marriage (Truog and Sayeed 2011). In fact, it has been shown that the majority of neonatologists do ascribe to an incorporation of family interests into decision-making for incompetent infant patients (Hardart and Truog 2003). This is particularly true in decision-making for the extremely preterm infant (Marcello et al. 2011). Paris et al. point out that the President's Commission report (President's Commission 1983) and the American Academy of Pediatrics (AAP) 2007 guidelines both endorse that "great discretion is to be afforded to the parents" (Paris et al. 2007; Presidents' Commission 1983; AAP 2007). Thus family interests need to be incorporated into any decision, although the limits of this are not easily defined.

Prognostic Uncertainty There are many clinical situations in which the infant's best interests may be unclear due to uncertainty about their predictable developmental outcome. In acute situations, and pending clarification of the circumstances, the presumption should normally be in favour of life-saving or life-sustaining treatment (Canadian Paediatric Society Statement 2004). When it is possible to defer or delay acute treatment, such a delay is encouraged as further prognostic information is gathered to clarify the issues.

Racine and Shevell emphasize that the prognostic uncertainty of neurologically compromised newborns frequently assumes that an outcome is inevitable, but multiple intrinsic factors (individual resiliency, plasticity of response, personal motivation) and extrinsic factors (timing, availability and access to rehabilitation services, socioeconomic factors and social supports) exist that may modulate outcome (Racine and Shevell 2009). Uncertainty and variability are intrinsic to the process of prognostication, rendering best interests assessments inherently complex.

My personal conviction is that in situations of great degrees of prognostic uncertainty, greater moral weight be given to the wishes of parents, but within mutually defined parameters. In a similar vein "the threshold view" advanced by Wilkinson suggests that treatment may be withdrawn from infants if their future well-being is below a threshold that is close to, but *above* the zero-point of well-being (Wilkinson 2011). He suggests that "where a critically ill newborn infant is predicted to have a severe and irremediable impairment, such that they will not be able to take part in and realize many of the good features of life, it is permissible for parents and doctors to allow them to die or to continue treatment" (Wilkinson 2011). It is important to emphasize here that these are joint decisions to which both parties must agree; parents have to live with the consequences and both parents *and* physicians have to live with their consciences.

Absence of a Parental 'Sounding Board' Because of the subjectivity of the assessment of an infant's best interests, the absence of parents in this interaction creates great difficulty for the healthcare team. They are rightfully hesitant to make a decision to withdraw LSMT without appropriate parental participation. This is typified by a case described by Boss in which the infant was abandoned by his mother in the NICU. Despite his extremely poor prognosis, care was continued for many months in view of the absence of the mother or other family members, and in the light of the authorities' fear that withdrawing LSMT might be perceived as motivated by prejudice (Boss 2008). Boss states that "systems in place to support physicians, composed of other healthcare providers, medical departments, hospitals, hospital lawyers and ethics committees, typically favor maintaining current levels of medical therapy" (Boss 2008) and not withdrawal of LSMT. In my experience, it takes great moral courage to pursue a decision to withdraw LSMT on the basis of a best interests assessment in the face of such obstacles.

Despite these challenges to the best interests of the newborn standard of judgment, it is accepted as the guiding principle for decision-makers for it unites under one standard different meanings and demonstrates reasonableness, given the prevailing conditions (Kopelman and Kopelman 2007). Properly understood, it focuses attention on the infant and can serve as a powerful tool for settling issues about how to make good decisions for those who cannot decide for themselves. It serves as a useful concept in choosing treatment or non-treatment options, although it does not serve as well in overriding parental wishes, nor does it answer the question of when, if ever, family-centred interests should or could override those of the infant.

14.5 Respect for Parental Authority

The idealized expression of autonomy in *adult* patients is one in which mentally and emotionally capable patients choose voluntarily and intelligently from among various options whose relative risks and benefits have been fully explained to them by their physicians, i.e., self-determined choices via a truly informed consent process. Strict adherence to respecting the adult patient's autonomy in decision-making has generated much comment and critique (Gert and Clouser 1990; McGrath 1998). Autonomy has a very rationalistic and individualistic bias and is regarded by Callahan as an escape from the complexity of life which stops moral analysis at what might be termed the "external conditions of moral decision-making" without entering into serious discussion of the ethical implications of the different choices possible (Callahan 2003). Modifications of autonomy are described: (i) Quill and Brody describe an "enhanced autonomy" model which encourages patients and physicians to actively exchange ideas, explicitly negotiate differences, and share power and influence to serve the patient's best interest (Quill and Brody 1996). Here the doctor's power should be used in such a way that it enables patients to bring their own perspective to the fore, thereby empowering patients by creating an atmosphere that encourages participation and dialogue; (ii) In Savulescu's "rational non interventional paternalism" model medicine is regarded as a moral practice which involves making value judgments and requires doctors to reflect on what ought to be done: if physicians are to respect patient autonomy and to function as moral agents they must make evaluations on what is best for their patients and argue rationally with their patients about what they ought to do (Savulescu 1995). This resonates with the model of physician as moral agent espoused by Pellegrino (Pellegrino 1994).

In the context of neonatal care, the term parental autonomy is often used without distinguishing it from parental *authority*. I will use the latter terminology, as autonomy, arguably, ought to be used only when making decisions for oneself. However, some suggest that the use of parental *autonomy* may be appropriate when a parent has a vision of what it means to be a good parent and wishes to do so in a *self-determined* way (Harrison C). Where the term parental autonomy has been used in the literature, I will retain that terminology.

14.5.1 Parental Authority and the Newborn

Neonatal healthcare professionals demonstrate strong respect for parental authority for a number of reasons:

i) In most jurisdictions the state confers the responsibility for healthcare decision-making on the parents of newborn infants, as they have a unique relationship with their children, one of concern, obligation, responsibility and intimacy
ii) There is the presumption that parents will act to promote their child's best interests and make healthcare decisions in that light
iii) Respecting parental views is a means of equalizing the parental role in decisions with physicians who shape parental views, sometimes even without consciously intending to do so. Physicians need to be conscious of the impact of *their* authority and the power that derives from expertise or that follows from information given in a convincing manner. They also have to recognize that parents' ability to reason and to act in a self-directed way is often diminished in the presence of serious illness in their newborn child.

14.5.2 Three Concerns Arise when Parental Authority is Narrowly Interpreted

i) An over-fixation on parental authority in *decision-making* may lead to the assumption that there are no other parental-related values to consider in the care of the patient
ii) The notion that parents always want to be the decision-maker and should be *compelled* to make an EOL decision for their infant may be unrealistic. Paris et al. feel the emphasis on parental autonomy may have gone too far (Paris et al. 2006). They and Montello and Lantos, in their allusion to Dostoevsky's *The Brothers Karamazov*, suggest that basic human nature may be fundamentally at odds with the emphasis on rationality and autonomy, and that we should appreciate that "the desire of parents and sometimes of physicians to avoid responsibility for the death of a patient, particularly a newborn infant, can be overwhelming" (Montello and Lantos 2002)
iii) In light of the systemic validity attributed to parental authority some physicians become reluctant to express their own judgment, recommendations or views, particularly when known to be counter to the expressed or even non-expressed views of parents. Physicians' fear that their assessment of an infant's best interests is interpreted as paternalistic could result in a general hesitancy to engage in interaction about values and preferences with parents. This may lead to a

degree of 'hiding behind' parental authority or, even more troubling, 'abandoning' parents to make their own decisions.

The concerns of a narrow interpretation of parental autonomy are well shown in the study entitled "Autonomy gone awry" by Orfali and Gordon (Orfali and Gordon 2004). They sought to determine whether a system that emphasized parents' autonomy enabled them to cope better with decision-making in neonatal intensive care. An ethnographic study in three neonatal units, two in France and one in the United States, was conducted. The U.S. unit was seen as an autonomy-based model in which informed consent was explicit, and where physicians offered information and choices that parents would or would not choose to consent to (there were restrictions on parents' requests for withholding treatment). The French units were characterized as paternalistic, with the physicians making decisions with little decision-making responsibility entrusted to parents. Mothers in the French units were more satisfied than their counterparts in the U.S. Key factors were strong doctor-parent relationships, continuity of care, presentation of information with little ambiguity, demonstration of emotional empathy by the physicians, and the belief instilled in mothers that whatever was best for their baby was being done. In contrast, in the U.S. unit mothers perceived that much information was provided, but medical uncertainty was emphasized, there was little emotional support to deal with the information, there was a lack of continuity, communication was inconsistent, and strong relationships with the physicians did not develop. The study also showed that the French mothers coped better with the deaths of their infants, and that expressions of dissatisfaction or guilt were almost never mentioned. The authors' conclusion was that decision-making may have become too principle-driven (specifically autonomy) in the U.S. unit.

A number of other studies demonstrate that process and relationships may be more important than the principle of autonomy when narrowly interpreted.

i) A French study compared shared, medical, informed and no-decision models of parents' perceived role in the decision-making process. The findings suggested that the fully autonomous decision would be ill advised and, since parental involvement preferences can vary, they should be able to decide what role they want to play (Caeymaex et al. 2011).
ii) In a study of antenatal counseling provided to mothers with impending delivery of extremely low birthweight babies, Keenan showed that the majority of mothers were satisfied with the amount of influence they had in the decision-making process. They felt informed and included, despite the fact that the majority of mothers stated that the counselor had made a treatment recommendation and that they had no choice in how their infant would be treated (Keenan et al. 2005).
iii) In a Canadian study, Payot described two divergent models used by neonatologists regarding resuscitation decisions with parents at risk of premature delivery: (a) a neutral information model driven by respect for parental autonomy in which information is given and parents ultimately decide, after receiving information on mortality statistics and risk estimates of complications and sequelae.

In this model parents are expected to manage the probabilities and uncertainties through their own self-determined decision-making process, and (b) an assent model in which the neonatologist's preferences are clearly expressed, a decision is sought during the consultation, and where prognostic statistics are used as information to justify the suggested course of action (Payot et al. 2007). From the parents' perspective, neither model fully addressed their expectations, but parents clearly expressed the need for an individualized and humane relationship which could not be addressed by the neutral information model. Decisions were more acceptable with the assent model when parents' expectations were confirmed by the information provided and when they felt they received the support needed to make a decision.

14.6 Shared Decision-Making

At the point in a newborn's NICU course when the responsible physician, the healthcare team, or the parents, raise the issue of whether treatment should be continued, withheld or withdrawn, a shared model of decision-making is embarked upon. This aims to ensure that the principles of the best interests of the infant and respect for parental authority enable coherence in reasoning and an ethically sound guide to action when harmoniously integrated within a shared decision-making process.

Achieving shared decision-making depends on building a good relationship in the clinical encounter so that information is shared and patients (in this context, parents) are supported to deliberate and express their preferences and views during the decision-making process (Elwyn et al. 2012). Barry and Edgman-Levitan advocate that (adult) patients should be educated about the essential role they play in decision-making and be given effective tools to help them understand their options and the consequences of their decisions. They should also receive the emotional support they need to express their values and preferences and to be able to ask questions without censure from their clinicians (Barry and Edgman-Levitan 2012). They go on to state that "recognition of shared decision-making as the pinnacle of patient centred care is overdue." These views are directly applicable to the parents of critically ill newborn infants.

14.6.1 A Procedural Framework for Shared Decision-Making in the NICU Entails:

- Creating an optimal environment for discussion
- Ensuring that both parents (wherever possible), the responsible physician and other team members are present
- Establishing the relevant facts and exploring treatment options
- Discussing the preferences and consequences of all reasonable options

- Moving to a decision as fast as the slowest member of the group
- Negotiating towards consensus
- Erring on the side of life with continuation of care if consensus is not reached
- Agreeing to continue negotiation as more information is gathered

14.6.2 Content of Discussion

Payot suggests that by focusing *first* on the issues raised by parents; sharing two-way information; recognizing the importance of 'relational space'; and overcoming, to any degree, the different starting points for parents and physicians, physicians may achieve greater consensus and more acceptable decisions for all parties (Payot et al. 2007).

Thus a conscious effort should be made to address key factors in this encounter. These include:

- The starting points and the degree of divergence which need to be traversed
- The degree of recognition of the unevenness of the encounter
- The degree of social, cultural, and language divergence
- The degree of empowerment given to parents to express their views
- The actual content (whether generalized or specific)
- How information is contextualized and made personally meaningful and filtered through parents' own values and beliefs
- The degree of transparency throughout the process of thinking
- The flexibility of either party
- The involvement of other members of the team or parents' community and friends

The vast majority of complex and difficult ethical decisions result from negotiation, consensus-building and ultimately harmonious agreement between the parents and healthcare professionals as to what is in the child's best interests. The ideal shared decision is one in which neither party feels individually responsible for that decision.

14.6.3 Cultural, Religious and Spirtual Dimensions of the Family Pertaining to EOL Decision-Making

The broad acknowledgement of the child's place within a family often brings out the cultural, religious and spiritual dimensions of the family's lives, and, at times of EOL decision-making, these family characteristics may present healthcare providers with significant challenges. The major challenge is to recognize, understand and respect the family's cultural, religious and spiritual views and values. Misperceptions caused by a lack of sensitivity can lead to inappropriate care and poor clinical outcomes. Cultural competence is more than sensitivity to cultural norms different from one's own; it encompasses four components (Martin and Vaughn 2007):

i) Awareness of one's own cultural worldview
ii) One's attitude towards cultural differences
iii) The knowledge of different cultural practices and worldviews
iv) Cross-cultural skills

Cultural competence is the ability to understand, communicate and effectively interact with people across cultures. A particular cultural challenge at the time of EOL decision-making involves the degree of acceptance of a family's decision-making practice where, for example, a male-dominated, patriarchal hierarchy may be the family's norm, as opposed to the more culturally acceptable *shared* parental responsibility for decision-making. The degree of tolerance of this and other cultural differences is a test for HCPs in their interpretation of respecting FCC approaches to care (Kopelman and Kopelman 2007).

While the healthcare team should recognize how families' views are shaped by social, cultural and other contexts, it is also important not to stereotype the members of specific cultural, ethnic or religious groups. Their backgrounds may not be predictive of their beliefs and values in the care of their infant: each patient and family should be considered unique and their specific views and values acknowledged.

Religion and the more general concept of spirituality as a major determinant of culture, tradition and family values need to be addressed when end-of-life decision-making is undertaken. Four explicitly spiritual/religious themes emerged from a questionnaire study completed by parents after their child's death (Robinson et al. 2006): prayer, faith, access to and care by clergy, and belief in the transcendent quality of the parent-child relationship that endures beyond death. In a study by Boss, significant themes with a religious/spiritual dynamic were identified when parents made decisions regarding delivery room resuscitation for high-risk newborns. These included finding meaning, hope, trust and love (Boss et al. 2008). Healthcare teams need to consider whether their environment is supportive of religious or spiritual practice, and, on a deeper level, appreciate parents' religious and spiritual perspectives in end-of-life discussions.

When attentiveness to the views and values of families is difficult because of language barriers, professional interpreters should be used. This is advisable for a number of reasons: it ensures that parents' views are available to the healthcare team, it removes the burden on family members or friends for the transfer of information, and it limits the potential for miscommunication. In certain situations a *cultural* interpreter not only facilitates language comprehension but also provides useful information about cultural norms and traditions that may be unfamiliar to the healthcare team.

14.6.4 When Parents Say "Do Everything Possible"

At any point in the course of a critically ill patient's care, parents or families may state that they want "everything done" (Gillis 2008; Feudtner and Morrison 2012; Jecker and Schneiderman 1995). It is important to ask what this really means. Is it a failure to comprehend the prognosis, a lack of confidence in the medical diagnosis or prognosis, a religious belief (often expressed as a belief in miracles) (Kopelman AE 2006), an expression of frustration and alienation, a simple one-liner in a complex situation, or the only available gesture of love and devotion left for the family? Feudtner considers the phrase vague at best and vacuous at worst, and views this as a starting point for a discussion, not an end point (Feudtner and Morrison 2012). Without exploring what underlies a parental declaration of "do everything," physicians neglect the complexity of the situation and the unperceived and possibly unexpressed fear of 'abandonment' (Cassell 2005).

There are situations in which the best interests of the infant may become overridden by other, usually extreme factors. This is exemplified in the following scenario.

Baby R was preterm at 32 weeks and severely growth-restricted. He was the first child born to parents in their forties and was conceived via in-vitro fertilization. He had a complicated course with severe chronic lung disease, multiple infections, prolonged feeding intolerance, and severe osteopenia. At 3 months he was still unable to be weaned from the ventilator, and all attempts to treat his lung disease had failed. He was not considered a candidate for tracheostomy or lung transplant given his clinical instability, size, and multiple co-morbidities. His neurologic examination was markedly abnormal and the team consensus was that there was no reasonable hope for recovery.

The mother spent every day in the intensive care unit at her son's bedside. When the team raised the issue of withdrawal of LSMT the parents stated that they would never consider end-of-life care, and the mother said that she would not be able to live without her child and would kill herself if the team removed the ventilator. After extensive and repeated efforts at determining the reality of the suicide threat and despite numerous discussions regarding the infant's suffering, the team continued to provide life-sustaining care (and continued to experience great moral distress).

The option to refer the case to an impartial third party, such as the Ontario Consent and Capacity Board, to determine whether the surrogate decision-makers, i.e., the parents, were acting in the best interests of the baby, was never implemented for fear that this would have engendered great tension and exacerbated parental and team stress. In addition, the mother appeared rational in her thinking and displayed great devotion to her child at all times. She remained adamant in her insistence on continuing assisted ventilation until his death some months later.

These types of cases are unfortunately not infrequent and sorely test the commitment of the NICU team to serve the patient's best interests within a family-centred care philosophy. In such situations, when all efforts fail to serve the infant's best interests, a certain 'moral residue' may remain, with the physician and team feeling that their duties and values were compromised (Webster and Baylis 2000). Such moral residue requires team and organizational support for resolution. We do not know if or to what degree *parents* experience moral residue, although the studies of McHaffie suggest that in a shared decision-making model parents felt the decision was made without guilt, doubt or adverse consequences, and that in 98 % of cases the right decision had been made (McHaffie et al. 2001).

In these conflictual situations I feel it is important to appreciate that neonates and parents suffer in different ways, and that the physical suffering of neonates can generally be managed more readily than the emotional suffering that parents endure when their values, beliefs and views are explored, scrutinized and challenged. We also need to be aware that the death of a child has a far-reaching, long-term impact upon parents; it is my contention that how an EOL decision is derived has a strong bearing on parents' adjustment to the death of their child. Here the quote attributed to Dame Cicely Saunders, the founder of the hospice movement, is particularly pertinent: "How people die remains in the memory of those who live on" (Saunders 1984).

14.6.5 Attempting to Resolve Conflict

It is clear that minimizing the potential for conflict is preferable to resolving issues when they have become more divisive and seemingly intractable. The following suggestions are in accord with shared decision-making and conflict-resolution models (Charles et al. 1999; Spielman 1993). These include:

- Ensuring full parental comprehension of the medical information and clarifying any misconceptions and misunderstandings
- Allowing time for further clinical observation and continuing discussion
- Exploring the cultural complexity of the decision-making process
- Clarifying the ethical issues in conflict
- Emphasizing the preferred consensual nature of the decision-making process and shared burden of the decision
- Removing medically or system-imposed obstacles to achieving consensus such as frequent staff schedule changes
- Recognizing the failure of one physician to establish a therapeutic alliance with the parents
- Preventing the development of a 'contest of wills' between the physician and parents
- Exploring the degree of (dis)agreement between the parents themselves
- Broadening the parents' 'moral community' by the inclusion of additional family members, significant others, and religious or spiritual advisors in meetings
- Employing creative means to promote ongoing discussion and understanding, such as the use of visual images
- Fostering more private opportunities for the responsible physician to engage in discussion with the parents.

There may be benefit in involving an institutional ethics committee with experience in case consultation and review, a clinical ethics consultation team, or a clinical ethicist. Ethics committees and ethicists vary in their practice: from playing a facilitator role ensuring that the ethical issues have been elucidated and an appropriate procedure followed, with sufficient time, information and freedom from coercion; to a more directive, decision-making role for themselves. Experience suggests that consultation with individual bioethicists and smaller ethics consultation teams may be of greater benefit to decision-makers than the practice of "ethics by committee."

14.6.6 Parental Insistence on Their Role as Final Decision-Maker

Parental insistence on their role and right as final decision-maker may be a valid expression of their perceived role as a parent. I believe, however, that this is usually not their first response. It may be a reaction to ineffective or poor communication, the result of an overly rational approach with a failure to connect to the parental narrative or emotion, or the impact of parents' stress affecting their ability to absorb information. HCPs also need to recognize that parents often use coping mechanisms of "hope and denial" to interpret data (Cole 2000) and that their understanding may be based more on their own background and beliefs than on the information provided. Their insistence on being the final decision-maker may also result when respect for their role in a shared decision-making process has been disallowed or denigrated.

14.6.7 Intractable Differences

When differences are intractable and the degree of a physician's moral compromise significant, it may be appropriate to consider transferring responsibility of care for the infant to another, accepting physician. If the conflict between parents and the responsible physician remains unresolved and if transfer is not a realistic option, physicians may, following ethics consultation, consider seeking institutional or legal advice. However, the impact of such a legal recourse must be seriously considered as it invariably destroys the parent-physician relationship, undermines trust in the medical system, and may increase the anguish for everyone involved (especially the families). In addition, seeking such unilateral authority for decision-making, even in the most extreme circumstances, is regarded by some not only as a clinical failure, but unjustified in principle (Burt 2003). In the NICU at the Hospital for Sick Children, decisions to withdraw or withhold LSMT are not taken without parental agreement. On rare occasions this may only be tacit agreement in the form of parents stating they *understand* why the decision is being made, even if they cannot formally say "they agree" for fear of the burden of guilt in being party to such a decision.

14.7 A Final Word

Parental views should carry strong moral weight in decision-making for their vulnerable infants in the NICU. Yet this does not diminish the responsibility for decisions to be made in the best interests of the infant. Finding the right balance between respect for parental authority and the physicians' role and responsibility in the assessment of infant *and* family interests in a shared decision-making process requires both scientific *and* humane judgment, empathy, imagination, insight, and effective communication skills. What establishes moral acceptability is not the application of a moral theory or a purely rationalist argument establishing indubitable first principles, but the moral clarity arising out of a mutually derived decision in which all parties have been empowered, their preferences established, and where dialogue and collaboration have (usually) achieved resolution.

References

American Academy of Pediatrics Committee on Fetus and Newborn, and E. F. Bell. 2007. Non-initiation on withdrawal of intensive care for high-risk newborns. *Pediatrics* 119 (2): 401–403.

American Medical Association. 2010. The AMA code of medical ethics: Opinions on seriously ill newborns and do-not-resuscitate orders. *JAMA Ethics* 12 (7): 554–557.

Barry, M. J., and S. Edgman-Levitan. 2012. Shared decision making—the pinnacle of patient centered care. *The New England Journal of Medicine* 366 (9): 780–781.

Bliton, M. J., and S. G. Finder. 2002. Traversing boundaries: Clinical ethics, moral experience, and the withdrawal of life supports. *Theoretical Medicine* 23: 233–258.

Boss, R. D. 2008. End-of-life decision-making for infants abandoned in the neonatal intensive care unit. *Journal of Palliative Medicine* 11 (1): 109–111.

Boss, R. D., N. Hutton, and L. J. Sulpar, et al. 2008. Values parents apply to decision-making regarding delivery room resuscitation for high-risk newborns. *Pediatrics* 122 (3): 583–589.

Bowlby, J. 1969. Attachment. 2nd ed. *Attachment and loss*. 1 vol. London: Hogarth.

Bowlby, J. 1988. *A secure base: Clinical applications of attachment theory*. London: Routledge.

Burt, R. A. 2003. Resolving disputes between clinicians and family about "futility" of treatment. *Seminars in Perinatology* 27 (6): 495–502.

Caeymaex, L., M. Speranza, and C. Vasilescu, et al. 2011. Living with a crucial decision: A qualitative study of parental narratives three years after the loss of their newborn in the NICU. *PLos One* 6 (12): e28633.

Callahan, D. 2003. Principlism and communitarianism. *Journal of Medical Ethics* 29 (5): 287–291.

Canadian Paediatric Society (Position Statement B 2004-01). Treatment decisions regarding infants, children and adolescents. CPS Statement. *Pediatrics & Child Health* 9 (2): 99–103.

Cassell, E. J. 2005. Consent or obedience? Power and authority in medicine. *The New England Journal of Medicine* 352 (4): 328–330.

Charles, C., A. Gafni, and T. Whelan. 1999. Decision-making in the physician-patient encounter: Revisiting the shared treatment decision-making model. *Social Science & Medicine* 49: 651–661.

Clark, J. 2012. Balancing the tension. Parental authority and the fear of paternalism in end-of-life care. *Archives of Pediatrics and Adolescent Medicine* 166 (7): 594.

Cleveland, L. M. 2008. Parenting in the neonatal intensive care unit. *Journal of Obstetric, Gynecologic, and Neonatal Nursing* 37 (6): 666–691.

Cole, F. S. 2000. Extremely preterm birth-defining the limits of hope. *The New England Journal of Medicine* 343 (6): 429–430.
Conner, J. M., and E. C. Nelson. 1999. Neonatal intensive care: Satisfaction measured from a parent's perspective. *Pediatrics* 103 (1): 336–349.
Daneman, D., and M. Daneman. 2012. What has attachment theory got to do with diabetes care? *Diabetes Management* 2 (2): 85–87.
De Rouck, S., and M. Leys. 2009. Information needs of parents of children admitted to a neonatal intensive care unit: A review of the literature (1990–2008). *Patient Education and Counseling* 76 (2): 159–173.
Downie, R. S., and J. Macnaughton. 2000. *Clinical judgment: Evidence in practice* xii. New York: Oxford University Press.
Dyer, K. 2005. Identifying, understanding, and working with grieving parents in the NICU, part 1: Identifying and understanding loss and the grief response. *Neonatal Network* 24 (3): 35–46.
Elwyn, G., D. Frosch, and R. Thomson, et al. 2012. Shared decision making: A model for clinical practice. *Journal of General Internal Medicine* 27 (10): 1361–1367.
Emanuel, E. J., and L. L. Emanuel. 1992. Four models of the physician-patient relationship. *The Journal of the American Medical Association* 267 (16): 2221–2226.
Feltman, D. M., H. Du, and S. R. Leuthner. 2012. Survey of neonatologists' attitudes toward limiting life-sustaining treatments in the neonatal intensive care unit. *Journal of Perinatology* 32 (11): 886–892.
Feudtner, C., and W. Morrison. 2012. The darkening veil of do everything possible. *Archives Pediatrics & Adolescent Medicine* 166 (8): 694–695.
Franck, L. S., S. Cox, and A. Allen, et al. 2005. Measuring neonatal intensive care unit-related parental stress. *Journal of Advanced Nursing* 49 (6): 608–615.
Gert, B., and K. D. Clouser. 1990. A critique of principlism. *The Journal of Medicine and Philosophy* 15 (2): 219–236.
Gillis, J. 2008. We want everything done. *Archives of Disease in Childhood* 93 (3): 192–193.
Gooding, J. S., L. G. Cooper, and A. I. Blaine, et al. 2011. Family support and family-centred care in the neonatal intensive care unit: Origins, advances, impact. *Seminars in Perinatology* 35: 20–28.
Hafferty, F. W. 1998. Beyond curriculum reform: Confronting medicine's hidden curriculum. *Academic Medicine* 73: 403–407.
Hardart, G.E., and R. D. Truog. 2003. Attitudes and preferences of intensivists regarding the role of family interests in medical decision making for incompetent patients. *Critical Care Medicine* 31 (7): 1895–1900.
Hardwig, J. 1990. What about the family? *The Hastings Center Report* 20 (2): 5–10.
Canadian Paediatric Society (Position Statement B 2004-01). Treatment decisions regarding infants, children and adolescents. CPS Statement. *Pediatrics & Child Health* 9 (2): 99–103.
Harrison C. 2005. Personal communication. Toronto.
Harrison, H. 1993. The principles for family-centered neonatal care. *Pediatrics* 92 (5): 643–650.
Hester, D. M. 2007. Interests and neonates: There is more to the story than we explicitly acknowledge. *Theoretical Medicine and Bioethics* 28: 357–372.
Heyland, D. K., G. M. Rocker, and P. M. Dodek, et al. 2002. Family satisfaction with care in the intensive care unit: Results of a multiple center study. *Critical Care Medicine* 30 (7): 1413–1418.
Holm, S., and A. Edgar. 2008. Best interests: A philosophical critique. *Health Care Analysis* 16: 197–207.
Hrdy, S. B. (1999) *Mother nature: A history of mothers, infants and natural selection* 387. New York: Pantheon Books.
Humikowski, C.A. 2012. Confession. *Archives of Pediatrics and Adolescent Medicine* 166 (7): 593–594.
Isaacs, D.. 2011. Controversial end-of-life issues in the neonatal intensive care unit. *The American Journal of Bioethics* 11 (2): 43–44.
Jecker, N. S., and L. J. Schneiderman. 1995. When families request that 'everything possible' be done. *The Journal of Medicine and Philosophy* 20: 145–163.

Keenan, H. T., M. W. Doron, and B. A. Seyda, et al. 2005. Comparison of mothers' and counselors' perceptions of predelivery counseling for extremely premature infants. *Pediatrics* 116: 104–111.

King, N.M. 1992. Transparency in neonatal intensive care. *The Hastings Center Report* 22 (3): 18–25.

Kopelman, A. E. 2006. Understanding, avoiding, and resolving end-of-life conflicts in the NICU. *The Mount Sinai Journal of Medicine* 73 (3): 580–586.

Kopelman, LM. 2005. Rejecting the Baby Doe rules and defending a "negative" analysis of the best interest standard. *The Journal of Medicine and Philosophy* 30: 331–352.

Kopelman, L. M., and A. E. Kopelman. 2007. Using a new analysis of the best interests standard to address cultural disputes: Whose data, which values? *Theoretical Medicine and Bioethics* 28 (5): 373–391.

Ladd, R. E., and M. R. Mercurio. 2003. Deciding for newborns: Whose authority? Whose interests? *Seminars in Perinatology* 27 (6): 488–494.

Marcello, K. R., J. L. Stefano, and K. Lampron, et al. 2011. The influence of family characteristics on perinatal decision making. *Pediatrics* 127: e934–e939.

Martin, M., and B. Vaughn. 2007. Strategic diversity & inclusion management magazine 31–36. San Francisco: DTUI Publications Division.

McGrath, P. 1998. Autonomy, discourse, and power: A postmodern reflection on principlism and bioethics. *The Journal of Medicine and Philosophy* 23 (5): 516–532.

McHaffie, H. E., I. A. Laing, and M. Parker, et al. 2001. Deciding for imperiled newborns: Medical authority or parental autonomy? *Journal of Medical Ethics* 27: 104–109.

Meaney, M. J., M. Szyf, and J. R. Seckl. 2007. Epigenetic mechanisms of perinatal programming of hypothalamic-pituitary-adrenal function and health. *Trends in Molecular Medicine* 13: 269–277.

Milgrom, J., C. Newnham, and P. J. Anderson, et al. 2010. Early sensitivity training for parents of preterm infants: Impact on the developing brain. *Pediatric Research* 67 (3): 330–335.

Mitchell, C. 1984. Care of severely impaired infant raises ethical issues. *The American Nurse* 16: 9.

Montello, M., and J. Lantos. 2002. The Karamazov complex: Dostoevsky and DNR orders. *Perspectives in Biology and Medicine* 45: 190–199.

Moore, K. A. C., K. Coker, and A. B. DuBuisson, et al. 2003. Implementing potentially better practices for improving family-centered care in neonatal intensive care units: Successes and challenges. *Pediatrics* 111 (4): e450–e460.

Orfali, K., and E. J. Gordon. 2004. Autonomy gone awry: A cross-cultural study of parents' experiences in neonatal intensive care units. *Theoretical Medicine* 25: 329–365.

Ortenstrand, A., B. Westrup, and E. B. Brostrom, et al. 2010. The Stockholm neonatal family centered care study: Effects on length of stay and infant morbidity. *Pediatrics* 125 (2): e278–285.

Paris, J. J., N. Graham, and M. D. Schreiber, et al. 2006. Has the emphasis on autonomy gone too far? Insights from Dostoevsky on parental decision-making in the NICU. *Cambridge Quarterly of Healthcare Ethics* 15: 147–151.

Paris, J. J., M. D. Schreiber, and M. P. Moreland. 2007. Parental refusal of medical treatment for a newborn. *Theoretical Medicine and Bioethics* 28: 427–444.

Payot, A., S. Gendron, and F. Lefebvre, et al. 2007. Deciding to resuscitate extremely premature babies: How do parents and neonatologists engage in the decision? *Social Science & Medicine* 64: 1487–1500.

Pellegrino, E. D. 1994. Patient autonomy and the physician's ethics. *Annals (Royal College of Physicials and Surgeons of Canada* 27 (3): 171–173.

President's Commission for the Study of Ethical Problems in Medicine and Biomedical and Behavioral Research. 1983. *Deciding to forego life-sustaining treatment* 218. Washington, DC: US printing office.

Preyde, M., and F. Ardal. 2003. Effectiveness of a parent "buddy" program for mothers of very preterm infants in a neonatal intensive care unit. *Canadian Medical Association Journal* 168 (8): 969–973.

Quill, T. E., and H. Brody. 1996. Physician recommendations and patient autonomy: Finding a balance between physician power and patient choice. *Annals of Internal Medicine* 125:763–769.
Racine, E., and M. I. Shevell. 2009. Ethics in neonatal neurology: When is enough, enough? *Pediatric Neurology* 40 (3): 147–155.
Robinson, M. R., M. M. Thiel, and M. M. Backus, et al. 2006. Matters of spirituality at the end of life in the pediatric intensive care unit. *Pediatrics* 118 (3): e719–e729.
Saunders, C. 1984. Pain and impending death. In *Textbook of pain,* eds. R. Wall and R. Melzack, 472–478. Edinburgh: Churchill Livingstone.
Savulescu, J. 1995. Rational non-interventional paternalism: Why doctors ought to make judgments of what is best for their patients. *Journal of Medical Ethics* 21: 327–331.
Scher, M. S., S. Ludington-Hoe, and F. Kaffashi, et al. 2009. Neurophysiologic assessment of brain maturation after an 8-week trial of skin-to-skin contact on preterm infants. *Clinical Neurophysiology* 120 (10): 1812–1818.
Shah, P. E., M. Clements, and J. Poehlmann. 2011. Maternal resolution of grief after preterm birth: Implications for infant attachment security. *Pediatrics* 127 (2): 284–292.
Shaw, R. J., L. Ikuta, and K. Ginzburg, et al. 2006. Acute stress disorder among parents of infants in the Neonatal Intensive Care nursery. *Psychosomatics* 47 (3): 206–212.
Spielman, B. J. 1993. Conflict in medical ethics cases: Seeking patterns of resolution. *The Journal of Clinical Ethics* 4 (3): 212–218.
Szyf, M.. 2009. Early life, the epigenome and human health. *Acta Paediatrica* 98 (7): 1082–1084.
Truog, R. D., and S. A. Sayeed. 2011. Neonatal decision making: Beyond the standard of best interests. *The American Journal of Bioethics: AJOB* 11 (2): 44–45.
Wall, S. N., and J. C. Partridge. 1997. Death in the intensive care nursery: Physician practice of withdrawing and withholding life support. *Pediatrics* 99 (1): 64–70.
Webster, G., and F. Baylis. 2000. Moral residue. In *Margin of error: The ethics of mistakes in the practice of medicine*, eds. S. Rubin and L. Zoloth, 217. Hagerstown: University Publishing Group.
White, R. D. 2011. The newborn intensive care unit environment of care: How we got here, where we're headed, and why? *Seminars in Perinatology* 35 (1): 2–7.
Wilkinson, D. 2006. Is it in the best interests of an intellectually disabled infant to die? *Journal of Medical Ethics* 32: 454–459.
Wilkinson, D. J. 2011. A life worth giving? The threshold for permissible withdrawal of life support from disabled newborn infants. *The American Journal of Bioethics* 11 (2): 20–32.
Woodroffe, I. 2006. Multiple losses in neonatal intensive care units. *Journal of Neonatal Nursing* 12 (4): 144–147.

Chapter 15
Legal and Ethical Issues in Child-and Family-Centred Care: Transplantation

Rebecca Greenberg, Michael Campbell and Linda Wright

Abbreviations

AAP	American Academy of Pediatrics
AHRA	Assisted Human Reproduction Act
CAS	Children's Aid Society
CCSS	Children conceived to save a sibling
COI	Conflict of interest
DBA	Diamond-Blackfan Anaemia
DD	Deceased donor
ESRD	End Stage Renal Disease
HFEA	Human Fertilisation and Embryology Act
HFEAu	Human Fertilisation and Embryology Authority

R. Greenberg (✉)
Bioethics Department, The Hospital for Sick Children,
555 University Avenue, M5G 1X8, Toronto, Ontario, Canada
e-mail: rebecca.greenberg@sickkids.ca

R. Greenberg · M. Campbell · L. Wright
The Joint Centre for Bioethics, University of Toronto,
555 University Avenue, M5G 1X8, Toronto, Ontario, Canada

R. Greenberg
Department of Family and Community Medicine,
Faculty of Medicine, University of Toronto,
555 University Avenue, M5G 1X8, Toronto, Ontario, Canada

M. Campbell · L. Wright
Bioethics Programme, University Health Network,
555 University Avenue, M5G 1X8, Toronto, Ontario, Canada

M. Campbell
e-mail: michaeldavid.campbell@utoronto.ca

L. Wright
Department of Surgery, University of Toronto,
555 University Avenue, M5G 1X8, Toronto, Ontario, Canada
e-mail: Linda.Wright@uhn.ca

R. Zlotnik Shaul (ed.), *Paediatric Patient and Family-Centred Care: Ethical and Legal Issues,* International Library of Ethics, Law, and the New Medicine 57,
DOI 10.1007/978-1-4939-0323-8_15,

HLA	Human leukocyte antigen
HSC	Hematopoietic stem cell
HSCD	Hematopoietic stem cell donor
HSCT	Hematopoietic stem cell transplantation
IVF	In vitro fertilization
LD	Living organ donor
PGD	Pre-implantation genetic diagnosis
UNOS	United Network for Organ Sharing
YLD	Young living organ donor

15.1 Introduction

15.1.1 History of Transplantation in Paediatrics

Paediatrics figured early in the history of transplantation when, in 1958, Canada's first living donor kidney transplant was performed between15-year-old twin sisters (Dossetor 2005). The transplant was successful and opened the door for many others, often to child recipients.

Deceased donor kidney transplantation developed after Dr. Joseph Murray performed the first such transplant in 1962 (Dean et al. 2012). The first deceased donor liver and heart transplants were performed in 1967, by Dr. Thomas Starzl and Dr. Christiaan Barnard respectively (Cooper et al. 2001; Starzl 2011). Deceased donation has enabled the number of transplants to grow, has proven very successful, and is now the treatment of choice for many cases of end organ failure of livers, lungs and hearts (Goldstein et al. 2011; Kirk et al. 2012; Spada et al. 2009). Expansion into other organs and tissues continues. For 2001–2010, the Canadian Organ Replacement Registry reported 547 first kidney transplants and 34 kidney re-transplants to paediatric recipients. In the same period, among liver transplant recipients under the age of 18, Canada had 365 first-time liver transplants (of which 105 were from living donors) and 57 re-transplants. First-time heart transplants were performed in 290 paediatric recipients over the decade, whilst 40 Canadian children received a lung transplant.

The first successful hematopoietic stem cell transplant (HSCT) was performed in 1968, on a 5-month-old using his sibling's stem cells (MacLoed et al. 1997). The success of HSCT depends largely on the type of disease treated. In favorable circumstances the success rate is 50–70 %, whereas in less favorable circumstances the rate is 10–25 %. Siblings are the most likely compatible hematopoietic stem cell (HSC) donors. Therefore, the practice of using children as HSC donors is widespread.

15.1.2 History of Pre-implantation Genetic Diagnosis and Children Conceived to Save a Sibling

When a child has a life-limiting disorder (e.g., Beta Thalassaemia, Fanconi's Anaemia or Diamond Blackfan Anaemia) that can be treated though a HSCT, some parents will opt to conceive a child to serve as an umbilical cord blood stem cell, HSC, or organ donor. These children are conceived through the use of in vitro fertilization (IVF) and pre-implantation genetic diagnosis (PGD).[1]

Until the availability of PGD, parents hoping to conceive a HSC compatible donor often failed, as the chance of successful match without medical intervention is 25 % (Wiener et al. 2007), often resulting in termination of pregnancy (MacLoed et al. 1997). The first successful use of PGD to conceive a sibling was on August 29, 2000 with the birth of Adam Nash who was a HSC donor for his sibling with rapidly progressive bone marrow failure and myelodysplastic syndrome secondary to Fanconi Anemia (MacLoed et al. 1997).

15.1.3 Overview of Chapter

In this chapter we will review the ethical issues related to the tensions and synergies in child and family centred care related to transplantation in minors. For the purposes of this chapter, paediatric patients are defined as persons ranging from birth to age 18 years (Mandal 2013). Transplantation may be from deceased or living donors and of solid organs or from HSC. Ethical issues arising in transplant pertaining to the tensions between child-centred versus child and family-centred care often pertain to balancing the best interests of the child with that of the family unit. We will discuss issues pertaining to solid living organ transplantation, including youth as living donors, living related donation, and disclosure of misattributed paternity. Issues pertaining to solid deceased organ transplantation include navigating the disagreement between child and family regarding proceeding with transplantation. Lastly, we will discuss the ethical issues pertaining to using siblings as HSC donors and the ethics of children conceived to save a sibling. Conflicts of interest (COI) may arise on several fronts. There may be challenges for healthcare professionals in their approach to family members who are potential living donors for a child, as it is the best interests of the child which are the focus of medical intervention. Laws and regulations often prohibit minors from being living organ donors, as they are considered vulnerable persons in need of protection and in many cases unable to provide informed consent.

The situation is different for HSC, a procedure involving less bodily intrusiveness and risk than organ transplantation. In Canada, HSC donation is not subject

[1] PGD is an assisted reproductive technology that permits genetic screening of an embryo before it is implanted, with an obvious advantage given that a woman need not wait until birth to know whether her child suffers from a genetic disorder.

to the same laws as solid organ donation. It is legally permissible for children to donate HSCs to family members. In most cases, parents may provide consent on behalf of the child and have a COI when the recipient is either themselves or another child for whom they are consenting. Clearly there are both synergies and tensions in situations where children may need a life-saving transplant for which they may not be able to give informed consent.

15.2 Canadian Policies and Law

15.2.1 Legislation—Children as Organ Donors

Each Canadian province and territory has its own laws governing age of consent to medical treatments, organ donation and transplantation. The legal safeguards related to living organ donation highlight the important distinction between accepting medical treatments intended to benefit oneself, and those where the medical benefit hinges on the gift (i.e., a solid organ) from another person. Organ donation laws typically require living donors to be of a minimum age, mentally capable, and able to consent (Trillium Gift of Life Network Act 1990). In Canada, minimum living donor ages range by province from 16 to 19 years. Canadian laws also vary by province in permitting children and young adults to consent to becoming an organ donor after death from the ages of 16–19. Legal substitutes (e.g., parents) may give consent for deceased donation for donors who are under the minimum age.

Canadian case law, to our knowledge, is largely silent on the issue of organ transplant. Many Canadian provinces and other jurisdictions do not allow children to serve as living donors. For example, while the Canadian province of Alberta does not permit solid organ donation they do allow children under 16 years old to donate regenerative tissue provided that the donor is an immediate relative of the recipient, among other conditions (Human Tissue Gift Act 1987). The issue of minor donation for transplantation has been reviewed in the American courts on a number of occasions, most commonly regarding kidney donation. The determining factor in most cases has been the best interests of the donor, which are the psychological interests. In *Strunk v Strunk*, Jerry, a 27-year-old-man with the cognitive abilities of a 6-year-old, was deemed a suitable donor for his dying 28-year-old brother, despite lacking capacity to consent. The court ruled that it was in Jerry's best interests to donate a kidney to his brother, largely because his brother was his closest friend and could readily understand Jerry's speech. In *Hart v Brown*, a 7-year-old was permitted to donate a kidney to his twin because it would be a great loss to the child and cause distress if the sibling were to die from a preventable illness. In *Little v Little*, a 14-year-old girl with Down's Syndrome was permitted to donate her kidney to her brother. She was emotionally close to him and understood that she could be harmed emotionally if he was ill. In *Bosze v Curran*, a father requested the twin half-siblings of his son to be tested as possible HSC donors. The twins' mother objected because the children were not close and they did not regard the child as their sibling. The court supported her refusal. These cases demonstrate the importance of high likelihood of donor benefit in permitting a person who is unable to provide fully informed consent to donate an organ or HSC. The focus appears to be on the donor, reflecting patient-centred care.

15.2.2 Legislation—Children Conceived to Save a Sibling and Pre-Implantation Genetic Diagnosis

Conceiving a child to serve as an umbilical cord blood stem cell donor, HSC donor or organ donor is subject to some ethical debate. PGD permits embryo screening to ensure the absence of certain genetic disorders and is used to conceive children to save a sibling. The ethical issues pertaining to children conceived to save a sibling (CCSS) will be explored in Sect 3.0 of this chapter. In Canada, the most relevant legislation regarding CCSS is the *Assisted Human Reproduction Act* (AHRA), which does not specifically mention CCSS, but creates rules that affect its use. Section 5 (1) of the AHRA prohibits gender determination for social or family planning reasons. It does not otherwise speak to the issue of genetic analysis of embryos for implantation. Some countries explicitly prohibit the use of PGD, while others permit it. Canada does not ban it, but at the same time, has not addressed the issue head on. Some countries have more developed rules regarding the issue of CCSS. It is expressly permitted in the U.K. under the *Human Fertilisation and Embryology Act* (HFEA), while in Australia, the state of Victoria has created an oversight body under its *Infertility Treatment Act*, and that body makes decisions on a case by case basis regarding the use of PGD. The UK Human Fertilisation and Embryology Authority's role is to interpret the HFEA and regulate the use of PGD (Human Fertilization and Embryology Act). Some countries have legislation which specifically prohibits the use of PGD. For example, Italy enacted a law in 2004 on assisted reproduction which bans PGD in an attempt to prevent planned wastage of surplus embryos (Dickens 2005).

15.2.2.1 Legal Considerations

A handful of cases in the early 2000s paved the way for the practice of PGD and CCSS. In the United States, the Nash family from Colorado, whose 6-year-old daughter suffered from Fanconi anemia, received ethical approval for IVF in Colorado, PGD in Chicago, and transplant of a suitable embryo in Minnesota. On August 29, 2000, their son Adam Nash was born—the first successful use of PGD for the purpose of a HSCT (Dickens 2005). In the UK, The Hashmi family requested approval for IVF and PGD from the Human Fertilisation and Embryology Authority for tissue typing to birth a child who would not inherit Beta Thalassaemia major and produce compatible stem cells. The Hashmis were permitted, but unfortunately miscarried (Dickens 2005). Also in England, the Whitakers requested approval of IVF and PGD from Human Fertilisation and Embryology Authority to provide a HSCT to their son suffering from Diamond-Blackfan Anaemia (DBA). Their 3-year-old son required blood transfusions every three weeks, and a HSCT would provide the opportunity for him to live more than 30 years. Human Fertilisation and Embryology Authority refused approval as the Whitakers were not at risk of transmitting the nonhereditary condition to another child. They ultimately received treatment in Chicago. A similar case came before Human Fertilisation and Embryology Authority in 2004 and, in light of evidence of the safety of PGD to born children, they approved the use of IVF and PGD (Dickens 2005).

15.2.3 American and Canadian Pediatric Society Position Statements

The American Academy of Pediatrics (AAP) has a policy statement entitled Pediatric Organ Donation and Transplantation, which aims to raise awareness pertaining to ethical issues in the practice of organ donation and transplantation including but not limited to: when it is ethical to procure organs, how to allocate organs, payment for organs, and fairness of the distribution system. Fourteen recommendations are put forward to address paediatric organ donation and transplantation practices, and the ensuing ethical issues. Recommendations speak to the importance of providing family-centred care at the end-of-life and providing staff education on the ethical issues related to potential donor and recipient families (American Academy of Pediatrics Policy Statement, Pediatric Organ Donation and Transplantation 2010b). Recommendations for providing family-centred care include providing support to families during the donation process for families whose children will be serving as deceased donors, and the importance of long term follow-up with these families.

The AAP also has a policy statement entitled Children as Hematopoietic Stem Cell Donors, which provides a discussion of the ethical considerations regarding minors serving as stem cell donors (American Academy of Pediatrics Policy Statement, Children as Hematopoietic Stem Cell Donors 2010a). The policy discusses conditions under which a minor may participate as a HSC donor and the acceptability of conceiving children to save a sibling with a life-limiting illness through HSCT. Six recommendations are put forward to address the ensuing ethical issues. Five conditions are stipulated that, when met, would permit a child to serve as a HSC donor. Recommendations speak to the need to exhaust other options, minimize foreseeable harms, maximize benefits, and use donor advocates as standard practice in evaluating potential HSC donors. The policy statement considers CCSS as ethically justifiable. Also, the statement urges paediatricians to educate the potential parents of HSC donors about the risks and benefits of CCSS.

15.3 Solid Organ Transplantation

15.3.1 Solid Organ Transplantation from Living Donors

15.3.1.1 Young Living Donors

Consent and Capacity Informed consent requires an individual to exercise choice and judgment regarding the proposed procedure. Some jurisdictions permit children under 16 years to be living donors. It can be challenging to ensure that the necessary conditions of informed consent are met when evaluating potential young living donors (YLD)s. The elements of valid informed consent include: (1) competence to understand and decide, (2) disclosure of information that is relevant to the decision, (3) understanding of the information, and (4) voluntariness of the decision and finally the consent (Beauchamp and Childress 2001). To ethically consent to organ donation, a YLD must be able to (1) understand information relevant to this choice and (2) appreciate the reasonably foreseeable consequences of a decision to undergo the proposed surgery (O'Donnell et al. 2010). Some potential YLDs may exceed the minimum legal age for living donation but lack the maturity, judgment or life experience to appreciate the consequences of living organ donation. Potential YLDs should be excluded if their capacity to consent is uncertain.

Risk-Benefit Analysis for Paediatric Living Donor Advances in immunosuppression have enabled excellent outcomes of kidney transplants from living donors who are not biologically related to their recipients, (Cecka 2000) including spouses, friends and strangers. However, some believe that living organ donation from minors should be restricted to those with whom the young donor has a close familial relationship (Brierley and Larcher 2011). Minors have historically been restricted to donating to family members, as the likelihood of psychological benefit is high (Ross et al. 2008).

Consensus statements have attempted to address the issue of allowing minors (less than 18-years-old) to become living organ donors (LD) by suggesting that certain conditions must be met, i.e., the donor and recipient are highly likely to benefit, the surgical risk to the donor is extremely low, all other deceased and LD options have been exhausted, and the donor assents freely without coercion (Abecassis et al. 2000; Canadian Council for Donation and Transplantation 2006). The AAP has proposed, additionally, that the emotional and psychological risks to the donor should be minimized (Abecassis et al. 2000), as paediatric donors could experience guilt or grief if the recipient has a poor outcome. Improved recipient health may improve family function and reduce stress on family members, including the YLD. In circumstances where a child is donating to a parent, potential benefit is high, as the child may be relieved of care-giving responsibilities and will benefit by avoiding the premature death of the parent (Spital 1997).

Risks of living organ donation include medical morbidity and mortality. Living kidney donation has a major complication rate of 2.9 % (Mjoen et al. 2009) and a mortality rate of 0.03% (Segev et al. 2010). Living liver donation has an overall complication rate of 28–37% (Adcock et al. 2010; Shah et al. 2005) and a mortality rate of 0.03–0.23% (Hashikura et al. 2009; Muzaale et al. 2012; Trotter et al. 2006). However, we are not aware of any evidence that YLDs are more prone to adverse outcomes than older donors. Some research has shown a very small number of living kidney donors later required kidney transplantation. These donors ranged in age from 17 to 61 years at the time of donation (mean age = 31 years) (Ellison et al. 2002).

Donor Autonomy vs Relational Autonomy Autonomy refers to "the capacities to think, decide, and act on the basis of such thought and decision freely and independently…" (Gillon 1985, p. 1807). These capacities begin developing in childhood and become more sophisticated as the individual gains life experiences. Even when capable of making independent decisions, many people involve others in their deliberations, particularly when the decision may have life-altering consequences. The ethical principle of respect for autonomy means that it is generally wrong to interfere with an individual's decisions and actions, provided s/he is mentally capable, is not harming another person, and is acting free from undue influence. Children and adolescents are in the process of developing morals and values. Respect for autonomy implies that people should be free to choose their values and pursue moral projects that give meaning to their lives.

The concept of relational autonomy implies that individuals are not wholly separate from the people who are around them. While individuals have unique identities, there are also social and relational sources of values, goals, and commitments (Ho 2008). Capable patients may include others in medical decision-making to varying degrees. Some patients choose to make decisions without conferring with others, while some defer decisions to another person. Patients may include family or friends in the decision-making process, or seek the opinions of healthcare workers. Before a YLD gives an organ to a family member in need, the healthcare team must be confident that the donor's best interests are considered. There may be concerns that the paediatric donor secretly wishes not to donate, but volunteers to please family members. Conversely, autonomous donors might prioritize the wishes of their family members over their own.

Family Coercion: Challenges of Informed Consent because of Parental Bias to Donate Unduly influencing people to make decisions or perform actions that they do not wish to perform violates the principle of respect for autonomy. Coercion refers to a class of unduly influential behaviours that includes but is not limited to: statements that there are no alternatives to living donation (if alternatives exist), threats, use of force, mistruths or withholding information that is relevant to the decision to donate. YLDs may be more vulnerable than older donors to coercion (Shaw 2001). Young donors may have strong feelings about donating. However, strong emotional impetus is not necessarily an undue influence, nor are the intrinsic benefits that result from donating. To ensure that LDs give organs voluntarily, many transplant programmes require that the LD have the opportunity to meet alone with an appropriate member of the living donor team and that the possibility of non-donation is offered. A YLD may need help in communicating the wish not to donate to a parent. Potential YLDs should be excluded if they are ambivalent about donating.

15.3.2 Living Donation Within Families

15.3.2.1 Strategic Planning for Living Donation to a Child Recipient

Children are prioritized for transplantation in several jurisdictions. In the United States, where a centralized points system is used to list patients awaiting transplant, the United Network for Organ Sharing (UNOS) allots extra points to a child until the age of 18. The points are not lost gradually, but are lost on attaining the age of 18. Canadian provinces' waiting lists do not have points systems, but prioritize children through allocation algorithms. The underlying philosophy is not always explicit. However, supporting arguments for giving preference to children include the "fair innings" concept that a child deserves the opportunity to grow to adulthood and have some longevity. Also, it may be argued that illness impacts on a child's chance to have a normal maturation from physical, psychological and social viewpoints. A transplant is likely to seriously augment the child's opportunity to achieve these goals, whilst an adult who is listed has already had the chance to reach these objectives. This argument is sometimes criticized on the grounds that all people have intrinsic value and should have an equal opportunity to receive an organ. Also, this argument focuses on quantity of life over quality.

The preferential allocation of deceased donor organs to children may have an impact on the family's decision regarding living donor transplantation. A child with End Stage Renal Disease (ESRD) may require several sequential kidney transplants during a lifetime. Parents facing this option for their child may choose to keep the possibility of being a LD for later years when the child, as an adult, will not be eligible for priority points and will therefore wait longer on the list. This decision-making reflects a careful balancing of patient and family-centred elements. Parents or older siblings may consider the likelihood of their suitability as a LD when they are older, the impact on the family dynamics and on their income of undergoing an operation at different time points. A mother may choose to donate to a 16-year-old son for his first transplant, while an 18-year-old brother may agree to donate for the recipient's second transplant. It is less likely that the mother would be a suitable living donor several decades in the future, when the recipient may require a second transplant, however, the recipient's brother will be appropriately age-matched. This strategy allows the recipient to enjoy the benefits of multiple living donor transplants while avoiding a lengthy wait for a Deceased Donor (DD) organ.

The parent(s) acts as decision-maker for the child, but may have obligations to additional children and family members. The transplant team has obligations to put the needs of the child first, which will most likely mean seeking transplantation as soon as possible. Family-centred care, on the other hand, draws attention to the needs of the whole family system, including other minors whose needs are different. The best interests of the child patient include stable and optimal family functioning. Resolution requires a careful balancing of these factors and sharing of the likely risks and benefits of the treatment options for the child with the decision-makers. Some parents may choose not to donate to their child even when they may be an excellent biological match. The reasons are varied: fear of donation surgery, inability to take time away from work, care-giving or other activities to recover from surgery. While some of these reasons are easy to understand, there may be others that are more difficult to accept (e.g., avoiding a scar). This raises the question of whether parents have a moral duty to donate, rather than to rely on deceased donation (Ross et al. 2002; Spital 2002).

15.3.2.2 Implications of Living Donation from Parents to Children: Influence of Offer by Team

Living organ donors (LDs) help to fill the gap left by the insufficient number of deceased donors. While some LDs are anonymous, most are relatives or friends of recipients. Some relatives decline to be evaluated for LD eligibility, hoping that a suitable DD organ will become available before the recipient dies or becomes too ill for transplantation.

Living donation can be ethically justified by the principles of utility, justice and respect for autonomy. When LDs give to recipients, the next available DD organ can be allocated to another recipient, increasing opportunities for those on the wait-list. For parents to give informed consent on behalf of their child-recipients, they need to know that the option of living donation is available. Potential parent-donors should receive information regarding the expected wait-time, possible outcomes, and the possibility that their child may become too ill for transplantation or die on the wait-list.

15.3.2.3 Re-Listing for Transplant

Transplant teams evaluate the ability of the child and their family to care for a transplanted organ. Success of an organ transplant requires a lifetime commitment to a regimen of medications and hospital appointments to monitor and control possible rejection of the graft and transplant complications. Reasons for not taking medication may include unpleasant side effects, forgetfulness, inconvenience, or a mistaken belief that the body will not reject the organ if the medications are not taken.

Parents who agree to their child being transplanted sign on to ensure that the child adheres sufficiently to the post transplant treatment plan—no mean feat, as a child grows independent from them. An ethical issue arises when a child's non-adherence that may be ascribed to poorer parental management is used as a reason for non-listing for subsequent transplantations. It is arguable that, until they demonstrate the maturity to take care of a graft, paediatric patients whose kidney transplants have failed due to non-adherence should receive only dialysis, as opposed to a further transplant. Interventions to prevent graft loss may be most effective when designed to support parents and child-recipients to take care of the transplanted organ.

15.3.3 Misattributed Paternity

Misattributed paternity has arisen in living donor kidney transplant when the necessary Human Leukocyte Antigen (HLA) testing provides unsought information relating to paternity. Transplant professionals are then in possession of highly sensitive information that is unnecessary for the purposes of transplantation. Whether this information should be disclosed or not, and to which of the relevant parties, has been hotly debated. Ethical issues include the right of the patient to their own health history, the relevance of the information to the decision to donate or not, confidentiality of each patient's health information, the potential for disclosure to cause harm, and trust in the healthcare system. Wright et al. concluded that disclosure was warranted in the case of an 18-year-old who was evaluated as a living donor to a parent (Wright et al. 2002). Ross argues that HLA testing's findings regarding paternity are not certain enough to warrant disclosure, given the potential harms of doing so (Ross 2010). A family-centred approach would not support disclosure for fear of upsetting family relationships and harmony. A more patient-centred view could support disclosure on the grounds that people have the right to information on their genetic origin and health history. For a child whose father is a potential living donor, the information would be on the child's health records. Yet, the mother is involved as a person in her own right and as a substitute decision-maker for the child. Clearly these serious decisions become very complex and extremely difficult to resolve.

One solution is to inform parties prior to testing that whilst HLA testing can indicate paternity, it is not paternity testing per se, and the results will not be shared with the parties as they are not necessary for the purposes of living donation. Thus, the parties are aware that the health record may contain this information, which is treated confidentially and the transplant centre focuses on its mission of donation and transplantation.

15.4 Solid Organ Transplantation from Deceased Donors

15.4.1 Parent-Child Disagreement Regarding Consent for Transplantation

In the vast majority of scenarios where a child would benefit from and qualifies for a solid organ transplant, there is no disagreement between parent and child about whether to proceed with a transplant. In these cases, patient and family-centred care are synergistic. Seldom do we see the wishes of parents and child diverge. However, when this occurs, there is clear tension between these two models of care. In cases where a transplant is indicated and parents refuse, legal steps are often taken to override the parents as substitute decision-makers. In Ontario, with children who lack capacity to consent, before 16 years of age, the best interest standard is to guide decision-making and after the age of 16 prior expressed capable wishes are expected to guide decision-making for substitute decision-makers. In Ontario, best interests require consideration of whether the treatment (in this case, an organ transplant) would improve the incapable person's condition or well-being (Health Care Consent Act). In the event that the parents do not want to proceed with a transplant and physicians feel it is in the best interests of the child, the Children's Aid Society (CAS) will likely be notified, or, if above the age limit for which CAS has jurisdiction (in most provinces, 16 or 18 years of age), an application to a court or court tribunal will be made.

There are some circumstances in which it may be challenging to determine if parental refusal of transplantation should be reported. Cronin et al. discuss the tensions between patient- and family-centred care when this arises (Cronin et al. 2013). They err on reporting to CAS in the vast majority of cases, should all attempts to resolve the conflict fail. However, they also note that organs are an absolute scarcity and children die waiting. They propose that it could be ethically defensible, given the scarcity of organs, not to force a transplant on a family where parents are refusing the transplant and to make it available to a family who is eager and willing. This is in accordance with a family-centred, rather than patient-centred approach.

On rare occasion a child will not want to undergo a transplant. From our experience, this has occurred with children who have been chronically ill all their lives and have already had an organ transplant. These children are aware of both what life is like with a chronic illness and what can be the consequences of a transplant. If the children are capable and able to consent, their wishes must be respected from both ethical and legal perspectives. In the event that the child does not reach the legal threshold of capacity, their wishes still deserve some credence. These children have much lived experience to bring to the decision-making process. They, more than the family or treating team, know what it is like to be ill and to live post-transplant. Substitute decision-makers are required to consider wishes expressed by the incapable patient (O'Donnell et al. 2010). Furthermore, from a practical vantage point, organ transplantation requires indefinite follow-up and lifelong anti-rejection medication. Executing the transplant and plan of care would be challenging, if not impossible, with a child who did not want such a procedure. Transplantation would likely not be done as the consequences to the child emotionally and to the child-parent relationship would be significant and possibly irreparable.

Case 1 Tiffany is a 15-year-old with Cystic Fibrosis. She underwent her first lung transplant at the age of eight and has multiple co-morbidities and recent hospital admissions. Tiffany lives with her mother and 19-year-old sister Alana. Tiffany's mom works nights and has often been an absent parent as a result of her job. Tiffany's dad is described as alcoholic and not a part of her life. She has often complained that her illness and the maintenance of her transplant impairs her life to such a degree that it is not worth living. Tiffany is experiencing organ rejection, requiring a second lung transplant. She does not perform the physical therapy prescribed, hides her anti-rejection medication, and is refusing a transplant adamantly. Psychiatry has been following Tiffany for many years and concludes she is not depressed. Tiffany's mother is requesting the medical team do whatever is necessary to ensure Tiffany is compliant with her plan of care and receives a second transplant.

This case highlights the tensions between patient- and family-centred care—what may be in the best interests of the child is not that of the family. Losing a child is clearly not in the best interest of any family, nor is suffering for any child. This is the crux of the divergence between patient- and family-centred care in this case. Despite the tragic consequences of not undergoing a second transplant, a capable patient is able to make such a choice. Although the wishes of the family may not be upheld providing family-centred care is still possible by offering appropriate supports and resources to the family.

15.5 Transplantation of Hematopoietic Stem Cells

15.5.1 Children as Hematopoietic Stem Cell Donors

Case 1 An 18-month-old girl, Maria, with ADA-deficiency Severe Combined Immunodeficiency requires HSC transplantation. Maria's 7-year-old brother, Tony, is a full 10/10 HLA match. The children have a great relationship and are very close. The parents would like the sibling to be the donor. When Tony arrives at the transplant centre, it is clear he is unaware of what is to take place.

The case of Tony and Maria illustrates the common situation of a child who requires a life-saving HSCT, either for an immunological disease or cancer, where the sibling is the best possible HSC donor (HSCD). When children need a HSCT, their siblings are often amongst the first to be considered. Biological sibling donors are generally the preferred donor as they can be identical at the major histocompatibility complex sites, thus reducing the likelihood of graft-versus-host disease and other complications compared to unrelated donors (Gustafsson et al. 2000). The practice of using minors as HSCDs is not without controversy, as there are a number of ethical issues: (1) *Parental Consent and Conflict of Interest*—most donors are not able to provide their own consent and parents serve as substitute decision-makers putting them in a position of conflict of interest (when one child requires the transplant and the other is to be the donor), (2) *Assessing Best Interests*—the best interests of the donor are not always clear, and (3) *Commodification of the Child*—questions regarding the permissibility of using children as means. Many of these issues highlight the tensions between patient- and family-centred care. What may be in the best interests of the family unit is not always in the best interests of the individual child. For example, parents may desperately want to save one child, while donation for the other child is not in their child's best interests.

15.5.2 Parental Consent and Conflict of Interest

The American Academy of Pediatrics Policy Statement for Children as HSC donors raises concerns about the "conflict of interest created if [parents] consider authorizing one of their healthy children to serve as a hematopoietic stem cell donor" for their ill child. It also notes that "transplant teams…face a conflict in their primary responsibility is to the potential recipient; yet, the same physicians may advise, consent, and possibly take the potential donor to the operating room for stem cell procurement" (Joffe and Kodish 2011, p. 517).

This begs the question, are parents the best persons to consent? Can a child's best interest be separated from that of a family's? Sometimes, even if parents are consenting for the wrong reasons, it could still be the "right decision" as the motives may be misplaced but the ultimate decision may still be in the best interests of all parties.

15.5.2.1 Donor Advocates

A survey by the donor health safety committee of the Centre for International Blood and Marrow Transplant Research showed that 70% of paediatric and adult HSCT centres involved donor and recipients being cared for by the same healthcare professionals (O'Donnell et al. 2010). The study recommended that the donor and donee have separate care providers to evaluate the suitability of the donor. The American Academy of Pediatrics Policy Statement for Children as HSC donors recommends that a donor advocate be appointed for all individuals who have not reached the age of majority (American Academy of Pediatrics Policy Statement, Children as Hematopoietic Stem Cell Donors 2010a). The donor advocate is to help the potential donor and parents understand HSCT and promote and protect the interests and well-being of the donor. The donor advocate is not intended as a substitute for parents but rather a supplement. They can serve as a support for the family as parents can be torn as they want to make decisions that are in the best interests of all their children (Ross 2011). Acting as a substitute decision-maker for two children may be a conflict of interest, and an illustrative example of where the tensions between patient—and family-centred care can arise. A donor advocate is one way to help mitigate this tension.

Although it is ideal to have two separate physicians, one for the donor and one for the recipient, this is not feasible in most centres and has been shown not to occur in 70% of centres (Ross and Glannon 2006). Using two separate teams is challenging because (1) most siblings share the same paediatrician, and (2) moving consent and education outside the HSCT program raises concerns of quality control

(Joffe and Kodish 2011). Joffe and Kodish propose that it is sufficient to have two separate physicians from the same transplant program, instead of utilizing a donor advocate in all circumstances. The advantages include having physicians both of whom are familiar with HSCT and encouraging communication between the teams (Jansen 2004). Separate teams for each child can provide a second opinion, put the best interests of each child at the forefront, and provide a safe harbour for the child to feel free to dissent. A donor advocate could be used as a supplement but not as standard practice.

A study exploring the attitudes of clinicians regarding ethical and legal issues in using minors as HSCDs in 70 paediatric HSCT centres across North America showed that most paediatricians endorse the validity of parental consent, even in potentially controversial situations (Chan et al. 1996). Most are prepared to have donors from as young as 6 months of age and are willing to use the same donor more than once. Consent was received primarily from parents alone, while, throughout the consent process, few used child advocates (9 of 56 centres), child welfare (4 of 56 centres) or ethics committees (1 of 56 centres) (Chan et al. 1996).

15.5.2.2 Intrafamilial Justification

Although we can try to separate the interests of one family member from another, the well-being of one family member is intimately tied to that of the others (Ross and Glannon 2006). Self-interests cannot easily be distinguished from those of the family's collective interests. The interests of the family itself carries moral weight and so it can be acceptable for children to put themselves at some risk, commensurate with the relationship with the prospective benefit, for the benefit of a family member (Pentz et al. 2004). When a child donates HSC to a sibling, with whom they have a relationship, their life is often significantly improved by a successful transplant. The intimate attachment principle explains that when a family member dies, with whom the child has a relationship, the child's interests and quality of life are not being served (Jansen 2004).

The interests of one family member are directly tied to another. Correspondingly some have argued that family members, including children, have duties to each other. It is argued that there is a heightened sense of obligations because of the family relationship and its associated indissoluble bond (Weijer 2001). Despite a duty-bound relationship between family members, there remain both age appropriate duties and limits to such duties (Snethen and Broome 2001). Is HSC donation beyond the call of duty? The risks and benefits of each individual situation should be assessed. For example, in a family in which the siblings are close and there is a high likelihood of a successful HSCT and low risk to the donor, this would be considered a circumstance in which the duty to donate is appropriate. Despite potential duties, voluntariness to donate should trump a duty to donate. In other words, if a child did not want to donate the HSC donation should not be mandatory.

15.5.3 *Assessing Best Interests*

15.5.3.1 Best Interests Moving Beyond the Medical

Best Interests: Risks

The medical risks of serving as a HSCD are relatively minor at 1 % for all complications. Common risks include that of anesthesia and nerve, bone or tissue injury (Grupp et al. 2006). The most common adverse effects in children include pain and joint pain (Pulsipher et al. 2006). Psychological risks are less known and can include distress to the donor pertaining to the donation, perception that they did not have a choice about whether to donate, and a sense of responsibility about the outcome of the transplant (MacLeod et al. 2003). There is some research that suggests that children donors are not adequately prepared for what to expect (Shama 2002; Heiney 1998). Siblings have reported feeling responsible for an unfavourable outcome. Research also suggests that minor sibling donors may feel neglected after the HSCT as the recipient still requires much support from parents. This, however is also true of non-donor siblings (Weiner et al. 2007). Studies have also shown that donor siblings report increased behavioural problems, higher levels of anxiety and self-esteem than non-donor siblings, and moderate levels of post-traumatic stress were found in some donors (American Academy of Pediatrics Policy Statement, Children as Hematopoietic Stem Cell Donors 2010a).

Best Interests: Benefit

Psychological benefit has been studied extensively in adult HSC donors (mostly with unrelated donors). Studies show child HSCDs experience increased self-esteem, pride, increased meaning, worth in life, and feel like a better person (Pentz et al. 2008). Only a handful of small studies have looked at minor HSCDs. Despite possible risks to sibling donors, research has shown they agreed the psychological benefits of donating outweighed the harms (Burgio and Locatelli 1997; Packman et al. 1997).

Requiring a strong personal and positive relationship between children donors and recipients has come to be expected for HSCT (American Academy of Pediatrics Policy Statement, Children as Hematopoietic Stem Cell Donors 2010a). There must also be reasonable likelihood that the recipient will benefit, particularly as research has shown that the adverse psychological effects on the donor are felt especially when the recipient dies (Butterworth et al. 1993). Efforts should be taken to considerably reduce the possibility of any harm to the HSCD. This includes the use of a separate medical team, donor advocates, ethics consultation, and a robust consent or assent process.

Best Interests: Non-Maleficence

Emotional harm and a sense of guilt can ensue for children when they are able to understand the consequences of not being allowed to donate (Burgio and Locatelli 1997; Pentz et al. 2008). Research exploring the experience of siblings of children with illness (which included a subset of siblings of children requiring a HSCT) revealed that siblings felt they were outside of the "circle of care" and wanted to help their siblings. Serving as HSCD was regarded as one way to be of help and join the circle of care (Snethen and Broome 2001). In keeping with the values that lie behind patient-centred care, one must be cognizant not to excessively limit the opportunities for children. Patient-centred care supports a capable individual to make informed decisions and for those who are not capable, for substitute-decision-makers to consider all their best interests broadly, beyond that of possible medical harm. Preventing children from donating could result in long term harm resulting from the ensuing guilt.

The American Academy of Pediatrics has put forward recommendations for HSCT (American Academy of Pediatrics Policy Statement, Children as Hematopoietic Stem Cell Donors 2010a). Criteria for HSCD include: (1) there is no medically equivalent histo-compatible adult relative who is willing and able to donate, (2) there is a strong personal and emotionally positive relationship between the donor and recipient, (3) there is some likelihood that the recipient will benefit from transplantation, (4) the clinical, emotional and psychosocial risks to the donor are minimized and are reasonable in relation to the benefits expected to accrue to the donor and to the recipient, and (5) parental permission and, where appropriate, child assent have been obtained. Other recommendations include the use of donor advocates and ensuring children and adolescents are included in all stages of the decision-making process to the extent that they are capable.

15.5.4 Commodification of the Child

Concerns regarding using a child HSCD as a means to assist another raises ethical considerations. It can be acceptable to use a child as a means if the parents' motive is not only to save the sick child, but also to love and care for the child acting as the HSCD for his/her own sake. Further exploration of this issue will be discussed in Sect 3.1.1.

Case 1 The case of the siblings Tony and Maria illustrates the need for a donor advocate and/or separate teams for the HSCD and the potential donor. It also demonstrates the importance of adequately preparing the HSCD about the process of HSCT and associated risks and benefits. Although there is a perceived tension between what is in the best interests of Maria and Tony (as Tony has not until this point been privy to the decisions made by is parents for him to be the HSCD), their interests are intimately intertwined (it is in the best interest of both siblings for Maria to get a transplant) and likely a HSCT would be in the best interests of both.

15.6 Children Conceived to Save a Sibling

In 2001, the President's Council on Bioethics accepted that savior siblings were loved by their parents for their own sake. They questioned the name of "savior sibling" and asked if it truly is the right name and suggested a change of language (Collins 2002). Accordingly, although the language of "savior siblings" is widely known, we have opted to use the term Children Conceived to Save a Sibling (CCSS).

Case 1—Isaac Joanne and Paul have a 3-year-old son, Isaac, who suffers from the blood disorder Beta Thalassaemia, which requires regular blood transfusions. Without a HSCT he may die. BT is hereditary and both Joanne and Paul are carriers, which mean that any child they have carries a 25 % chance of having BT. A match in the HSC registry has not been found. Joanne and Paul would like to have another child, who could be a suitable HSC donor. They would like to undergo pre-implantation genetic diagnosis (PGD) on embryos to ensure their child is born free of the disease, and also to conduct tissue typing for an HLA-matched HSCD. A matched sibling would allow that umbilical cord blood to be used for a HSCT.

15.6.1 Children Conceived to Save a Sibling—Ethical Issues

The case of Isaac is illustrative of the ethical tensions in providing family-centred and patient-centred care. In Canada, although an embryo has no legal rights (Dickens 1997), morally and ethically bringing a child into the world through PGD requires consideration of the ethical issues. Issues include commodification of HSCD, CCSS leading to a slippery slope toward "designer babies", best interests of CCSS, and parental conflict of interest. The aforementioned issues highlight the tensions between patient- and family-centred care. Are the best interests of one child in alignment with other children or the entire family? Can these interests be determined? Do they align or diverge?

15.6.1.1 Commodification

The first argument against CCSS is that they are 'a commodity rather than a person.' This concern is that children could be conceived as a means to an end. It is morally unacceptable if a child is being used *solely* as a means. However, it may be acceptable if the parents' motive is not only to save the sick child, but also to love and care for this child for its own sake. The concern regarding commodification conflates two distinct concerns: (1) having children for the "wrong" reasons and (2) how a child will be treated by his/her parents. Determining the right and wrong reasons for having a child is fraught with problems. People have children for a myriad of reasons. Extremes such as having a child to serve as a servant are clearly wrong, however, the less extreme examples are challenging. Having a child to serve as a means to an end may not be ethically problematic if the child is also wanted for other reasons. Knowing how a child will be treated is challenging to determine. However, parents who have gone to such lengths to save the life of their sick child argue for believing that they are committed parents. In 2001, the President's Council on Bioethics accepted that savior siblings were loved by their parents for their own sake (Collins 2002).

15.6.1.2 Slippery Slope

A second concern is that permitting the creation of children for their HSC encourages a slippery slope toward "designer babies." Claims include that, "the new technique is a dangerous first step towards allowing parents to use embryo testing to choose other characteristics of the baby, such as eye colour and sex." (Sheldon and Wilkinson 2004, p. 534). However, there is a lack of evidence to support concern for the slippery slope (Sheldon and Wilkinson 2004). Another objection to this argument is that CCSS are morally different than "designer babies." Selecting an embryo for the purpose of saving a life is distinctly different from selecting an embryo for trivial reasons such as hair colour. The former is a more weighty and defensible reason for the use of PGD than the latter. Careful regulation can prevent improper use of this technique and avoid the "slide down the slope." For example, England's regulatory body Human Fertilisation and Embryology Authority determines the appropriateness of this technique.

15.6.1.3 Best Interests

An area of concern is the child welfare argument—is it in the best interest of the unborn child to be born? Concerns have been raised that children born to provide HSC will on average have worse lives than children conceived naturally (Collins 2002). This argument hinges largely on the psychological harm that would ensue because they would grow up knowing that they had been created for this particular function and this knowledge would be psychologically damaging (Weijer 2001). The possible adverse effects could also impact the emotional welfare of the sibling recipient who could feel a lifelong debt to the donor sibling. There could be psychological damage to other siblings who are not eligible to donate, creating feelings of inadequacy. The medical risks are minor or non-existent as cord blood is often used as the source for HSC donation.

Alternatively, the donor sibling could feel pride and contentment in the knowledge that he or she is responsible for saving the life of a sibling. There is an absence of research on the long-term psychological impact of being conceived to serve as an HSC donor and save a sibling's life (Pentz et al. 2004). The mere fact that parents are willing to go to such lengths for one child may demonstrate their commitment to their children. There is no compelling reason to believe that they would love any other children less. This argument implies that if in fact the donor sibling were to have a less happy life than other children, their life would not be worthwhile and the effects of being a child conceived through PGD are severe enough to make his or her existence worse than non-existence. The implications are that the life of individuals who are less happy than others is not worthwhile (Weijer 2001).

15.6.1.4 Parental Conflict of Interest

Parents have a conflict of interest in creating a child to save another child. Making decisions on behalf of one child brings challenges as it is nearly impossible for the parent not to consider the other child's interests (Wolf et al. 2003). A conflict of interest by no means precludes both interests being achieved in any one decision. Parents may well have the interests of the child in need of a HSCT but they can still have just as much regard for the CCSS. Families have children for many reasons; despite the reasons, these children can still be loved, cared for and happy.

There are many reasons to support the practice of conceiving children for HSCT. The use of IVF has become widespread. The idea that families can select which embryos they will implant, and then discard the rest is not prohibited. It follows that families have no reason not to choose an embryo on humanitarian grounds, for example, to provide an HSCT to a sick child. Banning the creation of a savior sibling could lead to avoidable deaths of existing children and an increase in abortion rates (MacLoed et al. 1997). Should there be a need for HSCT after a donation for the CCSS cord blood, guidance can be drawn from the literature on HSCD which is discussed in Sect 2.0.

The Case—Isaac The case of Isaac, a 3-year-old requiring a HSCT with no suitable match, and his family highlights the ethical dilemmas that arise in providing patient- and family-centred care together. CCSS is ethically defensible and should be limited to children who have a medical condition that is life-threatening or seriously disabling. Healthcare professionals should educate parents about the risks and benefits of attempting to conceive a child who will serve as a HSCD. If cord blood is not sufficient or a HSCT is required at a later date, the American Academy of Pediatrics recommendations for HSC donation should be employed. Emotional support should be offered to parents, child donors (if appropriate) and recipients.

15.7 Conclusion

Children have been both recipients and donors of organs throughout the history of transplantation. This area of medicine provides rich examples of tensions between patient- and family-centred care and of circumstances where they may work synergistically. The challenge for healthcare workers is to address these issues so as to extend the life-saving opportunities to children when it is medically and ethically appropriate.

References

Abecassis, M., M. Adams, P. Adams, R. M. Arnold, C. R. Atkins, and M. L. Barr, et al. 2000. Consensus statement on the live organ donor. *Journal of the American Medical Association* 284 (22): 2919–2926.

Adcock, L., C. Macleod, D. Dubay, P. D. Greig, M. S. Cattral, and I. McGilvray, et al. 2010. Adult living liver donors have excellent long-term medical outcomes: The University of Toronto liver transplant experience. *American Journal of Transplantation: Official Journal of The American Society of Transplantation and The American Society of Transplant Surgeons* 10: 364–371.

American Academy of Pediatrics. 2010a. Policy Statement, Children as hematopoietic stem cell donors. *Pediatrics* 125 (2): 392.

American Academy of Pediatrics. 2010b. Policy statement, pediatric organ donation and transplantation, committee on hospital care, section on surgery, and section on critical care pediatrics. *Pediatrics* 125 (4): 822–82.

Assisted Human Reproduction Act SC. 2004. c 2. Canada.

Beauchamp, T. L., and J. F. Childress. 2001. Respect for autonomy. In *Principles of biomedical ethics*, 57–112. 5th ed. New York: Oxford University Press.

Brierley, J., and V. Larcher. 2011. Organ donation from children: Time for legal ethical and cultural change. *Acta Paediatrica* 100 (9): 1175–1179.

Burgio, G. R., F. Locatelli. 1997. Transplant of bone marrow and cord blood hematopoietic stem cells in pediatric practice, revisited according to the fundamental principles of bioethics. *Bone Marrow Transplantation*. 19: 1163–1168.

Butterworth, V. A., R. G. Simmons, M. Schimmel. 1993. When altruism fails: reactions of unrelated bone marrow dnors when the recipient dies. *Omega* 26 (3): 161–173.

Canadian Council for Donation and Transplantation. 2006. Enhancing living donation: A Canadian forum. Report and recommendations. http://publications.gc.ca/collections/collection_2007/hc-sc/H14-15-2006E.pdf. Accessed 22 Feb 2013.

Cecka, J. M. 2000. Kidney transplantation from living unrelated donors. *Annual Review of Medicine* 51: 393–406.

Chan, K. W., J. L. Gajewski, D. Supkis, R. Pentz, R. Champlin, and A. Bleyer. 1996. Use of minors as bone marrow donors: Current attitude and management. A survey of 56 pediatrics transplantation centres. *Journal of Pediatics* 128 (5): 644–8.

Collins, F. 2002. Presentation at the meeting of the President's Council on Bioethics, Washington, D. C. www.bioethics.gov. Accessed 13 Dec 2002.

Cooper, D. K., and C. Barnaard. 2001. and his contributions to heart transplantation. *Journal of Heart and Lung Transplantation* 20 (6): 599–610.

Cronin, D. C., J. Squires, G. Mazariegos, and J. D. Lantos. 2013. Parental Refusal of a Liver Transplant for a Child With Biliary Atresia. *Pediatrics* 131 (1): 141–146.

Curran v Bosze. 1990. 566 NE2d 1319(III).

Dean, C., and J. E. Murray. 2012. Transplant doctor and Nobel Prize winner, dies at 93. *New York Times,* 27. Nov. http://www.nytimes.com/2012/11/28/health/dr-joseph-e-murray-transplant-doctor-and-nobel-winner-dies-at-93.html?pagewanted=all&_r=0. Accessed 22 Feb 2012.

Dickens, B. 1997. Interfaces of Assisted Reproduction Ethics and the Law. In *Ethical Dilemmas in Assisted Reproduction,* eds. F. Shendfield and C. Sureau, 77–81. Pantheon.

Dickens, B. M. 2005. Preimplantation genetic diagnosis and 'savior siblings'. *International Journal of Gynecology and Obstetrics*. 88: 91–96.

Dossetor, J. B. 2005. *Beyond the hippocratic oath.* Alberta: University of Alberta Press.

Ellison, M. D., M. A. McBride, S. E. Taranto, F. L. Delmonico, and H. M. Kauffman. 2002. Living kidney donors in need of kidney transplants: a report from the organ procurement and transplantation network. *Transplantation* 74 (9): 1349–1351.

Gillon, R. 1985. Autonomy and the principle of respect for autonomy. *British Medical Journal* 290 (6484): 1806–1808.

Goldstein, B. S., S. C. Sweet, J. Mao, C. B. Huddleston, and R. M. Grady. 2011. Lung transplantation in children with idiopathic arterial hypertension: An 18 year experience. *Journal of Heart and Lung Transplantation* 30 (10): 1148–1152.
Grupp, A. S., H. Frangoul, D. Wall, M. A. Pusiphe, J. E. Levine, and J. R. Schultz. 2006. Use of G-CSF in matched sibling donor pediatric allogenic transplantation: a consensus statement from the Children's Oncology Group (COG) Transplant Discipline Committee and Pediatric Blood and Bone Marrow Transplant Consortium (PBMTC) Executive Committee. *Pediatric Blood Cancer* 46 (4): 414–421.
Gustafsson, A., M. Remberger, J. Winiarski, and O. Ringden. 2000. Unrelated bone marrow transplantation in children: outcome and a comparison with sibling donor grafting. *Bone Marrow Transplant* 25 (10): 1059–1065.
Hart v Brown. 1972. 289 A. 2d, 386, Conn.
Hashikura, Y., T. Ichida, K. Umeshita, S. Kawasaki, M. Mizokami, and S. Mochida, et al. 2009. Donor complications associated with living donor liver transplantation in Japan. *Transplantation* 88: 110–114.
Heiney, S. P, L. K. Bryant, K. Godder, and J. Michaels. 1998. Preparing children to be bone marrow donors. *Oncology Nursing Forum* 29 (10): 1485–1499.
Ho, A. 2008. Relational autonomy or undue pressure? Family's role in medical decision-making. *Scandinavian Journal of Caring Sciences* 22: 128–135.
Human Fertilisation and Embryology Act. 1990.
Human Fertilization and Embryology Act., United Kingdom. 2008.
Human Tissue Gift Act (Alberta). 1987. C.C.S.M. c. H180, Sect. 11(1) (e).
Jansen, L. A. 2004. Child organ donation, family autonomy, and intimate attachments. *Cambridge Quarterly Healthcare Ethics* 13: 133–142.
Joffe, S., and E. Kodish. 2011. Protecting the Rights and L. Wright, and S. MacRae et al.: Disclosure of Misattributed Paternity–a case study of issues involved in the discovery of unsought information. Seminars in Dialysis: 2002: 15 (3): 202–206. Interests of Pediatric Stem Cell Donors. *Pediatric Blood Cancers* 56: 517–519.
Kirk, R., A. I. Dipchand, L. B. Edwads, A. Y. Kucheryavaya, C. Benden and J. D. Christie, et al. 2012. The registry of the International Society for Health and Lung Transplantation: Fifteenth pediatric heart transplantation report–2012. *The Journal of Heart and Lung Transplantation* 31(10): 1065–1072.
Little v Little. 1979. 576 S. W. 2d. 493, Tex.
MacLoed, K. D., S. F. Whitsett, E. J. Mash, and W. Pelletier. 1997. Pediatric sibling donors of successful and unsuccessful hematopoietics stem cell transplants: a qualitative study of their psychosocial experience. *Journal of Pediatric Psychology* 28 (4): 223–231.
MacLeod, K. D, S. F. Whitsett, E. J. Mash, and W. Pelletier. 2003. Pediatric sibling donors of successful and unsuccessful hematopoietic stem cell transplants (HSCT): a qualitative study of their psychosocial experiences. *Journal of Pediatric Psychology* 28 (4): 223–231.
Mandal, A. What is pediatrics? News medical. http://www.news-medical.net/health/Pediatrics-What-is-Pediatrics.aspx Accessed 22 Feb 2013.
Mjoen, G., O. Oyen, H. Holdaas, K. Midtvedt, and P. D. Line. 2009. Morbidity and mortality in 1022 consecutive living donor nephrectomies: Benefits of a living donor registry. *Transplantation* 88 (11): 1273–1279.
Muzaale, A. D., N. N. Dagher, R. A. Montgomery, S. E. Taranto, M. A. McBride, and D. L. Segev. 2012. Estimates of early death, acute liver failure, and long-term mortality among live liver donors. *Gastroenterology* 142: 273–280.
O'Donnell, P. V., T. L. Pedersen, and D. L. Confer, et al. 2010. Practice patterns for evaluation, consent and care of related donors an recipients at hematopoietic cell transplantation centres in the United States, *Blood* 115: 5097–5101.
Packman, W. L., M. R. Crittenden, E. Schaeffer, B. Bongar, J. B. Fischer, and M. J. Cowan. 1997. Psychosocial consequences of bone marrow transplantation in donor and nondonor siblings. *Journal of Developmental and Behavioural Pediatrics* 18 (4): 244–253.

Pentz, R. D., K. W. Chan, J. L. Neumann, R. E. Canplin, and M. Korbling. 2004. Designing an ethical policy for bone marrow donation by minors and others lacking capacity. *Cambridge Quarterly of Healthcare Ethics* 113: 149–155.

Pulsipher, M. A., A. Nagler, R. Iannone, and R. M. Nelson. 2006. Weighing the risks of G-CSF administration, leukopheresis, and standard marrow harvest: ethical and safety considerations for normal pediatric hematopoietic cell donors. *Pediatric Blood Cancer* 46 (4): 422–433.

Ross, L. F., W. Glannon, M. A. Josephson, and J. R. Thistlethwaite. 2002. Should all living donors be treated equally? *Transplantation* 74 (3): 418–421.

Ross, L. F., J. R. Thistlethwaite, and Committee on Bioethics. 2008. Minors as solid-organ donors. *Pediatrics* 122 (2): 454–461.

Ross, L. F. 2010. Good Ethics Requires Good Science: Why Transplant Programs Should NOT Disclose Misattributed Parentage *Am J Transplant* 10 (4): 742– 746.

Segev, D., A. D. Muzaale, B. S. Caffo, S. H. Mehat, A. L. Singer and S. E. Taranto, et al. 2010. Perioperative mortality and long-term survival following live kidney donation. *Journal of the American Medical Association* 303 (10): 959–966.

Shah, S. A., D. R. Grant, P. D. Greig, I. D. MacGilvray, L. D. Adcock, and N. Girgrah, et al. 2005. Analysis and outcomes of right lobe hepatectomy in 101 consecutive living donors. *American Journal of Transplantation* 5: 2764–2769.

Shama, W. I. 2002. The experience and preparation of pediatric sibling bone marrow donors. *Social Work Health Care* 29 (10): 1485–1499.

Shaw, M. 2001. Competence and consent to treatment in children and adolescents. *Advances in Psychiatric Treatment* 7 (2): 150–159.

Sheldon, S., S. Wilkinson. 2004. Should Selecting Saviour Siblings be Banned? *Journal of Medical Ethics* 230: 533–537.

Snethen, J. A., and M. E. Broome. 2001. Children in Research: The Experiences of Siblings in Research Is a Family Affair. *Journal of Family Nursing* 7: 92.

Spada, M., S. Riva, G. Maggiore, D. Cintorino, and B. Gridelli. 2009. Pediatric liver transplantation. *World Journal of Gastroenterology* 15 (6): 648–674.

Spital, A. 1997. Should children ever donate kidneys? Views of US transplant centres. *Transplantation* 64 (2): 232–236.

Spital, A. 2002. Justification of living organ donation requires benefit for the donor that balances risk: Commentary on Ross et al. *Transplantation* 74 (3): 423–424.

Starzl, T. E. 2011. Academy of achievement. http://www.achievement.org/autodoc/page/sta0bio-1.

Strunk v Strunk. 1969. 445 S. W. 2d. 145 Ky.

Trillium Gift of Life Network Act (Ontario). RSO. 1990. Chapter H.20, Part 1, Sect. 3.(1).

Trotter, J. F., R. Adam, C. Mau Lo and J. Kenison. 2006. Documented deaths of hepatic lobe donors for living donor liver transplantation. *Liver Transplantation* 12: 1485–1488.

Wolf, S. M., J. P. Kahn, and J. E. Wagner. 2003. Using Preimplantation Genetic Diagnosis to Create Stem Cell Donor: Issues, Guidelines and Limits. *Journal of Law, Medicine and Ethics* 31: 327–339.

Weijer, C. 2001. Family duty is more important than rights. *The Western Journal of Medicine* 174: 342–343.

Wiener, L. S., E. Steffen-Smith, T. Fry, and A. S. Wayne. 2007. Hematopoietic stem cell donation in children: a review of the sibling donor experience. *Journal of Psychosocial Oncology* 25 (1): 45–66.

Chapter 16
Patient and Family-Centred Care: Critical Partnerships when Treating Adolescent Anorexia Nervosa

Lori d'Agincourt-Canning, Lucinda Kunkel and Karen Dixon

16.1 Introduction

The principles of patient and family-centred care have long served as the organizing framework for treatment of children and adolescents with mental health disorders. For one, access to mental health services by youth often begins with the recognition and actions of family. Second, but more importantly, recognizing the centrality of family in a child's life has promoted an understanding of family as the primary resource for advancing ongoing therapeutic change and child improvement (Hoagwood 2005). Treatment of eating disorders is an area where collaboration with family is especially important, as the family is more than a decision-making partner, but a co-therapist in their child's treatment (Lock et al. 2001).

Yet, while mental health providers readily endorse patient and family-centred care,[1] the ethical obligations it invokes have only recently begun to be explored (Fiester 2012). Providers must manage the tensions between being patient-centred, supporting parents and families, and making medical recommendations families may not agree with (Paul 2007). One of the most difficult situations in which to

[1] Patient and family-centred care is open to different interpretations. We use the term to mean that the child's family members, namely his or her parents, are partners in the therapeutic process. Family-centred care includes open communication between the family and health care providers, respect for cultural differences, and respect for the family's decision-making authority (Kuo et al. 2012).

L. d'Agincourt-Canning (✉)
Ethics Services, BC Children's Hospital & BC Women's Hospital, K4-161, Oak Street, 4500, Vancouver, BC, Canada
e-mail: ldagincourt@cw.bc.ca

L. Kunkel · K. Dixon
Provincial Specialized Eating Disorders Program for Children and Adolescents, BC Children's Hospital, Vancouver, BC, Canada
e-mail: lucinda.kunkel@unb.ca

K. Dixon
e-mail: Kdixon2@cw.bc.ca

R. Zlotnik Shaul (ed.), *Paediatric Patient and Family-Centred Care: Ethical and Legal Issues,* International Library of Ethics, Law, and the New Medicine 57,
DOI 10.1007/978-1-4939-0323-8_16, © Springer Science+Business Media Dordrecht 2014

provide patient and family-centred care is when there is conflict between the values and priorities of the patient and/or family and that of the treatment team. This is especially problematic when the youth is very ill and the choices she or he or their family makes contradict recommended or evidence-based practices about how treatment is best delivered.

In this chapter, we explore patient and family-centred care in the treatment of adolescents with anorexia nervosa (AN). We begin with a case study to illustrate some of the ethical challenges that can arise. This is followed by an overview of AN, best practices for treatment in children and adolescents, and the role of parents. Emphasis is given to the importance of treating children and adolescents with parental involvement as soon as possible once an eating disorder is discovered. Next, we discuss ethical issues related to treatment refusal, both by patients and their parents. We conclude with an analysis of the case study, and brief description of its practical implications for clinical practice generally.

16.2 Background Case Study and the Disorder

Kelly[2] age 16, has been struggling with anorexia nervosa for the past three years. This is her second admission to hospital for treatment and weight restoration. Her first admission did not go well. She resisted treatment by refusing to participate in check-ins or group therapy; speaking rudely to staff; pacing in her room for hours for exercise and purging food after meals. She left hospital against medical advice (at 84% her ideal body weight), insisting that she would eat and gain weight at home. Her parents supported this plan.

Ten months later her parents bring her back to hospital. Kelly is now at 63% of her healthy weight, exhibiting signs of bradycardia, hypothermia and kidney failure brought about by severe malnutrition. She is admitted involuntarily to the eating disorders unit. Initially, she says that the hospital is the best place for her and she engages in therapy. She gains weight over the early course of her admission. Then, feeling stronger, she begins to change her mind. She becomes argumentative with staff and refuses to attend group activities. Her parents similarly show ambivalence about the treatment plans. They begin to investigate alternative therapies and often return her to hospital late after weekend passes. At hospital, they actively side with Kelly when she is dismissive or rude to staff who are trying to engage her in therapy. Kelly's weight gain slows and she convinces her parents that she will no longer improve while in hospital. Her parents inform the treating team they want her discharged, despite Kelly not reaching the desired weight. They say they know their daughter and that home, not the hospital, is the best place for Kelly to recover.

Anorexia Nervosa (AN) is an illness that can affect people of any age, but tends to present during adolescence and teenage years (Hudson et al. 2007). It is currently

[2] The patient's name has been changed and details of the case altered in order to preserve confidentiality.

defined by the Diagnostic and Statistical Manual of Mental Disorders as: (i) refusal to maintain weight more than 85 % of that expected for age and height by weight loss or failure to make expected weight gain during period of growth, (ii) Intense fear of gaining weight or becoming "fat" despite being underweight, (iii) Disturbed perception of one's body, undue influence of body shape on self-evaluation, or denial of seriousness of being underweight, and (iv) Amenorrhea (absence of at least three consecutive menstrual cycles) in post-menarcheal females (4th ed., text rev.; DSM–IV–TR; American Psychiatric Association 2000).

Statistically, AN effects about 0.9 % of all females in the United States (Hudson et al. 2007) and tends to be approximately ten times more common in females than males (4th ed., text rev.; DSM–IV–TR; American Psychiatric Association 2000). While studies are lacking, it is thought that AN occurs at similar rates in Canada as well.

16.2.1 Medical Sequelae of Anorexia Nervosa

People with AN often experience adverse symptoms associated with the effects of malnutrition in every organ of their body. As there is less available energy, the body starts to prioritize organ systems, with the brain having the highest priority to receive glucose energy over every other body organ. The brain consumes approximately 20 % of calorie intake, resulting in severe compromise of itself and other body systems if calorie intake is consistently too low.

A person who is underweight or loses weight quickly will have difficulty regulating body temperature, becoming intolerant of cold and ultimately hypothermic (Yager and Anderson 2008). As the body tries to conserve heat, a downy, soft lanugo hair may grow all over, similar to that seen on a newborn to conserve heat (Yager and Anderson 2008). There is general weakness and fatigue, with dizziness on standing as the body cannot respond quickly to increase blood pressure and return blood to the brain. There is often hair loss and poor healing response due to inability to produce enough functioning white blood cells.

Estrogen levels become low in females, contributing to a loss of menses and infertility if not reversed, while testosterone is low in males. The gastrointestinal system begins to have delayed emptying, which can lead to early satiety, reflux and constipation, compounding difficulties to eat (Kamel et al. 1991). Starvation can also lead to heart changes including bradycardia (a slow heart arrhythmia), hypotension, ECG abnormalities, decreased myocardial tissue mass and cardiac output (Mehler et al. 2010). These changes are a significant contributor to sudden death, especially if interacting with dehydration or electrolyte disturbances often seen with starvation such as low potassium, magnesium, or phosphate (Mehler et al. 2010). Edema (swelling) of extremities may also occur (Mehler et al. 2010). In some with severe malnourishment, acute or chronic renal disease may develop (Takakura et al. 2006).

Despite the body's attempts to spare the brain from starvation effects, deleterious changes occur in this organ as well. Structure is changed with increased "space" in the form of enlarged ventricles and decreased volume of white and gray matter (Katzman et al. 1997, 1998; Ellison and Fong 1998). Some studies have shown reversibility of changes with weight restoration after long term recovery (Mainz et al. 2012; Wagner et al. 2006), however others have indicated abnormalities persist (Katzmann et al. 1997).

Mood and behavior changes such as irritability, depression and anxiety are known to be a direct result of malnutrition (Pollice et al. 1997). Insomnia, hyperactivity, decreased concentration, memory loss, and obsessive thoughts can be a direct result of the brain not having enough energy to make normal levels of various neurotransmitters, including serotonin (Haleem 2012; Strüder and Weicker 2001; Kaye et al. 2009). A person in distress with an eating disorder may begin to harm themselves and/or have thoughts of suicide (Skårderud and Sommerfeldt 2009). Completed suicide is unfortunately a major contributor to mortality in AN (Rosling et al. 2011). AN has the highest mortality rate of any psychiatric illness, with studies showing up to 16.7 % of people diagnosed die from starvation effects on the body or suicide (Papadopoulos et al. 2009).

Importantly, the dangers do not disappear once someone malnourished begins to eat. Re-feeding must be regulated in those who are severely malnourished to prevent re-feeding syndrome, a potentially lethal disturbance of electrolytes including phosphorous, potassium, magnesium and calcium that can cause cardiovascular collapse (Mehler et al. 2010).

16.2.2 Additional Medical Sequelae Specific to Children and Adolescents

Because AN strikes at a time when children and adolescents are undergoing physical, cognitive and social development, they can experience all of the above medical sequellae, while being at risk for further compromises that could affect them the rest of their lives.

For example, malnutrition contributes to growth stunting in those who have yet to achieve their adult height, which may be permanent if the window of opportunity to grow is missed (Misra 2008). In female adolescents low estrogen levels contribute not only to loss of menses, but also to loss of bone density at a time in development that is normally devoted to bone formation (Mehler et al. 2010). If malnutrition is not reversed to allow estrogen to return to appropriate levels, a young person is likely not to achieve their bone density potential, causing them to be at risk for osteopenia or osteoporosis and ultimately fractures (Mehler et al. 2010; Misra 2008; Winston et al. 2008; Teng 2011).

Normal adolescent behaviour is also arrested by AN. Social and interpersonal development is disrupted as the adolescent withdraws from friends and family in order to focus on eating disorder worries, food restriction and exercise. Studies are beginning to show that the disease may also affect the normal sequence of neurohormonal development as it can strike at a time in an adolescent's life when the brain has not matured fully. Neural restructuring continues at an accelerated rate during puberty (Herpertz-Dahlmann et al. 2011). Thus, disrupting regular brain development with malnutrition may have long term consequences, in addition to the structural effects of malnutrition described above (Herpertz-Dahlmann et al. 2011).

16.2.3 Importance of Early Treatment

Once an eating disorder has developed, early diagnosis and treatment give the best chance for recovery (Steinhausen 2002; Von Holle et al. 2008). However, early treatment is often not possible. Psychological changes occur initially, which may not be readily discerned, or may be perceived as "normal" in our culture. Those with AN tend to not recognize the severity of their illness, and are ambivalent about and can be resistant to receiving treatment. AN can cause an individual to minimize, rationalize and hide symptoms, while providing a feeling of comfort, sense of achievement and identity. A person can become fearful of giving up this familiarity for an unknown future of recovery and therefore feel alone but aligned with the illness (Boachie and Jasper 2011). For these reasons it is relatively rare for someone with AN to voluntarily seek treatment initially. Treatment may be sought for medical symptoms related to their eating disorder or interference with life, however it takes an average of 7–13 years for someone to seek medical assistance for AN itself (Boachie and Jasper 2011, p. 51). By that time, AN is more likely to have evolved into a severe, life threatening chronic illness from which it is much harder to recover.

Weight restoration through food remains the primary treatment for those recovering from AN. Achievement of healthy weight improves physical, psychological, social and emotional functioning and is correlated with good outcomes, even if eating disordered thoughts and disturbed body image persist (Bodell and Mayer 2011). Failure to fully restore and maintain a healthy weight is correlated with worse recovery outcomes (Steinhausen 2002; Von Holle et al. 2008). It is therefore a medical priority to safely restore the weight of children and adolescents malnourished from an eating disorder as soon as possible, to prevent and possibly reverse medical complications and further progression of the illness. Weight restoration with family involvement will not only help the adolescent to recover, but also form trusting relationships with those who care about them.

16.3 The Role of the Family

The earliest accounts of AN treatment were focused on the medical treatment of the individual, and implicated the family as the cause of AN in youth. For example, Sir William Gull, an English physician who gave the illness its name in 1874, advocated nutritional rehabilitation as treatment, but removed the child from the family since he perceived that they were "the worst attendants" to help the child to eat (Gull 1874, as cited in Silverman 1997). Similarly, French neurologist Charles Lasègue advocated for complete removal of the patient from family and friends and recommended visitation only after the patient began making progress toward recovery (Silverman 1997). Finally, French neurologist Jean-Martin Charcot described the family's influence on the child with the eating disorder as "particularly pernicious" (Charcot 1889, as cited in le Grange and Eisler 2008). This isolating treatment approach came to be known pejoratively as a "parentectomy" (Brumberg 2000) and unfortunately, still survives in some sectors even today (i.e., F.E.A.S.T. 2012).

Over time, professionals began to examine the contextual factors of the family unit that they believed to play a role in the development of the illness. For example, Bruch (1973) posited that AN arose as an attempt to individuate from controlling parents (usually the mother), who did not permit development of ideas or a personality of their own. Minuchin et al. (1978) hypothesized that there were interactional patterns common to families in which there is a child with AN. These patterns included "enmeshment, over-protectiveness, rigidity, and lack of conflict resolution" (Minuchin et al. 1978, p. 30). Professionals came to believe that if these family mechanisms could be targeted in treatment and the family could come to function in a different way, the anorexic adolescent could be helped to recover. Thus, the family became actively involved in the treatment process (le Grange and Eisler 2008). However, one problem with these early attempts at treatment lay in the discovery that there are no patterned family interactions—there are as many different types of family interactions as there are youth with AN.

Nonetheless, since these early conceptual understandings of family treatment, evidence has established that family therapy is an effective way to support adolescent recovery of AN. Other family therapy models have come to be employed in the treatment of adolescent AN, such as narrative therapy (White and Epston 1990) and systemic family therapy (Palazolli 1996). The work of Dare et al. at the Maudsley Institute in England, as well as that by Lock, le Grange, Agras, and Dare in the United States, created the foundation for the family treatment with the most empirical support. This model, referred to here as Family Based Therapy (FBT), has consistently been shown to lead to clinically meaningful weight gain and psychological improvement in adolescents with anorexia (Dare 1985; Lock et al. 2001; Bulik et al 2007).

FBT is based on the premise that: (a) the adolescent is not in control of the illness, rather the illness is in control of the adolescent; (b) parents were able to instinctively support their child to eat prior to the onset of the eating disorder and (c) the family's structure and functioning has become reorganized around the illness, and these altered patterns may be maladaptive (Eisler 2005). For example, parents often feel disempowered in their attempts to support their anorexic child to eat, creating stress and tension in the home. As such, "the field has moved away from understanding family therapy as a treatment of families ... to a treatment with families" (Eisler et al. 2010, p. 151). Rather than perceiving the family as somehow to blame for the illness, the family is instead perceived as a powerful resource in the recovery of the adolescent. Thus, the treatment aims to provide support to the (out of control) adolescent by recruiting the guidance and love of their parents, and the therapist helps the parents to become re-empowered to support their child to eat. The family is regarded as a key resource for treatment since they know their child best.

The therapist helps the family in learning to externalize the illness, meaning that the adolescent is separated conceptually from the illness and that the battle is with the disease not the adolescent. This permits the parents to focus on fighting the illness with a caring but firm stance while at the same time creating an environment of unconditional love for their child. Externalization alters the dynamic of conflict that may have been present prior to the intervention, since parents are better able to ally themselves with the child against the illness.

The application of the FBT model in an inpatient setting (such as the tertiary level inpatient program) can be challenging. First, while FBT is intended to foster the expertise of the parents through their own decision-making, decisions in the inpatient setting are made by professionals such as psychiatrists, dieticians, psychologists, nurses, and social workers (Lock et al. 2001). Second, an unintended consequence of inpatient admission is that it provides respite to the remainder of the family, while a key component of FBT is the raising of anxiety of the parents in order to mobilize them to act. Third, it is crucial in FBT treatment that all family members eat together on a regular basis, a process made more challenging in the inpatient setting, particularly if family members do not live in proximity to the hospital. Fourth, the patients admitted may have significant co-morbidities or a longer duration of AN illness, parents may have mental health issues of their own, or there might be a high degree of conflict among parents. Indeed, Federici and Wisniewski (2012) have noted that when the presentation of the adolescent with AN is complex, the success rates of FBT are much lower.

16.4 Treatment of Anorexia Nervosa: Ethical Considerations

Ethical issues concerning eating disorders have frequently focused on decision-making capacity, consent, and compulsory treatment. In this section, we use these concepts to explore the question: how much should an adolescent's wishes count in making medical decisions, in particular refusing treatment for AN? Does it make a difference that the young person is undergoing treatment for a psychiatric condition? We also expand the discussion to examine best interests and parental or surrogate decision making in the context of providing family-centred care.

16.4.1 Capacity

AN is a perplexing and difficult condition to treat. As previously stated, patients frequently deny that they are ill and that treatment is needed, even when they are near death. Further, while there appears to be greater success in youth than in adults, treatment efficacy remains limited. Relapses—as this case study shows—are frequent and treatment can continue for years.

A key ethical issue to either treatment acceptance or refusal in AN is competence, that is the ability of patients to make their own treatment decisions (Tan 2008). The ability to make treatment decisions is viewed largely as a cognitive skill, based on elements of understanding (nature of health problem, treatment and non-treatment options), reasoning (weighing the pros and cons of each option), appreciation (considering consequences of each option) and expressing choice (Grisso and Appelbaum 1998, 2006; Tan et al 2006; Paul 2004). The value placed on the capacity and in turn, liberty to make decisions, is reflected in the principle of respect for autonomy and the right to self-determination.

In adult medicine, a competent patient has the right to accept or refuse any treatment, including life-saving treatment. Physicians must respect the competent patient's wishes based on the presumption that the competent patient knows what is best for him or herself. A competent decision is considered to be one that has a good chance of promoting the patient's values and goals, even if the physician disagrees (Dickenson and Jones 1995). This does not mean, as Ross (2009) points out, that the physician should accept all refusals as final, but after attempting to convince the patient to change his or her mind, the physician must ultimately respect the patent's refusal.

> A competent adult's decision is respected because, even if the physicians are sure that a medical treatment would serve the patient's medical best interest, physicians do not know what is best for any particular patient, all things considered (Ross 2009, p. 302).

With this view, the legal right of medical autonomy is only challenged if the patient shows significant impairment (physical, psychiatric and/or cognitive) such that their ability to make decisions regarding their own welfare is compromised.

Traditionally, minors have had little say about their medical care, but over the past two decades greater weight has been given to an adolescent's right to participate and made decisions about his or her own health care. This stems, in part, from studies indicating that most neurologically normal adolescents, starting at around age 14 or 15, have approximately the same capacity to make informed decisions as adults (Weithorn and Campbell 1982). The "mature minor doctrine" is now well embedded in current thinking about adolescent development, reminding us that parental authority declines in accordance with the minor's evolution into adulthood (Duncan and Sawyer 2010; Dickenson and Jones 1995). In many countries, legislation has been revised giving children and adolescents the authority to consent to treatment, provided the minor is capable of understanding the proposed treatment and expressing her wishes, although the same right may not always apply to refusal of treatment.

Yet, while there is broad recognition in the literature of the importance of cognitive skills, most accounts of capacity also refer to the possession of life goals and values as critical to medical decision-making. Some have described this aspect of competence as the capacity to appreciate the consequences of a decision, implying a deeper comprehension and stability in that person's values and decisions (Grisso and Applebaum 1998). Others refer to the notion of authenticity, an authentic choice being one that makes sense within a stable framework of beliefs and values (Brudney and Lantos 2011).[3] Because respect for autonomy recognizes a person's right to make unsound or even irrational decisions, what we really want to know here is that "whatever decision a person makes, it is truly his decision: a decision for which he can finally be held accountable" (Elliott 1997, p. 114).

Treatment refusal by an adolescent with AN begs the question: is the decision authentic to the patient's true self, meaning is it based on a stable set of personal beliefs and values? There are two aspects to consider here in relation to adolescent capacity and accordingly, patient-centred care: (1) To what degree can we say choices made by adolescents in general are an expression of a mature, stable or authentic self? (2) Can certain disorders, such as AN, distort a person's values so much so that is reasonable to say the decisions she or he makes are no longer hers/his?

16.4.2 Adolescents: Developing Autonomy

While there is broad agreement that many adolescents possess the cognitive skills to make complex decisions, others caution this does not automatically translate into having the necessary maturity or independence of judgment to make truly autonomous choices.

Mark Cherry (2010) argues that from a scientific point of view, mature minors may not be mature decision-makers. He points to a large body of neuro-biological evidence that demonstrates the parts of the brain (prefrontal lobes) responsible for judgment, problem-solving, planning, mental flexibility, inhibition and behavioural self-regulation are still developing during adolescence. Because the prefrontal cortex does not reach full maturity until the mid-20s, adolescents, when compared to adults, tend to be more impulsive, reactive to stress, more likely to take risks, vulnerable to peer pressure, and prone to focus on short-term rewards over longer-term consequences of actions (Gilbert and Burgess 2008; Casey et al. 2008). This is not to say that adolescents mature in a linear fashion, or that all adolescents develop capacities at the same rate, but simply that age and developmental stage are important considerations when asking whether an adolescent's choice is reflective of a mature and stable self.

Dickenson and Jones (1995) explore notions of personal identity to ask whether an adolescent's wishes can be said to consistent with self-determination and autonomous decision-making. While acknowledging that the concept of identity is open to many philosophical interpretations, they define it as that which characterizes us as unique individuals (Dickenson and Jones 1995). A hallmark of identity is knowing who one is, one's values, commitments and goals. Understanding of self and

[3] Meyers' (2000) definition of authenticity is also helpful here. She describes authenticity as a way of living a life that is distinctly one's own.

personal identity is secured through experience and development of skills and capabilities that enable individuals to realize who they are. Yet, adolescence is a time of exploration and enormous psychological and emotional change, when youth are just setting out to establish an identity for themselves. Attention to identity then begs us to ask whether the values the adolescent claims and the choices she or he makes are consistent with a fully formed identity. Dickenson and Jones draw on the work of developmental psychologist Erik Erikson (1968) to say this is likely not the case.

> Perhaps, then, the relevant distinction between young people and adults is not that they are less rational, but that their identities are less securely formed. Their wishes and intentions, their motives, their emotions, the whole structure of values within which they make choices, may be less secure, less stable, than those of adults. In particular, children may have less sense of connection, of responsibility, and of a socially constructed identity. Erikson, for example, views the essence of adolescence as the attempt to establish a self that can be seen as continuous and unified (1968). Perhaps it is misleading to talk of self-determination when a young person's sense of self is perishable and transient? (Dickenson and Jones 1995, p. 295)

Blustein and Moreno (1999) make a similar argument. They argue that the rationale for drawing a distinction between decision-making of adults and adolescents has less to do with cognitive ability than the development of a stable moral self. They write:

> There is no doubt that the normal adolescent is capable of recognizing selfhood: a teenager has long since developed the ability to say 'I' with meaning. However, it is not nearly so clear that or when the normal adolescent develops a moral self, so that he or she can say with meaning, 'I value this or that,' so that a particular decision is authentic, an expression of who I am in a moral sense, of what kind of person I am, of what is really important to me. When an adolescent has not yet developed a moral self, his choices would not be his in the sense that demands respect as an expression of his autonomy. There might still be grounds for respecting his choices, but this would not be because his choices reflect values that are authentically his own (Blustein and Moreno1999, p. 101)

Yet, valid consent in medical ethics relies on an understanding of persons and their values as being relatively stable. We accept a normal adult's decision to make choices that are contrary to the person's best interest because we believe them to be based on values that reflect what the person genuinely believes and truly wants. In contrast, Blustein and Moreno argue the developing moral self of the adolescent can render the authenticity of some choices questionable. Thus an adolescent's choice, while acceptable from a purely cognitive stance, may not to be based on deeply held values that can be attributed to "a stable and recognizable moral self" (1999, p. 104). This is not to say that adolescents are incapable of making mature and responsible decisions. However, Blustein (2009) asserts that this lack of a distinctive moral self—rather than best interests or parental rights—is why we should think carefully about an adolescent's capacity to refuse efficacious life-saving treatments.

Although writing about adults, Brudney and Lantos (2011) also appeal to notions of authenticity to argue that "mere decisional competence does not reflect a sufficiently robust value to justify the refusal of lifesaving treatment" (2011, p. 222). The point about authenticity, they write, is not just to make a choice but to make a choice that accords with one's distinctive life and values. According to Brudney and Lantos: "The value that we look for in an authentic life is that it is a life that I have made and made in a specific way because it is the life that I believe fits me. An authentic choice is one that makes sense within the framework of the beliefs and values that I affirm" (2011, p. 221).

Brudney and Lantos claim that the "moral justification for accepting a patient's decision to refuse lifesaving treatment is lessened to the extent that this decision, though made by a decisionally competent patient, is less than authentic" (2011, p. 223). They argue that in situations where the medical team questions the authenticity of a patient's decision to refuse life-saving treatment, additional efforts should made to talk to the patient, understand her or his reasons for refusing treatment and try to see if those reasons fit into an account of a patient's life journey and who she/he say they are. They also suggest consultation with the patient's family and friends may be required. If a patient's decision to refuse life-saving treatment is deemed sufficiently inauthentic, they conclude that it would be wrong to accept refusal. Their conclusion may even have greater weight when applied to adolescent refusal of life-saving treatment if adolescence is understood as a time of developing autonomy and emerging authenticity.

Brudney and Lantos' discussion of authenticity pertains primarily to adults in later stages of their life. Accordingly, their depiction of authenticity is a relatively static conception of the self. In contrast, others write about authentic self as an evolving self (Meyers 2000). Authenticity emerges through a process of self-knowledge and self-definition. Authenticity is not innate, but develops as individuals come to know what they truly believe in, desire and care about through life experiences. It is this concept of authenticity that we see applying to adolescents. We do not claim that adolescents have no authenticity, but rather as they are developing the skills for autonomy, they are developing authenticity.

16.4.2.1 AN and Self-Interest

Having a mental disorder such as AN can add a further layer of complexity to treatment refusal by adolescents. Research conducted by Tan et al. (2003, 2006) shows that young women suffering from AN experience difficulties with making decisions to accept treatment because of shifts in value systems, incorporation of the mental disorder into the patient's sense of personal identity, and issues of control. They interviewed 10 patients (age range 13–21) who met DSM-IV criteria for AN. In addition to interviews, all the young women were tested using the MacArthur Competence Assessment Tool—Treatment (MacCAT-T) test of competence. Interestingly, 8 of 10 participants scored well on the MacCAT-T test of competence. Only two participants did not show full appreciation (e.g., apply the facts to themselves) that they had the disorder.

As typical of the disorder, patients accorded enormous value to being thin. All the participants described having value systems in which being fat was perceived as not only highly undesirable, but directly connected to their identity and self-worth. Patients viewed themselves being fat as a failure, being unlovable, or being an indictment of their entire personalities, and as such a was state to be avoided at all costs (Tan et al. 2003, 2006). The fixation on thinness increased to the extent it superseded most other aspects of their lives, including health, family, friendships and academic achievement. From the point of view of treatment refusal, the value given to being thin (and the corresponding anxiety and fear around food and gaining weight), "affected the way participants viewed the risk of death and disability, with a significant loss of the motivation to avoid these outcomes" (Tan et al. 2003, p. 274). As a result, a patient with AN may be aware that treatment refusal threatens her well-being, and even survival, but simply does not care about those risks compared to being thin. Thinness becomes overvalued with AN, and the overvaluing of thinness a medical abnormality that is both caused and reinforced by the disease. According to Tan et al., this lack of self-interest has implications for how we think about capacity to refuse treatment. A competent evaluation of the risks and benefits of treatment involves taking into account one's well being. If a particular disorder affects a person (by distorting values) such that she/he fails to take her own welfare into account, then one must ask whether she/he is accountable for making decisions whether to accept treatment or not. Elliott makes a similar point with respect to depression: "the assumption that other people ordinarily both have some minimal degree of self-interest and are in the best position to judge their own interests lies at the heart of the institution of informed consent" (Elliott 1997, p. 115). Further, notions of authenticity ask us to explore whether a person who is in the grips of AN can be said to make a choice that is constitutive of her true self, if one of the effects of AN is a distortion of values.

16.4.2.2 Anorexia Nervosa, Adolescents and Reasons for Caution

Several conclusions emerge from our analysis of adolescence and AN. First, certain features of the developmental period of the adolescence, namely impulsivity, short-term thinking and high risk behaviour, justifies our treating adolescents as less responsible for their decisions than adults.[4] This is because the prefrontal cortex, which is associated with judgment and the ability to control emotions, is not fully developed, until the mid 20s. Second, because adolescence is a time of change when a youth's moral self has not yet crystallized, questions arise as to whether the choices they make can be said to be authentically their own. The concerns that adolescents articulate as central to their lives are likely to change with experience and as they mature. Third, while AN leaves a patient's cognitive abilities intact, it can distort a person's values to the extent that interest in personal welfare is overridden by the drive to be thin. The impact of AN on values also raises questions as to whether an adolescent's refusal of necessary treatment is an expression of her true wishes (Dickenson and Jones 1995). These considerations, together with cognitive ability, are relevant to assessment of adolescents' capacity to consent to or refuse treatment.

[4] At the same time, we are not making the claim that all adolescents make bad or immature decisions.

Clearly, there are many factors to consider in deciding whether to accept or override an adolescent's refusal of treatment, including the patient's presentation and stage in their treatment course. If treatment is not yet critical to the welfare of the patient, then efforts should be made to continue engagement and persuade the patient to accept therapy (Tan et al. 2007). However, in situations where treatment would be life-saving or preventative of irrevocable medical damage, it is a physician's fiduciary responsibility to act in the patient's benefit. In practice, this means physicians are obligated to seek involvement of others (typically parents) who are entrusted to support the patient and make decisions on his or her behalf (Rhodes and Holzman 2004).

16.4.3 Parental Dissent

As stated previously, family plays a central role in the treatment of adolescents with AN. Support and assistance from family members is a key part of recovery, especially initially while the youth is in the throes of the disorder and significantly malnourished. With family involvement in re-feeding, the goal is to break the young person's alliance with the eating disorder and strengthen bonds with family and people who care about them (Steinhausen 2002). This means that therapeutic decision-making needs to take into account family choice and preference as much as possible.

Yet, despite the focus on family-based therapy and family-centred care, inevitably situations arise where collaboration does not seem possible. For example, treatment for AN often involves a re-feeding program implemented under strict supervision, enforcement of a prescribed diet, prevention of exercise or purging, and sometimes nasogastric feeding (Lock et al. 2001). These kinds of interventions may be viewed as unnecessarily harsh by patients and their parents. Alternatively, parents without understanding of the illness may think they are helping by advocating for another chance for the young person to recover on their own, giving the eating disorder more opportunity to establish itself (Boachie and Jasper 2011). Some may fear the stigma of a mental illness. A major challenge occurs, as our case demonstrates, when the youth is severely malnourished and she or he together with parents rejects medical recommendations for in-hospital treatment.

In most countries, parents have a legal right to make health care decisions for their children. This reflects a normative presumption that parents know the child best, and affection and close family ties make parents most likely to reach decisions based on the child's best interest (Forman and Ladd 1995). Yet, while parents are viewed by society as the most appropriate decision-makers for their child's health, "the state has the authority to intervene when the parents' decision falls below some threshold that qualifies their decision as abusive or neglectful" (Ross 2009, p. 302). Thus, if parents refuse to consent to life-saving treatment for their child, it is the physician's duty to seek intervention through legal measures. Parental decision-making authority deserves respect, but ultimately it is the patient who is the focus of care and to whom professional duties of care are owed.

It is widely recognized, however, that physicians should seek court intervention only as the last resort (Bryden et al 2010; Diekema 2004; Tan et al. 2007). Promoting parental authority and shared decision-making requires that the health care provider engage in open dialogue, inform families about therapeutic options and chances for success, explore patient/family's values, as well as their own and make recommendations that take into account—if not necessarily follow—each party's values, experience and knowledge. Quill and Brody (1996) call this model "relationship-centered" rather than exclusively patient- or family-centred. It does not deny the inherent hierarchical nature of the patient/family-provider relationship, but emphasizes the importance of therapeutic engagement when there is disagreement. They state:

> [T]o accept a patient's choice when it flies in the face of strong recommendations, without a full exploration and vigorous exchange of ideas and perspectives, can be tantamount to abandonment. This exchange between two persons who disagree but who both care deeply about what happens to the patient often yields better decisions than those that would have been made by either the physician or the patient independently. Sometimes the decision itself does not change, but the meaning of the decision to both participants is more fully appreciated. At other times, exploration leads to a better decision, one that can embrace the best of both positions. (Quill and Brody 1996, pp. 766–767)

Patient and family-centred care is a commitment to meet patients and families on their own terms. When conflicts arise, every effort should be made to negotiate with family. Partnership means that both patients and providers must change, and share responsibilities, as well as information and decision-making (BMJ 1999). Short-term involuntary treatment may be necessary if the patient is critically ill, but recovery from AN is a long-term process. Obviously, a trusting relationship between providers and family is critical to the patient's ongoing recovery (Kelly 2010). Taking decision-making authority away from parents risks doing significant harm by fracturing the trust that is necessary for long-term follow-up and patient care. Health care providers have a duty of care to make difficult decisions if the welfare of the adolescent patient is being significantly compromised, but also needed is sensitivity to the potential harmful consequences of these actions.

16.4.4 Case Review

By the usual standards applied, Kelly was not incompetent to make medical decisions. She could understand the information provided, assess the consequences of her refusal to stay in hospital, and explain her decision in a coherent and consistent way. Likewise, her parents felt clearly they had her best interests at heart. Although they brought her to hospital for initial care, they believed she had stabilized and they could achieve greater success with further weight gain at home. Yet, given Kelly's past history and current weight status (84 % of her ideal body weight, the same weight at which she left AMA on her first admission), the team knew that early discharge would put Kelly at high risk for relapse, reducing her chance of ever recovering from her disorder. They thought Kelly's capacity to make a truly informed decision about treatment was compromised by her current physiological state and a disease that caused her to value thinness above everything else. The team respected the parents' authority to make medical decisions for their daughter, but felt they did not understand the gravity of the situation and poor prognosis for Kelly if she was allowed to return home early and at such a low weight. However, they also knew that if they involved Child Protection Services, the parents would consider this a grave violation and affront to themselves. The providers also worried that court intervention would be viewed so serious a betrayal of trust that any chance of maintaining a long-term therapeutic relationship with the family would be lost.

Several meetings were held with the parents, psychiatrist, family therapist and hospital ethicist. The team recognized the need to build a stronger partnership with the family and that this would take time. They began by demonstrating a willingness to negotiate over weekend passes. The parents requested that Kelly be given three and sometimes four day passes so that she could meet with the alternative practitioners they had found. The health care team offered to arrange videoconferences to assess Kelly's health status so that the family did not need to physically return to the hospital for assessment during these passes. They also invited the alternative practitioners to meet with them to discuss Kelly's current status and projected needs.

Further conversations focused on the centrality of the parents' roles in helping Kelly to overcome AN. The team explained that hospitalization would be a short-term solution to the acute physical and medical dangers that Kelly faced, but long-term recovery required that the parents become co-therapists with them in fighting the disorder. Emphasis was placed on externalizing the disease and how they must take necessary steps to help Kelly to gain weight, as physiological malnutrition compromised her ability to look at her disorder differently. The team also acknowledged the strong relationship Kelly had with her parents, and emphasized that this bond was critical to her recovery. Discussed at length were the reasons for the hospitalization. The team provided her parents with further information about how AN had already affected Kelly's bone health, putting her at risk for osteoporosis, bone fractures and other life long complications. They told the parents about other situations in medicine where adolescents' preferences would be over-ridden if their choices put them at significant risk (e.g., adolescent refusal of chemotherapy for a curable cancer). It was their position that while it may be hard for Kelly to go through this now, re-feeding must be the first priority, to reduce current and future physical health risk. Finally, the team was forthright about its professional, legal and ethical obligations. They explained that while they respected the parents' authority to make decisions on Kelly's behalf, they had a duty to seek outside intervention if they thought Kelly's well-being would be significantly harmed.

Despite some tense moments, the therapeutic alliance was maintained. After a number of extended weekend passes, it became apparent that Kelly was still deeply affected by her illness. At home, she would exercise compulsively and did not participate in meals as promised. This, together with ongoing team discussions, helped Kelly's parents to understand the necessity of in-patient care and they agreed to ongoing treatment. What about Kelly? Both her parents and providers told her that they would not let her starve to death, that they cared for and respected her, and that all together they would fight the disease.

16.4.5 Conclusions

Anorexia nervosa is a life threatening illness that involves a lengthy course of treatment. When adolescents fall prey to this illness, research indicates that involving the family in treatment affords the best chance for successful recovery. Ideally, decisions to accept treatment are made by adolescents and their parents together, and treatment plans are shaped by their values, goals and preferences. However, ethical tensions may arise when patients and/or their parents refuse recommended therapy and the adolescent is severely ill. While adolescents may demonstrate the cognitive capacity to refuse treatment, the authenticity of their choices requires careful examination. Further, the nature of the illness itself may interfere with the adolescent's cognitive functioning and judgment, making consent and staying the course of treatment even more problematic. When the parents of the adolescent also struggle with the treatment recommendations, we argue that the risk to the adolescent is high. It requires the treating team to focus on maintaining the therapeutic alliance, since the partnership between parents and the team is critical to support the youth through the weight recovery process and beyond. As discussed with the case study, preserving partnerships with families often requires flexibility and creativity in treatment approaches, in hopes of finding a solution that is acceptable to all (e.g., allowing extended passes with video check-ins). At the same time, though, the healthcare professional's duty of care is to the patient and ensuring that his or her welfare is protected. In those exceptional cases where collaboration is not possible and the situation is of sufficient risk and urgency, providers may have to consider whether legal measures are warranted.

References

American Psychiatric Association. 2000. *Diagnostic and statistical manual of mental disorders*. 4th ed. (text rev). Washington, DC: American Psychiatric Association.

Blustein, J., and J. Moreno. 1999. Valid consent to treatment and the unsupervised adolescent. In *The adolescent alone blustein*, ed. J. Blustein, C. Levine and N. Dubler, 100–110. New York: Cambridge University Press.

Blustein, J. 2009. Response. *Cambridge Quarterly Healthcare Ethics* 8 (3): 315–320.

BMJ. 1999. Editor's choice, Take your partners for the dance. *BMJ* 319, 18. Sept.

Boachie, A., and K. Jasper. 2011. *A parent's guide to defeating eating disorders*. Philadelphia: Jessica Kingsley.

Bodell, L., and L. Mayer. 2011. Percent body fat is a risk factor for relapse in anorexia nervosa: a replication study. *International Journal of Eating Disorder* 44:118–123.

Brudney, D., and J. Lantos. 2011. Agency and authenticity: Which value grounds patient choice? *Theoretical Medicine and Bioethics* 32:217–227.

Bruch, H. 1973. *Eating disorders: Obesity, anorexia nervosa, and the person within*. New York: Basic Books.
Brumberg, J. 2000. *Fasting girls*. New York: Vintage Books.
Bryden, P., C. Steinegger, and J. Jarvis. 2010. The Ontario experience of involuntary treatment of pediatric patients with eating disorders. *International Journal of Law and Psychiatry* 33:138–143.
Bulik, C. M., N. D. Berkman, K. A. Brownley, et al. 2007. Anorexia nervosa treatment: A systematic review of randomized controlled trials. *The International Journal of Eating Disorders* 40:310–320.
Casey, B., R. Jones, and T. Hare. 2008. The adolescent brain. *Annals of the New York Academy of Sciences* 1124:111–26
Cherry, M. 2010. Parental authority and pediatric bioethical decision making. *The Journal of Medicine and Philosophy* 35:553–572.
Dare, C. 1985. The family therapy of anorexia nervosa. *Journal of Psychiatric Research* 19:435–453.
Dickenson, D., and D. Jones. 1995. True wishes: The philosophy and developmental psychology of children's informed consent. *Philosophy, Psychiatry, & Psychology* 2 (4): 287–303.
Diekema, D. 2004. Parents refusal of medical treatment: The harm principle as threshold for state intervention. *Theoretical Medicine and Bioethics* 25:243–264.
Duncan, R., and S. Sawyer. 2010. Respecting adolescents' autonomy (as long as they make the right choice). *The Journal of Adolescent Health: Official Publication of The Society for Adolescent* 47:113–114.
Eisler, I. 2005. The empirical and theoretical base of family therapy and multiple family day therapy for adolescent anorexia nervosa. *Journal of Family Therapy* 27:104–131.
Eisler, I., J. Lock, and D. le Grange. 2010. Family-based treatment for adolescents with anorexia nervosa: Single-family and multifamily approaches. In *The treatment of eating disorders: A clinical handbook*, ed. C. Grilo and J. Michell. NewYork: The Guilford Press.
Elliott, C. 1997. Caring about risks: Are severely depressed patients competent to consent to research? *Archives of General Psychiatry* 54:113–111.
Ellison, A., and J. Fong. 1998. Neuroimaging in eating disorders. In *Neurobiology in the treatment of eating disorders*, ed. H. Hoek, J. Treasure, and M. Katzman, 255–269. Chichester: Wiley.
Erikson, E. H. 1968. Identity, youth, and crisis. London: Faben.
F.E.A.S.T. 2012. http://feast-ed.org/TheFacts/DoParentsCauseEatingDisorders.aspx. Accessed 23 Aug 2012.
Federici, A., and L. Wisniewski. 2012. Integrating dialectical behavior therapy and family-based treatment for multidiagnostic adolescent patients. In *A collaborative approach to eating disorders*, ed. J. Alexander, J. Treasure. New York: Routledge.
Fiester, A. 2012. What "Patient-Centered Care" requires in serious cultural conflict. *Academic Medicine: Journal of the Association of American Medical Colleges* 87:20–24.
Forman, E., and R. Ladd. 1995. *Ethical dilemmas in pediatrics: A case study approach*. University Press of America
Gilbert, S., and P. Burgess. 2008. Executive function. *Current Biology* 18:R110–114
Grisso, T., and P. Appelbaum. 1998. Assessing competence to consent to treatment: A guide for physicians and other health professionals. London: Oxford University Press.
Grisso, T., and P. Appelbaum. 2006. Appreciating anorexia: Decisional capacity and the role of values. *Philosophy, Psychiatry, and Psychology* 13 (4): 293–297.
Haleem, D. 2012. Serotonin neurotransmission in anorexia nervosa. *Behavioural Pharmacology* 5:478–495.
Herpertz-Dahlmann, B., J. Seitz, and K. Konrad. 2011. Aetiology of anorexia nervosa: from a "psychosomatic family model" to a neuropsychiatric disorder? *European Archives of Psychiatry and Clinical Neuroscience* 261 (Suppl. 2): S177–181.
Hoagwood, K. 2005. Family-based services in children's mental health: A research review and synthesis. *Journal of Child Psychology and Psychiatry, and Allied Disciplines* 46 (7): 690–713
Hudson, J., E. Hiripi, H. Pope, et al. 2007. The prevalence and correlates of eating disorders in the National Comorbidity Survey Replication. *Biological Psychiatry* 61:348–358.

Kamel, N., T. Chami, A. Anderson, et al. 1991. Delayed gastrointestinal transit times in AN and bulimia nervosa. *Gastroenterology* 1011:11320–11324
Katzman, D., E. Lambe, D. Mikulis, et al. 1998. Cerebral gray matter and white matter volume deficits in adolescent girls with anorexia nervosa. *The Journal of Pediatrics* 129:794–803.
Katzmann, D., R. Zipursky, E. Lambe, et al. 1997. A longitudinal magnetic resonance imaging study of brain changes in adolescents with anorexia nervosa. *Archives of Pediatrics & Adolescent Medicine* 151:793–797.
Kaye, W., J. Fudge, and M. Paulus. 2009. New insights into symptoms and neurocircuit function of anorexia nervosa. *Neuroscience* 10:573–584.
Kelly, M. 2010. Case study commentary. *Hastings Center Report* 40 (6):12.
Le Grange, D., and I. Eisler. 2008. Family interventions in adolescent anorexia nervosa. *Child and Adolescent Psychiatric Clinics of North America* 18:159–173.
Lock, J., D. Le Grange, W. Agras, and C. Care. 2001. *Treatment manual for anorexia nervosa: A family-based approach*. New York: Guilford Press
Mainz, V., M. Schulte-Ruther, G. Fink, et al. 2012. Structural brain abnormalities in adolescent anorexia nervosa before and after weight recovery and associated hormonal changes. *Psychosomatic Medicine* 74:574–582.
Mehler, P., L. Birmingham, S. Crow, et al. 2010. Medical complications of eating disorders. In *The treatment of eating disorders: A clinical handbook*, ed. C. Grilo and J. Mitchell, 66–82. New York: The Guilford Press.
Meyers, D. 2000. Intersectional identity and the authentic self? Opposites attract! In *Relational autonomy: Feminist perspectives on autonomy, agency and the social self*, ed. C. MacKenzie and N. Stoljar, 151–180. London: Oxford University Press.
Minuchin, S., B. Rosman, and L. Baker. 1978. *Psychosomatic families: Anorexia nervosa in context*. Cambridge: Harvard University Press.
Misra, M. 2008. Long-term skeletal effects of eating disorders with onset in adolescence. *Annals of the New York Academy of Sciences* 1135:212–218.
PalazolliSelvini, M. 1996. *Self-starvation: From individual to family therapy in the treatment of anorexia nervosa* (*The master work series*). New Jersey: Jason Aronson Inc.
Papadopoulos, F., A. Ekbom, L. Brandt, et al. 2009. Excess mortality, causes of death and prognostic factors in anorexia nervosa. *The British Journal of Psychiatry: The Journal of Mental Science* 194:10–17.
Paul, M. 2004. Decision-making about children's mental health care: Ethical challenges. *Advances in Psychiatric Treatment* 10:301–311.
Paul, M. 2007. Rights. *Archives of Disease in Childhood* 92:720–725.
Pollice, C., W. Kaye, C. Greeno, et al. 1997. Relationship of depression, anxiety, and obsessionality to state of illness in anorexia nervosa. *International Journal of Eating Disorder* 21:367–376.
Quill, T., H. Brody. 1996. Physician recommendations and patient autonomy: Finding a balance between physician power and patient choice. *Annals of Internal Medicine* 125 (9): 763–369.
Rhodes, R., and I. Holzman. 2004. The not unreasonable standard for assessment of surrogates and surrogate decisions. *Theoretical Medicine* 25:367–385.
Rosling, A., P. Sparén, C. Norring, et al. 2011. Mortality of eating disorders: a follow-up study of treatment in a specialist unit 1974–2000. *International Journal of Eating Disorder* 44:304–310.
Ross, L. 1997. Health care decision making by children: Is it in their best interest? *The Hastings Center Report* 27 (6): 41–45.
Ross, L. 2009. Against the tide: Arguments against respecting a minor's refusal of efficacious life-saving treatment. *Cambridge Quarterly of Health Care Ethics* 18 (3): 302–315
Silverman, J. 1997. Anorexia nervosa: Historical perspectives on treatment. In *Handbook of treatment for eating disorders*, ed. P. Garner and P. Garfinkel. 2nd ed. New York: The Guilford Press.
Skårderud, F., and B. Sommerfeldt. 2009. Self-harm and eating disorders. *Tidsskrift for den Norske laegeforening* 129:877–881.
Steinhausen, H. 2002. The outcome of anorexia nervosa in the 20th Century. *The American Journal of Psychiatry* 159:1284–1293.

Strüder, H. K., and H. Weicker. 2001. Physiology and pathophysiology of the serotonergic system and its implications on mental and physical performance. Part I. *International Journal of Sports Medicine* 22:467–481.

Takakura, S., T. Nozaki, Y. Nomura, et al. 2006. Factors related to renal dysfunction in patients with anorexia nervosa. *Eating and Weight Disorders: EWD* 11:73–77.

Tan, J., T. Hope, and A. Stewart. 2003. Competence to refuse treatment in anorexia nervosa. *International Journal of Law and Psychiatry* 26 (6): 697–707

Tan, J., A. Stewart, R. Fitzpatrick, and R. Hope. 2006. Competence to make treatment decisions in anorexia nervosa: Thinking processes and values. *Philosophy, Psychiatry, and Psychology* 13 (4): 267–282.

Tan, J., G. Passerini, and A. Stewart. 2007. Consent and confidentiality in clinical work with young people. *Clinical Child Psychology and Psychiatry* 12 (2): 191–210

Teng, K. 2011. Premenopausal osteoporosis, an overlooked consequence of anorexia nervosa. *Cleveland Clinic Journal of Medicine* 78:50–58.

Von Holle, A., A. Pinheiro, L. Thornton, et al. 2008. Temporal patterns of recovery across eating disorder subtypes. *The Australian and New Zealand Journal of Psychiatry* 42:108–117.

Wagner, A., P. Greer, U. Bailer, et al. 2006. Normal brain tissue volumes after long-term recovery in anorexia and bulimia nervosa. *Biological Psychiatry* 59:291–293.

Weithorn, L., and S. Campbell. 1982. The competency of children and adolescents to make informed treatment decisions. *Childhood Development* 53 (6): 1589–1598

White, M., and D. Epston. 1990. *Narrative means to therapeutic ends*. Adelaide: Dulwich Centre.

Winston, A., A. Alwazeer, and M. Bankart. 2008. Screening for osteoporosis in anorexia nervosa: Prevalence and predictors of reduced bone mineral density. *International Journal of Eating Disorder* 41:284–287.

Yager, J., and A. Anderson. 2008. Clinical practice: Anorexia nervosa. *New England Journal of Medicine* 353:1481–1488.

Zuo, D., A. Houtrow, P. Arrango, K. Kuhlthau, J. Simmons, and J. Neff. 2012. Family-centered care: Current applications and future directions in pediatric health care. *Maternal and Child Health Journal* 16:297–305.

Chapter 17
Lost in Transition: Legal and Ethical Approaches When Applying Patient- and Family-Centred Care to Adolescents with Disabilities

Maria L. Z. McDonald and Lucie Wade

17.1 Introduction

One noticeable weakness when applying a patient- and family-centred care model (P&FCC) is the difficulty in finding a resolution when there is a direct conflict between the wishes of a child or adolescent and the wishes of the parent(s).

In this chapter we examine the legal and ethical obligations of health practitioners, adolescent patients and the family to support decision-making in times of conflict. We propose that health practitioners should understand human rights and healthcare law in order to assist adolescents in protecting their rights to decision-making and privacy. Further, we seek to build upon the understanding of clinical practice as ethically complex (Langlois 2012) and explicitly discuss another layer of complexity that is present in the rehabilitation setting. Specifically, health practitioners, patients, and families who work and receive care in the rehabilitation setting do so within a broader social context that requires them to recognize and address important ethical issues related to living with a disability. It is important to consider contextual factors that impact life with a disability as an integral part of ethical decision-making, not only because these factors affect almost every individual who will receive care in the rehabilitation setting, but because bioethics has a history of failing to adequately recognize, address, and integrate disability perspectives in decision-making and argumentation (Kuczewski 2001).

To facilitate consideration of these factors, this chapter explores a composite case arising from our experience working in the paediatric rehabilitation setting. This scenario provides context for our discussion of how a model of P&FCC should be applied to the benefit of patients when tension exists between values. First we provide background on P&FCC and the law, specifically considering implications

M. L. Z. McDonald (✉) · L. Wade
University of Toronto Joint Centre for Bioethics, Toronto, ON, Canada
e-mail: mzmcdonald@rogers.com

L. Wade
Jewish General Hospital, Montreal, QC, Canada
e-mail: lwade@jgh.mcgill.ca

R. Zlotnik Shaul (ed.), *Paediatric Patient and Family-Centred Care: Ethical and Legal Issues,* International Library of Ethics, Law, and the New Medicine 57,
DOI 10.1007/978-1-4939-0323-8_17,

for child and adolescent decision-making. Next we explore the impact of social context on the rehabilitation setting and propose more explicit integration of disability studies perspectives in P&FCC. Finally, we return to the case, relying on an application of a robust model of P&FCC that incorporates legal and human rights standards as well as disability studies perspectives to facilitate who makes the decision and how the decision is made.

17.1.1 Case Scenario

Sasha, a 14-year-old girl who has recently immigrated to Toronto with her family, has received a referral for an assessment for a portable communication device. Sasha is able to make sounds but has difficulty forming words, a symptom attributable to cerebral palsy. Sasha uses a power wheelchair most of the time because she finds it tiring to walk everywhere.

The speech language pathologist (SLP) has recommended a large communication device that is easy for Sasha to use and sits on the tray of her wheelchair. The device allows Sasha to track with her eyes to locate words because her hand tremors prevent her from using a keyboard when she is tired. Unfortunately, while the size of the communication device enables efficient eye tracking, it blocks her facial expressions and does not allow her to have eye contact with her peers when communicating.

Sasha's mother, Diana, takes great care to ensure Sasha fits in with her social circle. She helps her dress nicely in recent fashions and jewelry, and paints her nails. Sasha's mother is concerned that Sasha's classmates and people in their community will ostracize her because this device blocks her face. In a discussion with the SLP she stated: "Sasha is not retarded![1] She doesn't need this big device. It just takes her longer to say things and with more practice she will be faster. What she needs is to have friends!"

Diana has asked the SLP to recommend government funding for a computer tablet for Sasha instead of the larger communication device. The tablet won't block her face, won't interfere with driving her power chair and would better help her fit in with the kids at school and in the community.

The SLP was caught off guard by Diana's impassioned reaction to the device. She hasn't worked with Sasha's family before, and feels it is her responsibility to do what the family asks. Many families see the tablet as a solution, and many kids

[1] We recognize and appreciate that this is a derogatory and violent term and do not endorse its use. For the purposes of our discussion regarding the impact of social context on the experience of health practitioners, patients and families we feel it is necessary to include this term here as an explicit example of how disability discrimination infiltrates clinical settings and decision-making. Following Magasi (2008a), elsewhere in our chapter we have made the conscious decision to use both 'person-first' terminology, e.g., people with disabilities and the term 'disabled people,' to reflect the important work of disability activists and scholars. Further explanation of why both terms are considered respectful can be found in 17.3.1.

request it, but the SLP knows that it is not designed to address Sasha's specific communication concerns.

Further, Sasha has conveyed to the SLP during a private moment that she doesn't care so much about her face being seen; rather she is worried that the words on the communication device are 'for babies.' After some probing, the SLP discovers that the kids in Sasha's class use swear words frequently and often talk about sex. Sasha wants to be able to join in the conversations and wants to use words that her friends use, for example, awesome, wicked, sweet, as well as a few swear words. She doesn't want her mother to know about all these new words, as her mother would be upset.

17.2 Patient- and Family-Centred Care and the Law

When we refer to P&FCC, we generally invoke the core concepts of dignity and respect; information sharing; participation; and collaboration as outlined by the Institute for Patient- and Family-Centered Care.[2] Importantly, P&FCC emphasizes the concepts of respect for the adolescent as well as partnering with the adolescent and family to provide care. Many adolescents still rely on family members to remember details of their medical history, to assist in getting to appointments, and to remember to take medications or do therapies. The concept of partnering also reminds us that the adolescent is influenced by family well-being when making decisions (Sine and Sharpe 2011; Harrison 2010). The complexity of the day to day triadic relationship between adolescent, family and healthcare team (Harrison et al. 1997) often results in confusion and losing focus on the spirit of P&FCC, especially on the important concepts of respect for the adolescent and partnering with adolescents and families.

While these core concepts allow a rich consideration of ethical principles, the practice of upholding all these values is challenging. A P&FCC approach becomes especially difficult when the adolescent and the parent(s) (or other family members) disagree on what to do. To deal expeditiously with such challenges, health practitioners may feel they should do everything the adolescent or family wishes, or feel pressured to do what the most strident voice in the room demands (Fine 2010). Kenny et al. (2008, p. 123) warn that:

> Simply complying with the parents' wishes in such cases is inadequate. Furthermore, the family-centred approach must not be taken to allow family members' interests to trump the child's interests. Rather it must be seen as recognizing the fact that children are embedded in their families and the interests of the child can be seen as bound up with the interests of the other family members. The child's interests must always be the basis for a decision to be followed by the healthcare team. This approach does not discount the parents' concerns and authority but it does recognize the child (albeit as a member of a family) as the particular patient to whom the healthcare team has a primary duty of care.

[2] A more extensive discussion of P&FCC is found in Part 1 of this book.

Thus, it is important for health practitioners to understand the distinctions between P&FCC and patient- and family-*directed* care. P&FCC does not change the fact that the continuing responsibility of the health practitioner is to recommend the most appropriate treatments and alternatives to the appropriate decision-maker, to seek to harmonize the values of everyone involved in the decision (Kenny et al. 2008) and to keep in mind the rights of the adolescent.

Neither case law nor legislation provides any assistance in understanding or applying P&FCC as a model, as we are not aware of any cases dealing with P&FCC specifically. Should a challenge be brought to court by a patient or family, the court will however be influenced in its considerations of P&FCC by any policies, practices and guidelines developed to interpret and implement this model (Dickens 1997, 1994). There is a wealth of jurisprudence relating to the obligation to respect persons and their healthcare decisions. However, health practitioners often express dissatisfaction with the law, saying that it does not help them resolve challenging situations nor does it provide a clear answer on what to do next. We consider the law to represent the minimal ethic "meaning that there is an ethical obligation to keep the law but that many decisions, such as whether or not to invoke a legal power, are based on ethics and not simply on law" (Dickens 1994, p. 306). We outline the law relating to decision-making here to provide the legal framework for respecting the autonomy of adolescents, and protecting their human rights in healthcare decision-making. By integrating legal standards and ethical principles within a model of P&FCC we strive to raise the 'minimal ethic' of common law and legislation, and suggest options that advance the rights and values of adolescents.

17.2.1 General Legal Concepts Related to Decision-Making

The law relating to decision-making in healthcare is supported by the underlying ethical principles of respect for persons, beneficence, non-maleficence and protection of human rights (Dickens 1994) and forms the basis for the P&FCC concepts of dignity and respect.

Many countries, including Canada and the United States, recognize that each person has the right to life, liberty and security and have adopted the United Nations Universal Declaration of Human Rights when it was first drafted in 1948 (UDHR 1948, article 3). Adopting countries accept their obligations under this declaration, and reference these obligations in subsequent international treaties, regional human rights instruments, national constitutions and laws to promote respect for these rights and freedoms. The UDHR is the driving force behind the *Canadian Charter of Rights and Freedoms* (Charter 1982), as well as the *Health Care Consent Act* of Ontario (HCCA 1996) which relates specifically to decisions about healthcare and includes the topics of consent and capacity and substitute decision-making.[3] The

[3] In this chapter we will use the term 'capacity' to remain consistent with the terminology used in the Health Care Consent Act although the term 'competence' is also used in literature and statute law to refer to the ability to make healthcare decisions.

HCCA is formulated from common law, reflects human rights obligations, and is representative of the approach taken by many provinces across Canada and by other Commonwealth countries.

Both case law and legislation clearly set out that only one person makes the health-care decision (that is either the patient or the substitute decision-maker(s)) (HCCA 1996; Gillick v West Norfolk 1986; C et al v Wren 1986; Van Mol v Ashmore 1999). However "[l]aws tend not to address the peculiarities of particular circumstances as sensitively as ethical assessments and judgements may" (Dickens 1994, p. 307). In practice, health practitioners will refer to the law to confirm who is the ultimate, legal decision-maker but prefer to facilitate a shared decision between the adolescent and parent(s) to harmonize the values of all participants (Harrison et al. 1997; Kenny et al. 2008) and to ensure cooperation and good relations with all. There are times, however, when there remains a discord between the adolescent and parent(s), and health practitioners need to be attentive as to who is the person in law who makes the treatment decision and to respectfully, clearly communicate this with all participants.

Foundationally, the HCCA asserts that treatment may not be administered unless the person has given consent and the health practitioner is of the opinion that the person is capable. Or, if the person is incapable, his or her substitute decision-maker must provide consent to the treatment (HCCA 1996, s. 10(1)).

Consent Consent is valid if all the following elements are fulfilled: (1) consent must relate to the treatment; (2) consent must be informed; (3) consent must be given voluntarily; and (4) consent must not be obtained through misrepresentation or fraud.

The consequences of not carefully following this law could include a civil action against the health practitioner. If the consent of a capable person or the substitute decision-maker is not obtained prior to treatment, the health practitioners could be found to have committed the civil tort of battery (Malette v Shulman 1990). Providing treatment to a patient after obtaining inadequately informed consent constitutes negligence (Reibi v Hughes 1980).

Informed decision-making is more than just giving a piece of paper to the patient and family to explain the treatment or service. It is the sharing of information and discussion that helps a person, often with the support of family, make an informed decision. A person makes an informed decision (to consent or refuse) following his or her own beliefs, values and standards. At times a decision is criticized as a poor medical decision; however these are personal decisions whether made by an adult or an adolescent (Dickens 1994; Harrison et al. 2004). So long as a person is capable, the person maintains the right to make a poor decision for his or her own reasons, and bears the consequences of such a decision (C (Re) 2009).

Criteria for Capacity It is important to note that laws around capacity focus on decisional capacity rather than functional capacity. That is, the test for capacity is a cognitive one (Gilmour 2002), and not one based on whether or not the person will have the ability to follow through on a decision.

Ontario legislation does not set an age requirement for decision-making about healthcare treatment decisions, rather a person may make a decision if he or she is 'capable,' that is if the person is (HCCA 1996, s. 4):

1. able to understand the information provided relating to the treatment, *and*
2. able to appreciate the consequences of making or not making a decision.

Health practitioners are permitted to presume that a person, regardless of age, is capable to make the specific decision at hand unless there are reasonable grounds for suspecting that a person is not capable, for example, a very young age, severe intoxication, etc. (HCCA 1996, s. 4(2)). In addition, a person may be incapable with respect to some treatments and capable with respect to others as well as incapable with respect to a treatment at one time and capable at another time (HCCA 1996, s. 15). It is not reasonable to presume a person is incapable merely by reason of:

- age
- a physical or intellectual disability
- a psychiatric or neurological diagnosis
- a difficulty in communicating or
- a difficulty in determining capacity.

The definition explicitly requires the 'ability' to understand and appreciate. Health practitioners are not required to be prepared to convince a tribunal or court that the person 'actually' understood the information and 'actually' appreciated the consequences, rather it is sufficient to establish that the person had the ability to do so. The legal burden of establishing incapacity is on the health practitioner when challenged during a court or tribunal proceeding, and the level of proof is that of the civil standard of 'balance of probabilities' (G H (Re) 2012).[4]

Determining Capacity When it is unreasonable to presume capacity, then capacity must be evaluated (HCCA 1996, s. 4(3); Etchells et al. 1996b). The determination of capacity is part of the consent process, and does not occur in a linear fashion after the consent discussion is completed because a health practitioner is continually evaluating what information was understood, what has to be re-explained and whether the person has the ability to understand and appreciate the consequences of a particular decision. Since the capacity of a person may fluctuate depending on the time of day, the types of medications received, the illness, etc., a person may become capable after being determined to be incapable. Thus, health practitioners should re-evaluate the person to confirm whether the person has capacity (HCCA 1996, s. 15) whenever there are reasonable grounds to suspect that there has been a change in capacity.

While the law sets out the criteria for capacity, there is no single test for determining whether capacity exists; evaluation is based largely on clinical judgment. To our knowledge, professional schools do not teach or discuss this skill in a detailed

---

[4] The civil standard of proof is also described as 'more likely than not,' 'greater than 50%', or 'probably not possibly.' The civil standard of proof is much lower than the criminal standard of 'beyond a reasonable doubt.'

manner; professional regulatory bodies provide very little guidance to members; and hospital practices rarely include assistance on which tools are useful in determining capacity. How to perform a good evaluation of capacity is still vigorously debated in the literature and the consensus amongst authors is that more research is needed to develop a standardized tool for quick and reliable evaluations of capacity to increase the consistency and accuracy of health practitioners' judgments (Larcher and Hutchinson 2010; Sessums et al. 2011). After reviewing 19 different instruments for evaluating capacity, Sessums et al. concluded that the Aid to Capacity Evaluation (ACE) (Etchells and Darzins 1999) was the best available instrument to assist health practitioners in ensuring a consistent, defensible and accountable evaluation.[5]

Process If Found Incapable In cases where the person is determined to be incapable, consent is sought from a designated substitute decision-maker (Etchells et al. 1996b; HCCA 1996, s. 20).

Ontario legislation also sets out procedural steps which must be fulfilled to ensure that the human rights of a person are protected (HCCA 1996, s. 17, 18; Hesson and Bakal 1993; B (Re) 2008). These steps include notice that the person has been found incapable, information about who has been determined to be the substitute decision-maker, the right of the person to request a review of the decision of incapacity or the determination of the substitute decision-maker, and the right of the person to request a different substitute decision-maker. If the person proposes to appeal a decision of incapacity or challenge the determination of the substitute decision-maker, members of the healthcare team have an ethical obligation to assist the person since the person may not have the energy and ability to challenge the decisions.

No treatment may begin until the hearing has been held or 48 hours elapse from the time the healthcare team became aware that the person wished to challenge the decision. In Ontario, this review is performed by a specialized tribunal called the Consent and Capacity Board.[6] In other jurisdictions the courts hear these cases.

Role of the Substitute Decision-Maker The role of the substitute decision-maker (SDM) is to take into consideration the expressed capable wishes of a person 16 years old or older, if they exist. If there is no prior expressed capable wish relevant to the decision at hand, or the person is not yet 16 years old, the SDM must act in the person's best interests (HCCA 1996, s. 21). The SDM is usually a parent or other family member (Etchells et al. 1996b; see HCCA 1996, s. 20 for the hierarchy of substitute decision-makers).

---

[5] Sessums et al. (2011, p. 426) found that the ACE was validated in the largest study, "was the only instrument evaluated against a gold standard, with an acceptable Rational Clinical Examination level-of-evidence score and robust test characteristics, that can be performed in less than 30 minutes, is available for free online, and includes training materials. Moreover, the ACE is based on the actual decision the patient is facing. Most of the other instruments use a clinical vignette, violating the tenet that capacity assessment is specific rather than generic."

[6] See the Consent and Capacity Board's website for detailed information about the tribunal and its mandate, at www.ccboard.on.ca.

17.2.2 Law Specific to Adolescents

International declarations form the foundation for the interpretation of laws related to adolescents and advance their human rights. The United Nations Convention on the Rights of the Child (UCRC 1989), article 12 requires us to assure "the child who is capable of forming his or her own views the right to express those views freely in all matters affecting the child."

More recent international advocacy on behalf of persons with disabilities has resulted in the United Nations Convention on the Rights of Persons with Disabilities (UCRPD 2006, art 7) which confirms that signatory countries:

> shall ensure that children with disabilities have the right to express their views freely on all matters affecting them, their views being given due weight in accordance with their age and maturity, on an equal basis with other children, and to be provided with disability and age-appropriate assistance to realize that right.

This declaration also reaffirms that every human being has the inherent right to life (UCRPD 2006, art 10). The emphasis on right to life is reflected in the Canadian Charter of Rights and Freedoms (Charter 1982), Sect. 7 and is discussed in more detail below.

Adolescents[7] are in transition between relying on others to make decisions and becoming leaders of their own healthcare (Kieckhefer and Trahms 2000; Gall et al. 2006). As an adolescent becomes more mature, he or she is better able to understand the possible harms, benefits and outcomes and may be able to make more medical decisions. Some adolescents need or desire help from their family and healthcare team to make a decision. Adolescents experience fluctuations in their developing decision-making capacities depending on the complexity of the decision at hand, and healthcare teams need to be prepared to re-assess capacity frequently (Gilmour 2002).

In addition, adolescents often struggle against their parents who may mistrust the adolescent's ability to make sound choices or may feel obligated to continue making the decisions as part of fulfilling their parental duties (Dickens and Cook 2005). Some jurisdictions outside Ontario may have a specified age of consent but all apply the 'mature minor' rule to recognize the right of capable adolescents to make decisions about healthcare (Dickens and Cook 2005; Harrison et al. 2004; Gilmour 2002; Etchells et al. 1996b). The 'mature minor' rule refers to the common law's development of a concept which recognizes situations where the adolescent is capable to make healthcare decisions independently, for example, when minors are no longer dependent on parents, and working and living on their own, and more prevalently when the minors' decisional capacity is similar to that of a mature person. Thus, in every Canadian common law jurisdiction, a capable adolescent has the right to make his or her own decisions about healthcare.

---

[7] The World Health Organization defines 'adolescents' as young people aged 10–19 years (WHO 2002).

The evolving capacities of the adolescent are also recognized in the U.N. Convention on the Rights of the Child, which specifically points out in article 5 the limitations on parental responsibilities and rights (Dickens and Cook 2005). The article states:

> States Parties shall respect the responsibilities, rights and duties of parents or, where applicable, the members of the extended family or community as provided for by local custom, legal guardians or other persons legally responsible for the child, to provide, in a manner consistent with the evolving capacities of the child, appropriate direction and guidance *in the exercise by the child* of the rights recognized in the present Convention. [Emphasis added]

Together, these international declarations provide the foundation for how we should behave to protect the human rights, including healthcare decision-making rights, of adolescents even when statutes fail to give adequate guidance.

Potential Charter Challenges An adolescent seeking to protect his or her healthcare decision-making rights may consider a legal challenge to the validity of the application of one or more provisions of the HCCA. While the HCCA incorporates the human rights obligations set out in international declarations, interpretations of how to apply the HCCA may vary amongst institutions and health practitioners. Inaccurately interpreting or applying the HCCA could lead to two types of Charter challenges by adolescents: under either Sect. 7 (security of the person) or Sect. 15 (equality rights). The adolescent would also have to establish that the limitation to the right is not justified under Sect. 1 of the *Charter* in order to convince the court that the relevant parts of the statute are unconstitutional and should be struck down as invalid.

A Charter argument under Sect. 7 would begin with the proposition that every capable person, irrespective of age, has the right to determine what shall, or shall not, be done with one's own body, and to be free from non-consensual medical treatment. Fleming v Reid (1991) identified that this right to bodily integrity and autonomy is deeply rooted in our common law. The court in Fleming v Reid observed this right warrants the 'highest order' of constitutional protection under Sect. 7 of the Charter as a right to security of the person, only to be denied according to the principles of fundamental justice under Sect. 1 of the Charter (Garton 2005, p. 7). The case involved a patient with mental illness who successfully challenged the provision in the Ontario Mental Health Act authorizing a review board to override an involuntary patient's competent refusal to take anti-psychotic drugs, as expressed by the patient through his or her substitute decision-maker.

The Sect. 7 Charter challenge would then assert that everyone involved in the healthcare of adolescents has a duty pursuant to the HCCA to protect their rights and freedoms as they do for adults. The right to 'security of the person' set out in Sect. 7 may be used to enforce the fundamental rights and freedoms of adolescents relating to medical decision-making (Hesson and Bakal 1993). While there is no case law on this point relating to the HCCA, there are compelling arguments which could be made in the right factual scenario. This possibility is reinforced by the recent case of Rasouli v Sunnybrook (2011 at para. 36) where the Ontario Court of

Appeal, referring to the HCCA, acknowledged, "as per this court's decision in the Act should be construed in a manner consistent with the Charter." It could be argued that health practitioners employed by a hospital (which receives substantial government funds) are agents of the government, must abide by the Charter, and are under a duty to ensure that capable adolescents have the opportunity to decide about the treatments or services proposed. If health practitioners fail to interpret the HCCA in a way that allows capable adolescents to have a say about the treatment they receive, it may be argued that the HCCA has *not* been construed in a manner consistent with the Charter. The courts considering this challenge to the constitutionality of the HCCA would then decide whether or not the provision is saved under Sect. 1 of the Charter which allows a limitation on an individual's right if, among other things, the limitation is reasonable and justifiable (Garton 2005, p. 1).

The second way the Charter may arguably be infringed is where delegated decision-makers (for example, a hospital) use government funds in a discriminatory way contrary to the Sect. 15 of the Charter. The case of Eldridge v. Attorney General British Columbia (1997) involved an application under Sect. 15 (equality rights) by patients who are deaf challenging the failure to provide sign language interpretation during the provision of healthcare to allow effective communication. The court found that there are positive obligations to allocate resources reasonably; that is, positive steps must be taken to ensure that disadvantaged groups benefit equally from services offered to the general public subject to reasonable accommodation. The unanimous Supreme Court of Canada found that the violation of the Charter occurred not in the legislative provisions but in the exercise of discretionary decision-making by government agents. The exercise of discretionary decision-making could be challenged if adolescents are not supported in the same way as adults in hospitals. For example, hospitals developing informational brochures and making available supports aimed at assisting adults to pursue their right to make healthcare decisions may also be under a duty to provide similar information and supports to adolescents. These hospitals may be perceived as using public health funds to help adults to the exclusion of adolescents. Further, hospitals may also be under a positive duty to assist adolescents with disabilities to make their own healthcare decisions, to the extent it is reasonable to do so.

No Legal Duties to Parents Health practitioners may be concerned that they have a duty to parents and should not provide care that parents oppose. However, there is no legal requirement to obtain parental consent or advise parents prior to treating capable adolescents, even if the adolescent's choice is not supported by the medical team (HCCA 1996). Both "adults and adolescents capable of autonomy must also bear the consequences of choices they make that are not in their best interests, and may frustrate or damage their best interests" (Dickens and Cook 2005, p. 180). In addition, offering care only on the condition that parents are notified creates the risk that the confidentiality of the adolescent's personal health information may be breached pursuant to the Personal Health Information Protection Act of Ontario (PHIPA 2004). Adolescents capable of making choices about medical care are likely capable of deciding with whom to share their personal health information because the capacity needed to understand how their information is used and shared is usually lower than that needed to make healthcare decisions (Gillick v West Norfolk 1986; PHIPA 2004, s. 21).

Capacity Determination of Adolescents Health practitioners find it especially challenging to determine capacity of adolescents. Ross' concerns (2003, p. 194) remain valid:

> … since there are no criteria on which to base maturity or decision-making capacity, the decision of whether or not a child has decision-making capacity is dependent upon the judgement of the particular pediatrician—a judgement for which he or she has no training.

The Appendix sets out a framework for determining capacity in adolescents, largely based on the ACE, and developed at Holland Bloorview Kids Rehabilitation Hospital in Toronto (McDonald et al. 2010; Etchells and Darzins 1999). Larcher and Hutchinson (2010) and MacKean et al. (2012) suggest health practitioners also have an ethical duty to enhance capacity by using appropriate techniques, like: facilitating complex decision-making by breaking the process down into smaller but linked choices or considering inviting a third party, such as a trusted adult or peer, to help and support the adolescent.

Like adults, whether an adolescent has capacity to make a decision depends on how complicated the treatment plan is, and the kinds and amounts of benefit and harm.[8] For many routine treatments, an adolescent will have the capacity to decide when information is given in an understandable way, and he or she is helped in the decision-making process. It may also depend on how the adolescent is feeling, whether the adolescent is on medications, whether the adolescent is in pain, etc. When in doubt, capacity should be evaluated.

Adolescents may have a more developed decision-making capacity than functional capacity (the ability to organize, remember and carry out daily activities). Health practitioners need to differentiate between these two abilities, and remember that decision-making capacity is the only criterion considered in law for healthcare decisions (HCCA 1996; Dickens and Cook 2005; Gilmour 2002).

In Canadian and English law, the definition of capacity does not include a demonstration of rationality (Larcher and Hutchinson 2010; HCCA 1996). However, in practice, health practitioners often strive to understand whether the reasons for a decision by an adolescent appear rational, thought out and compatible with the adolescent's values, especially when the adolescent refuses treatment. This technique may enhance the adolescent's feeling that whatever the outcome, he or she is being heard and is valued for him or herself (Larcher and Hutchinson 2010).

Even when the adolescent is not capable, health practitioners should continue to show respect for the adolescent in the decision-making process by including the adolescent as much as appropriate in the discussions (Kenny et al. 2008; Harrison et al. 2004). Health practitioners should also remain attentive to the possibility that the adolescent may be capable to decide which substitute decision-maker he or she

[8] The courts have distinguished cases where the adolescent refuses life-saving treatment. In such cases, a higher level of understanding and appreciation is required. See discussion by Gilmour (2002, 214 et seq.).

prefers even if not capable to decide about treatment. Under the Substitute Decisions Act of Ontario (SDA 1992), the criteria for capacity are set out in Sect. 47(1):

A person is capable of giving a power of attorney for personal care if the person,

a. has the ability to understand whether the proposed attorney has a genuine concern for the person's welfare; and
b. appreciates that the person may need to have the proposed attorney make decisions for the person.

Since these criteria are easier to meet in most circumstances, the adolescent will often have capacity to appoint someone to make personal care, including healthcare, decisions on his or her behalf, enabling some control over decision-making.

In summary, the important concepts of dignity and respect in P&FCC are based on the ethical principle of autonomy and must be honoured whether the person is an adolescent or adult, and whether the person is disabled or typically developing. Understanding the legal duties owed to an adolescent with disabilities when dealing with complex consent and capacity issues will assist all participants in better knowing how to protect human rights even if the adolescent and family do not agree on what to do. Furthermore, this increased knowledge and sensitivity will assist health practitioners to better follow and demonstrate the core concepts of P&FCC: dignity and respect; information sharing; participation; and collaboration to all participants involved in healthcare decisions.

17.3 Added Complexity in the Rehabilitation Setting

While P&FCC is well-suited to the paediatric setting, the rehabilitation setting presents an added layer of complexity that is not explicitly acknowledged or addressed in P&FCC. In the rehabilitation setting there is a need to explicitly address the influence that disability—specifically the adolescent and family's experience with and understanding of disability—may have on treatment decisions and goals.[9] Explicitly acknowledging that families are affected by dominant social norms related to disability complements the P&FCC mission to improve care by acknowledging that adolescents are affected by their family (Harrison 2010): if adolescents are affected by their family, they are also affected by their family's experience and understanding of disability. Acknowledging the impact that social context related to disability may have on the family, and hence the adolescent, can enable health practitioners

[9] The influence of social norms on decision-making may arise as a potential contextual factor during a consideration of diversity and cultural values within P&FCC, yet the possibility that a dominant view of disability may negatively impact goals of care cuts across cultures. Due to its breath and potential for negative influence on the well-being of the child and his or her family, P&FCC applied within the rehabilitation setting should include a mechanism to help health practitioners recognize and address issues related to disability and social norms as they arise within the therapeutic setting.

to better understand the context informing decision-making and thus enable a more appropriate and tailored process of examining values and assumptions that may be affecting treatment goals and decisions and, ultimately, influencing the adolescent's well-being. The call to integrate an awareness of disability into the rehabilitation professions is not new (see Magasi 2008a, b; Oliver et al. 2012 for an overview). Here we extend this call specifically to P&FCC.

Our position on the importance of including an understanding and exploration of the influence of disability within P&FCC is informed by critical disability ethics scholarship. The emerging field of critical disability ethics reflects the interest of and need for bioethicists to attend to and learn from issues of disability studies, and to integrate this knowledge to inform health practitioners and change healthcare practice (Kuczewski 2001). Recently described, critical disability ethics "asserts that policies and practices responsible for systematically excluding disabled people from participating in social life are morally wrong and must be addressed on multiple fronts" (Gibson et al. 2012, p. 212). The following sections explore how disability studies scholarship, specifically recent work in critical disability studies (CDS), informs our recommendation that the impact of disability on decision-making should be considered within the context of P&FCC as an additional responsibility of the health practitioner.

17.3.1 Critical Disability Studies and the Rehabilitation Setting

CDS takes a critical stance on how disability is viewed and reproduced in society. As Goodley (2011) summarizes, it focuses on challenging the commonly accepted binary of dis/ablism to create space for new understandings of disability and improved well-being for disabled people. Following four decades of disability studies scholarship that considers how non-disabled persons are valued over disabled persons and how social practices uphold this hierarchy, CDS considers: "In what ways do disabled bodies rearticulate what qualifies as a body that matters?" (Goodley 2011, p. 159). CDS asks us to acknowledge the social context of oppression experienced by disabled people and to question how we can change our understanding of a 'good life,' or, in the case of treatment, a 'good outcome' based on this awareness.

To respond to this challenge health practitioners must first have an understanding of the complex nature of disability. In turn they must apply this knowledge to question dominant assumptions regarding what constitutes 'a body that matters' and consider how those assumptions are influencing their practice. A concise review of the work of disability studies scholars to model disability and promote appropriate language will clarify this point.

Understanding Disability Social context and what it means for our understanding of disability in the therapeutic setting has been highly theorized by disability studies scholars. Beginning circa 1970, scholars began to put forward interpretations of disability that challenged the 'dominant view' of disability. The dominant view (the view we see replicated widely across societies) individualizes disability, considering it a personal tragedy that harms the bodily integrity of an individual (Goodley 2011). This view asserts that any problem experienced by disabled persons is directly related to their bodily impairment such that any hope of a resolution is intimately connected to their ability to overcome their disability by way of treatment, rehabilitation or cure (Oliver et al. 2012).

Disability scholars have constructed models of disability to help us better understand why we we've come to view disability in this way, and to promote an alternative interpretation (Finkelstein 2001a). Comparing two prominent models, Goodley (2011) explores how disability is interpreted both as a moral condition (moral model), caused by sin, and a medical condition (medical model). This second model presents a medicalized view of disability, reflecting how disability is interpreted as pathology when it is defined "in medical terms, using medical language… adopting a medical framework… or using a medical intervention to 'treat' it" (Conrad 1992, p. 211). While the moral model of disability is prevalent globally, the medical model is considered to reflect majority views on disability, especially those which inform rehabilitation facilities and journals, as it includes the expectation that disabled people will seek medical intervention aimed at 'returning' the individual to a 'normal' state (Oliver et al. 2012; Goodley 2011).

Implicit in the concept of medicalization, and hence the medical model, is the idea that social factors contribute to the construction of disability as a medical problem, requiring medical intervention. "As identified in other areas of medicine, what we consider disease intervention and what we do not… will conform to what the culture or health practitioners see as the proper objects of medical intervention" (Wolpe 2002, p. 389). In their challenge to the 'dominant view,' disability activists and theorists have sought to overthrow the medicalized understanding of disability, and replace the medical model with a third model, the social model of disability. With the social model, disability is explicitly de-medicalized and explained as a form of social disadvantage that is imposed on top of bodily impairment (Oliver et al. 2012, p. 16):

> *Disability* is the disadvantage or restriction of activity caused by the political, economic and cultural norms of a society which takes little or no account of people who have impairments and thus excludes them from mainstream activity. (Therefore disability, like racism or sexism, is discrimination and social oppression).
> *Impairment* is a characteristic of the mind, body or senses within an individual which is long term and may, or may not, be the result of disease, genetics or injury. [Emphasis original]

In this way, a distinction is made between disability and impairment. Disability is not a problem of the individual in isolation, but is understood as the product of social factors contributing to the marginalization of individuals with impairments.

The social model has generated considerable support as it acknowledges the influence of social values on the lives of individuals, and asks society to change in order to support all people, rather than view those with impairments as possessing a distinct form of vulnerability that necessarily separates them from the rest of society (Finkelstein 2001a). Following the social model, any remedy for disability must address social factors to result in the "restructuring of society so that it is neither competitive nor disabling for all people" (Finkelstein 2001a, p. 5).

Recent disability studies scholarship (referred to by some as *postmodern*) further challenges the dominant view by continuing to question basic assumptions that are made about disability and impairment—even within the social model (see Corker and Shakespeare 2002 for further discussion). For example, the social theory of impairment sets out to explain a new understanding of impairment as also socially constituted, thus expanding the typical view of the social model that distinguishes between impairment and disability on the grounds that the latter is socially constituted while the former is not. As Tremain argues, while the social model presented above may be helpful in de-medicalizing disability, it effectively reduces the 'impaired' body to "the exclusive jurisdiction of medical interpretation" (Tremain 2002).

Thus, Tremain proposes that the same exercise that de-medicalized disability in the social model should be applied to impairment: the biophysical qualities of an individual can be understood as neither impairments nor enhancements until they have been acted upon by social factors (Tremain 2002). Biophysical factors *may* be problematic for an individual. Conversely, the individual may not experience them as problematic at all. In this way, the social theory of impairment extends the social model to critique the practice of treating the biophysical features of individuals without attending to social context. Such treatment may not be relevant to attending to the problem at hand, and may be addressing a 'problem' that the individual does not find problematic.

To better serve individuals and families seeking paediatric rehabilitation services, health practitioners must be aware of these divergent and developing views of disability and impairment and recognize that the way disability is understood can change which treatment decisions are considered appropriate for an adolescent and which are not.

In addition to an awareness of disability studies scholarship, knowledge of key social movements that have developed from it may empower practitioners to recognize when social context is influencing the therapeutic context. Understanding the role of language related to disability is one such area where practitioners can both gain heightened sensitivity to situations where social context may be implicitly affecting goal setting or impinging on a therapeutic request and also role-model positive valuation of life with a disability.

Using Appropriate Language when Discussing Disability For decades the United States Disability Rights Movement has been advocating for appropriate language-use related to disability due to an understanding that language has implications for both disabled people's self-perception as well as how they are viewed and valued within society (Haller et al. 2006; Ben-Moshe 2005; Linton 1998). One of the first coups on this front was to replace the term 'handicapped,' which many felt connoted charity and pity, with 'person first' terminology, such as 'persons with disabilities'(Haller et al. 2006), which asserts that the person is primary, while disability is but one part of their life experience. In the United Kingdom, where the social model of disability was first described, the preferred terminology is 'disabled person' as it reflects the influence of society acting upon an individual to cause their experience of disability. Understood in this way, the term implies oppression of the person, rather than defectiveness (Olkin 2002). Since both terms are accepted, they may be used interchangeably. The central goal is that people make a conscious decision to use language that reflects the decades of work that has gone into challenging oppression and leaving behind essentialist (e.g., "the Down syndrome adolescent") and medicalized (e.g., "the adolescent *suffers from/is a victim of* cerebral palsy") terms which reduce the individual to the biophysical and convey a helpless individual acted upon by their diagnosis (Olkin 2002; Linton 1998). Additional attention should be paid to the use of discriminatory descriptive terms such as 'wheel-chair bound' that perpetuate the limiting view that individuals are restricted and burdened by wheelchairs. Instead people should be recognized as using a wheel-chair. This turn of phrase acknowledges the user as the agent and the use of the chair as both assistive and time specific (i.e., people may use the chair, but they also use cars, beds, couches, etc.) (Linton 1998).[10]

Currently, Special Olympics is supporting a high profile campaign, *Spread the Word to End the Word*—which may be the most wide-spread and accessible movement yet—to reduce the use of the derogatory term 'retard(ed)' in formal and lay speech in order to increase the positive valuation of individuals with disabilities (Special Olympics 2013). This campaign has reached President Obama, inspiring him to sign a law to ensure harmful language is no longer used in American statutes (Rosa's Law 2010). All but seven US states have since taken a similar pledge, and many celebrities and public figures have used the *Spread the Word* website to openly apologize for their derogatory use of the term or to pledge to stop using it (Special Olympics 2013). Legislatures in Canada have been slower to adopt these practices, however Saskatchewan has passed a bill arguing that similar steps are needed (Bill No. 625 2013).

National and international organizations have also begun to respond to the challenge to use respectful language, integrating accepted terminology in their names, e.g., the American Association on Intellectual and Developmental Disabilities, International Association for the Scientific Study of Intellectual Disabilities, President's Committee for People with Intellectual Disabilities (Schalock et al. 2007). The commitment on the part of these organizations to replace 'mental retardation' with 'intellectual disability' reflects their broader understanding of disability; specifically that it is "the expression of limitations in individual functioning within a social context" (Schalock et al. 2007, p. 117). In addition, a recent change in the diagnostic label 'Mental Retardation' to 'Intellectual Disability (Intellectual Developmental Disorder)' in the recent DSM-5, demonstrates that disability activism is beginning to influence and be incorporated within the medical field (American Psychiatric Association 2013).

In the face of such high-profile efforts to reduce stigma and improve lives by changing language, health practitioners have an obligation to role-model these lan-

[10] Disability scholarship on the importance of language is extensive. For a more comprehensive review of how certain terms are seen to respect or oppress individuals with disabilities see Linton (1998).

guage preferences. Sensitivity to language preferences demonstrates both an awareness of the larger issues adolescents and families face and, more importantly, signifies that their needs will be met in ways that value them as contributing members of society.

17.3.2 Integrating Disability Studies into the P&FCC Process

Regulatory bodies have actively amended or added policy statements and guidelines to require their members to meet the minimum legal standards set out in the *Accessibility for Ontarians with Disabilities Act* (AODA 2005) to make healthcare services accessible to disabled people. However, no professional regulatory bodies that we know of have guidelines to support health practitioners in navigating the challenging position of providing medical treatment and intervention while recognizing that medicine itself has contributed to the oppression of disabled people.

This is not for a lack of effort on the part of disability studies scholars. Persistent attempts have been made by disability scholars to integrate an understanding of the social model of disability into professional training, as it is strongly felt that any attempt by a health practitioner to improve the well-being of disabled people without an explicit understanding of disability studies perspectives will fail (Oliver et al. 2012). Other disability scholars, disappointed in the seeming inability of health practitioners to extricate themselves from understanding disability through the medical model, have called for a new profession all together: 'professionals allied to the community' (PACs). They argue that PACs would be specifically trained in the social model and as a result would be better prepared to respond to the aspirations of disabled people (Finkelstein 2001b; see Goodley 2011, Table 10.1, 174 for an overview of how the role of PACs would differ from that of health practitioners).

It is not uncommon for discussion of the medical and social models to cast a mutually exclusive view of healthcare provision and disability activism: problems identified by disabled people are often reduced to either inherent problems of the individual to be addressed medically or ones of society to be addressed through social change. However, maintaining such a clean division outside of academia is unlikely, and moreover does not always benefit people with disabilities (Magasi 2008a). In practice, there is a need to anticipate the complex reality where disabled people may benefit from both medical treatment and social change. With an understanding of the complexity of disability, health practitioners can critically approach the challenge of determining when individual treatment options might be appropriate verses when the focus on technical intervention obscures the role that social factors play in preventing the individual from achieving their goals. The inherent risk of failing to recognize the impact of social factors on the therapeutic setting is the possibility that treatment in isolation may re/produce patterns of social oppression in unintended, and likely unrecognized, ways (Magasi 2008b).

Little (1998) asserts that health practitioners have a responsibility to move beyond the sole appraisal of the morality of an individual intervention to also appraise social context when they offer medical treatments or interventions for enhancement purposes. As one of the authors (LW) has argued elsewhere (Wade 2011), Little's argument provides meaningful insight into disability as it is based on the idea that when a medical technology is evoked to overcome a negative, or unjust, social norm—in this case using an individual intervention or treatment to counteract the marginalization experienced by disabled people in society—its use requires grave consideration as it risks being complicitous with the unjust norm itself. Little defines being complicitous as "to bear some improper moral relation to the evil of some practice or set of attitudes… when one endorses, promotes, or unduly benefits from norms and practices that are morally suspect" (Little 1998, p. 170). Importantly, a health practitioner may be complicitous with suspect norms not only by direct endorsement, but also if their actions end up reinforcing that norm (Little 1998). Arguably, the risk of complicity is more likely in an interaction involving a health practitioner because of medicine's high status; such that a well-intentioned intervention may unwittingly legitimize a suspect norm as "others see in [it] a legitimization of or pressure to meet norms" (Little 1998, p. 172). Thus, when a negative social norm is present, Little contends that health practitioners are not justified to provide a technical intervention by the claim that they are helping to relieve an individual of the distress caused by this norm. In so doing they may reinforce the norm and propagate the distress it causes.

Following Little, an important way that health practitioners can reduce complicity is to integrate disability studies scholarship into their practice, recognizing and attempting to minimize the fact that they have significant control over the lives of disabled people, and challenging themselves and colleagues to question whether social norms are influencing treatment goals or decisions. The recognition of a power imbalance is especially important for health practitioners working in the rehabilitation setting, as their work influences disabled people in all aspects of their social lives as well as in the healthcare setting—including education, transportation, employment, housing, and leisure activities (Goodley 2011). In these various settings, 'treatment' options take different forms. It is important to recognize that regardless of the intervention at hand rehabilitation practitioners may constrain the choice of disabled patients as they present and frame certain goals and their associated treatments as appropriate (Magasi 2008b).[11]

An understanding of disability studies prepares health practitioners to begin the difficult work of attempting to understand how an adolescent and family view disability to determine whether social context is playing a part in treatment decisions and goals. Recognizing the rationale driving an adolescent and family's interaction with the rehabilitation setting is critical to determining appropriate avenues for redress and understanding which treatment interventions will be viewed as having successful outcomes and why (Hjorngaard 2011; Olkin 2002). As Hjorngaard

[11] To this end, it is important for health practitioners to understand the perceived gap between current practice norms and standards and the ideal role of PACs as informed by disability studies scholars. This insight into 'ideal' care may encourage health practitioners to find ways to incorporate these standards into their daily practice.

(2011, p. 244) has deftly argued, this practice "involves a commitment to long-term multi-faceted development of a child at the center of her family in which disability exists on a continuum of care between the medical and social models of practice." Finally, an awareness of the complexity of disability prepares health practitioners to ask crucial questions such as: "what makes the proposed outcome a good outcome?", "who does this [treatment] really benefit?" and "what message to [the adolescent] am I endorsing by [offering] this [treatment]" (Hjorngaard 2011, p. 244). Such critical engagement enables health practitioners to be confident that the treatment options they recommend or decisions they support promote the best interests of the adolescent, rather than reflecting the hegemonic view of what it means to have 'a body that matters.'

17.4 Returning to the Case

This case presents a number of challenges for the Speech Language Pathologist (SLP). She is trying to provide treatment or service in the rehabilitation setting, specifically a device that addresses a particular need, to help her patient achieve her personal goals.

The SLP is confronted with a sense of legal and ethical uncertainty as well as conflicting professional obligations. Unpacking her uncertainty reveals a number of legal and ethical concepts which are important to clarify. Notably, this case presents the SLP with a challenging ethical dilemma related to social context, specifically disability discrimination and stigma, raising questions about the limits of her professional role. We explore the layers of complexity here to provide an example of what our understanding of an enhanced application of P&FCC would entail.

Identifying the Ethical Tensions At the end of the case scenario, the SLP is focused on her recommendation and Diana's strong emotional response.[12] She feels that she could justify following Sasha's directions as well as justify acting upon Diana's directions through an appeal to the 'patient- and family-centred' standards advanced by her institution; however she must decide which is more ethically acceptable, and why.

The SLP understands that if she provides the device Diana wants for Sasha, she has a better chance of maintaining Diana's trust and willingness to return with Sasha, which will ensure that she is able to continue to provide support to Sasha. However, if the SLP provides the device she strongly believes is best based on her assessment of Sasha's needs—which is also justified by an appeal to evidence-based care supported by the professional standards set out by her professional college—Sasha will be able to communicate more easily with her family and friends. This option also seems to be what Sasha wants. At first glance, both options seem

[12] At times, families attempt to sway the healthcare team by using intimidation or the threat of involving lawyers. Teams should discuss any such comments with the manager, risk manager or others to clarify what is happening and understand the extent of their legal obligations.

to offer different ways for the SLP to 'do good' and 'not harm' her patient, yet they appear to be mutually exclusive.

In her attempt to integrate Sasha's concerns without damaging her relationship with Diana, the SLP will struggle to optimize the ethical principles of practicing based on appropriate evidence; upholding professional standards of care; complying with the law and hospital policy; honouring the wishes of Sasha and thus respecting her as a person; honouring the wishes of Diana and respecting the role of the family; doing good; avoiding harm; protecting privacy and promoting truth-telling.

Situations like this one are very challenging. The SLP should not hesitate to contact her colleagues (e.g., other SLPs, the chaplain, professional practice leader, risk manager, privacy officer, patient representative, child life specialist, etc.) or her professional college, as they might be able to assist bringing clarity to the issues and help to address them. Sometimes a healthcare ethicist or members of the ethics committee are available to help with the specific tasks of clarifying and analyzing the ethical issues at stake and facilitating a resolution.

Who Decides? In working through this case from the standpoint of P&FCC, the question "How can we achieve a shared goal?" should be at the forefront of the SLP's mind. Though the SLP's training in P&FCC may not have emphasized the importance of integrating human rights and healthcare law into applications of P&FCC, it is imperative that her quest to achieve a shared goal includes ensuring she correctly identifies the appropriate decision-maker to ensure respect for the autonomy of the adolescent.

Sasha is a growing, developing teenager who is striving to be involved in decisions affecting her, and is learning to express her ideas and wishes. Even if Sasha has never explicitly been informed of her rights, she has some natural intuition that she should 'have a say in her life.' Sasha may even be testing her mother and healthcare team to see how much control she has over her treatment and choice of assistive technology, and the boundaries of this control.

A health practitioner offering treatment has an obligation to gain information from the adolescent about his or her values, and take instruction from the adolescent if he or she has decisional capacity. The SLP should begin by presuming that Sasha is capable, and evaluate her capacity only if there is reason for concern. In this case, though there is no express reason for the SLP to be concerned about Sasha's capacity, conducting a formal capacity evaluation to explicitly establish Sasha's decision-making capacity for this proposed treatment may help Diana recognize and value her daughter's wishes, and reassure Diana when the SLP takes instructions from Sasha. If the SLP feels a capacity evaluation is needed or helpful, she should clearly explain what this means to Sasha[13], restate her treatment recommendation and ensure she provides enough information to allow Sasha to make an informed decision after considering all reasonable options. Importantly, the SLP must ensure

[13] Under the HCCA (1996), consent is not required to undertake a capacity evaluation; however procedural fairness requires an explanation of what is being done and why (G H (Re) 2012). A capacity assessment for treatment is explicitly excluded from the definition of the term 'treatment' in s. 2(1), and consent is required for 'treatment', see s. 10(1).

there are no barriers to communication; namely that Sasha is comfortable using any communication devices available for the evaluation. As she conducts the evaluation, the SLP may consult or work with other members of the team, and may use the Adapted Aid to Capacity Evaluation tool (Appendix) to help perform and document the evaluation of Sasha's capacity.

The requirement that Sasha's consent be voluntary does not imply that the SLP must refrain from using persuasion to help Sasha accept the recommended device. Persuasion involves understanding what is important to Sasha, describing how it could achieve her goals and appealing to Sasha's reason in an attempt to convince her of the merits of a recommendation. This can be done in a way that leaves Sasha free to accept or reject the SLP's advice (Etchells et al. 1996a). However, ensuring the decision is voluntary does involve asking Diana to leave the room to ensure Sasha can speak openly about her preferences.

If Sasha is capable to make a decision, then she should be the principal decision-maker of her own care. Remembering the influence that family can have on adolescent decision-making and health outcomes, the SLP should rely on the P&FCC model and support Diana to play a role in the decision-making process. The way that the SLP communicates with Sasha and Diana will have a significant impact on the trusting relationship between them: being calm, pleasant, compassionate and genuine; maintaining good eye contact, body language and calm tone of voice are important techniques for effective communication. Statements and open-ended questions that encourage input from Sasha and Diana like "Tell me more about that…" and "Is there anything I can do for you?" assist in providing the time and space to communicate well about what is important to each of them. It will also be vital for the SLP to promote hope and optimism in appropriate ways as she interacts with Sasha and Diana (MacKean et al. 2012).

Ultimately, the SLP may need to make clear to Diana that the primary duty is to the adolescent patient Sasha. In cases where a facilitated solution between Sasha and Diana is not feasible, the SLP will support Sasha. This is not a situation where there is a risk of imminent harm to Sasha by following Sasha's wishes, thus the SLP should continue to steadfastly advocate on Sasha's behalf, even while working at resolving the tensions between Sasha and Diana. If Diana disagrees with the capacity evaluation or with Sasha's treatment decision, the team should point out the options in 'the road ahead', including facilitation, dispute resolution,[14] and assistance or direction from the Consent and Capacity Board. Everyone involved should have a good appreciation for the various options open to each person to escalate the issue or get assistance as needed (Handelman and Parke 2008).

If Sasha were determined to be incapable to make a decision about the device, Diana would most likely be the SDM. In fulfilling this role, Diana maintains a responsibility to consider Sasha's values beliefs and wishes, as well as assess the

[14] The literature in end-of-life care underscores the usefulness of having available methods of dispute resolution when the patient, family and teams are no longer communicating effectively and cannot come to agreement. Pope (2007) describes a six step process which could be adapted for this case scenario.

harms and benefits of the treatment (HCCA 1996, p. 21). It is important for Diana to also realize that her own values, beliefs and wishes are not relevant in her SDM role (E J G (Re) 2007). The team may be able to assist Diana in her role, and provide more detailed information about her role as needed.

Achieving a shared goal and plan respects Sasha's autonomy and ensures that she will be able to use the device and receive the support she desires from her mother and from the team.

Understanding the Rationale Once the appropriate decision-maker has been determined, it remains important to investigate the rationale behind the different treatment options. As discussed, the social context of disability creates an imperative for further consideration of potential harms, such as being complicit with a negative social norm.

There is evidence that both Sash and Diana are influenced by the strong social desire to be 'normal.' Sasha's interest in a 'teenage' vocabulary demonstrates that she is feeling set apart from her peers and desires to be part of their social group. Diana's use of derogatory language in her interaction with the SLP demonstrates her own insecurity, and perhaps fear, related to disability. Her 'horizontal hostility'—discriminating against one disadvantaged group (in this case individuals with intellectual disability) as a means of privileging another (those with physical disability)—provides insight into Diana's experience with disability and points to concerns she may be having that society does not value Sasha as it should. It is likely that she is battling the heavy influence of negative social norms, has no mechanism to consider an alternative, and is not feeling supported when she is vulnerable.

If the SLP does not address this limited perspective, there is a risk of undermining the focus on Sasha's best interests in favour of using medical technology to combat a negative social norm. For example, the SLP might be tempted to justify the smaller tablet device because Sasha is likely to encounter discrimination with the larger one. With this thinking, the SLP would be 'doing good' by reducing the harm Sasha experiences. However, this approach fails to challenge the dominant perception that it is not possible for Sasha to flourish if she lives according to her own standards of well-being and what is important to her. In this way, the SLP would be complicit with these norms, conveying to Sasha that she should be ashamed of her disability and she should want to cover it up to be as 'normal' as possible, rather than empowering her to select the method of communication she feels most helpful.

While Sasha's desire to use language that her mother and teachers may not condone can also be linked to a social norm, i.e., to be 'cool' you must swear and talk about sex, it is not an unjust social norm, and thus to participate in this norm does not involve complicity. To the contrary, providing Sasha with access to the vocabulary she wants respects her right to self-determination and empowers her to take on the responsibility of learning to respectfully engage with her peer group.

Sasha's request that the team keep secret her wish to have swear words and words about sex included on her device also needs to be further explored. Sasha specifically does not want her mother to know because she fears her mother will not approve and this will create a conflict between them. The SLP is concerned about this, as fostering trusting relationships with each of Sasha and her mother is part of providing good clinical care and she recognizes that truth-telling is considered a necessary condition for trust in North American culture (Harrison et al. 2004; Hébert et al. 1997). Furthermore, the SLP understands that losing that trust could be detrimental to her on-going involvement in Sasha's care.

Sasha has a right to privacy which is supported by underlying ethical and human rights principles, and PHIPA. If Sasha is able to "understand the information that is relevant to deciding whether to consent to the collection, use or disclosure, as the case may be; and to appreciate the reasonably foreseeable consequences of giving, not giving, withholding or withdrawing the consent," then she has the right to decide which parts of her personal information to share with her mother (PHIPA, s. 21(1)). The level of capacity required for making a decision about sharing personal health information is typically less than that needed for many healthcare decisions because healthcare decisions are often more complex. Health practitioners should be attentive to this difference and make a point of having separate conversations with adolescent patients about decisions relating to their information versus decisions about treatment.

This dilemma provides an explicit example of the kind of power with which health practitioners are entrusted. At a basic level, limiting someone's vocabulary is impossible unless there is a communication device involved, and to do so would generally be considered socially and ethically inappropriate, drawing parallels to the use of restraints.[15] The further action to allow an adolescent to determine her vocabulary, yet withhold her ability to keep her choice confidential is just as unacceptable. Attempts to limit one's right to privacy would have to be rigorously justified. In this case, the harm to Sasha of restricting her right to privacy is greater than the potential benefit of disclosure. The SLP should counsel Sasha about the importance of truth-telling, just as Sasha is counseled on the management of a medical issue (Hébert et al. 1997). The SLP and the other members of the team should work thoughtfully with Sasha to clarify why she wishes to keep information private, what options exist, and whether a staged and supported disclosure to Diana could be planned. Likewise, the SLP should work with Diana to help her understand that Sasha is successfully transitioning to be the leader of her own healthcare information and may not choose to share everything with her mother (Gall et al. 2006; Kieckhefer and Trahms 2000). Ultimately, as Sasha is capable to decide about sharing her own information, her desire to keep things from her mother should be respected.

Managing Conflict When challenging situations arise, members of the health care team rely on the core concepts of P&FCC, especially those of respect, information sharing and partnership. Careful attention to the values of transparency and accountability is also critically important to ensure effective information sharing by health practitioners.

Prior to a joint adolescent-family-team meeting, the healthcare team may wish to set aside a safe time and space to plan how to approach Diana and Sasha. Discuss-

[15] While communication devices are not traditionally defined as restraints, these limitations on the use of the device should be considered similar to a restraint. See Selekman and Snyder (1996) for a detailed discussion on the use of restraints in paediatrics.

ing together the challenging aspects of this scenario allows the team members to learn all the facts, distinguish whether assumptions have been made, contribute to a more accurate picture of all that they have heard from the family and friends, clarify their professional and legal responsibilities, brainstorm possible options, discuss their professional recommendations, consult with others as needed, and formulate a consistent team approach. Understanding that the family is at 'point A' even though some team members think the family should already be at 'point B' can be a revelation, and helps to clarify the unspoken expectations of some team members. Feeling like a cohesive team and well prepared for the meeting, the team may be more open to hear the adolescent's and the family's stories.

Adolescent-family-team meetings are often a stressful time for the adolescent and family because the healthcare team tends to control the agenda, the players and the time. Sometimes adolescents and families end up feeling defensive, and feel they will lose power if they give up any perspective or position. Inviting families to contribute to the creation of an agenda and to bring a support person to the meeting helps to increase the perception of strength in numbers for the family, and may help to balance out the feeling that the team has all of the power. Sometimes a cultural support person is helpful in helping the family understand the implicit values or beliefs that may be unique to different cultures as well as the differences between cultures.[16]

Ensuring transparency and consistency about the information being shared is often difficult. At times a concept or explanation is given verbally but the team is not confident the adolescent or family understands the information. One effective method of sharing information is to send a letter summarizing significant or important developments, recommendations and decisions. It is also important for the team to be accountable by being explicit in the analysis of the various options, the harms and benefits; how this analysis is being undertaken; and by allowing questions or requests for second opinions. Families often find a summary letter reassuring and helpful to confirm what they have heard when overwhelming amounts of information are shared. The written document can also be used to help recall all the various details when reviewing events with other family members or friends.

Throughout her involvement with Sasha and Diana, the SLP must remain gently professionally authoritative and continue to have evidence-based discussions to ensure that a consistent message is given. Trust is enhanced when the team is willing to be transparent and accountable, and is open to have decisions reviewed.

17.5 Conclusion

In summary, the challenges specific to adolescents with disabilities relate to their developing autonomy as well as their understanding, and that of others, of disability. We have used a case scenario to stimulate thinking about how P&FCC should be enhanced to recognize and address the ethically complex social context. We have

[16] Myser (2007) talks about the power of white culture and contends that each society has a culture but citizens can't see their own 'normal'.

proposed that health practitioners who work in rehabilitation must understand and protect the important human and legal rights of adolescents as well as thoughtfully interpret and apply the ethical principle of respect. Health practitioners should further explore concepts related to disability studies and learn to ask crucial questions about what is a good outcome, who benefits from the treatment, and what messages the adolescent hears when various treatments are offered. Armed with this increased knowledge and sensitivity, health practitioners will be well prepared to address and resolve challenges that arise in the paediatric rehabilitation setting, and adolescents and their families will feel better heard and be more involved in decision-making.

So, "let us put our minds together and see what future we can make for our children."[17] And by listening to our children and adolescents, we will keep on learning what is important to them.

Appendix—A Framework for Assessing Capacity for Treatment in Children and Adolescents (Adapted Aid to Capacity Evaluation)

A health practitioner in Ontario proposing a treatment uses professional judgment to determine whether a child or adolescent (hereinafter referred to as 'child' for simplicity) has the capacity to consent to a specific treatment or plan of treatment. There seem to be few tools to assist in the assessment of capacity, but the Adapted Aid to Capacity Evaluation, below, may be useful to health practitioners.

In Ontario, capacity (also referred to as competence) is set out in the Health Care Consent Act (1996) as the ability to understand the information provided, and the ability to appreciate the consequences of making or not making a decision.

More specifically, capacity can be described as:

The Ability to Understand—The child should be able to:

- Possess factual knowledge about his or her own health and functional status
- Know the available options
- Remember the information provided

The Ability to Appreciate—The child should be able to:

- Appraise the risks and benefits associated with the potential outcomes of the various options
- See how the facts apply to his or her own situation
- Have a reason for his or her decision

A person makes an informed decision (to consent or refuse) in accordance with one's own beliefs and values.

[17] Attributed to Sitting Bull (c. 1831—December 15, 1890) a *Hunkpapa* Sioux holy man who led his people as a tribal chief during years of resistance against United States incursion into Indian lands.

A Continual Consent and Capacity Process

The goal is to respect the child, help the child develop decision-making abilities, and help the child make decisions about his or her treatment as he or she is able. Capacity of a child can change with time, with development and with changes in medical condition. Every interaction with a child can be an opportunity to assess capacity, encourage understanding of the treatment, and review willingness to continue with the treatment.

Important Considerations During the Process

- **Communication**: are there barriers to communication (language, understanding or expression) which should be addressed with the child? With either of the parents?
- **Optimize Capacity**: what can be done to optimize the capacity of the child?
- **Relationship Building**: begin discussion with relationship-building and introduction to treatment being proposed. Begin with language and ideas appropriate to the child's cognitive level.
- **Reassess Professional Impressions**: is the child able to participate fully or partially? Are parents able to participate fully? (See fuller discussion below.)
- **Informed Discussion**: restart or continue the discussion regarding proposed treatment.
- **Opinion on Capacity**: by asking questions to probe the child's decision-making capacity (see questions suggested by the Adapted Aid to Capacity Evaluation, below) and, using professional impression, formulate an opinion about whether the child is capable to make this treatment decision. Or seek help from another clinician to assess capacity, or consult with psychiatrist or psychologist/psychological associate.
- **Review Often**

Health Practitioner's General Professional Impressions

During interactions the following may be helpful to guide impressions:

- **Attention**: Is the child able to sit still and listen to a discussion? Is the child running around the room and not paying attention to the discussion? Is the child able to make eye contact and listen?
- **Engagement**: How engaged is the child with his or her parent? With you as the clinician?
- **Basic Facts about Self**: Is the child able to tell you his or her name, age, brothers or sisters?

- **Proposed Treatment**: Is the child able to describe to you the basics about treatment? Can he or she tell you in simple terms about the treatment, risks, what you are going to do, etc.?
- **Firm Decision**: Is the person able to make a firm decision?[18]
- **Reasons**: The reason a child provides for explaining his or her preferences lends insight into decision-making capacity. If the child is able to articulate a plausible justification for his or her expressed decision, then that expression should be viewed as evidence of the child's capacity.
- **Realistic**: Are the child's expectations for outcome realistic?

Adapted Aid to Capacity Evaluation

If it is unclear whether a child has the capacity to consent, then the researcher could use a tool, such as the adapted Aid to Capacity Evaluation (adapted ACE). The ACE was developed by Etchells and Darzins (1999) and has been validated for use in adults. The "Adapted ACE" was developed at Holland Bloorview Rehabilitation Hospital for use with children, although its use in children has not been validated.

While explaining the treatment and answering any questions the child raises, these questions can be used to determine if the child is able to understand and able to appreciate the information. If the child cannot answer the question initially it should not automatically be assumed that the child is incapable. Rather the information should be re-explained and the question asked again. Determining capacity requires a judgment regarding the quality of the answers, i.e., does the child answer correctly, is the response a rote repetition of information or is he or she able to accurately discuss and consider the information in his or her own words?[19]

A. The following are sample questions to begin to address whether the child is "able to understand" information relevant to making a decision about medical treatment:

- What health problems are you having right now?
- Why are you here?
  Probe: Do you have [said health problems]?
- Do you know what we could do to help you with [your problem]?
- What else can we do to help you?
- Are there any other things you think we (or others) could do to help you?
- What could happen to you if you do not have [proposed treatment]?
- Can you say no to [proposed treatment]?

[18] College of Physicians and Surgeons of Ontario (2007) uses 'settled' decision.

[19] Etchells and Darzins1999) used a scoring system with his ACE tool for adults. This tool has not been validated with children and adolescents. It is difficult to quantify what is essentially a qualified impression or judgment. This Adapted ACE does not include a scoring system.

B. The following are sample questions that could be used to begin to address whether the child is "able to appreciate" the reasonably foreseeable consequences of a decision or lack of decision:

- What could happen to you if you have [proposed treatment]?
- Can [proposed treatment] cause more or other problems? [side effects]
- What could happen to you if you don't have [proposed treatment]?
  Probe: Could you get sicker/worse if you don't have [proposed treatment]?
- Can you help me understand why you do or do not want to have [proposed treatment]?

Acknowledgments The authors wish to thank the caring, thoughtful health practitioners and staff at Holland Bloorview Kids Rehabilitation Hospital who have shared with us their hopes, fears and dilemmas, thereby enriching our understanding of the ethical issues that arise in paediatric rehabilitation. We also thank our colleagues at the University of Toronto Joint Centre for Bioethics who generously help and challenge us to grow in our ethics knowledge and work.

References

Accessibility for Ontarians with Disabilities Act. 2005. SO 2005, c 11. http://www.e-laws.gov.on.ca/html/statutes/english/elaws_statutes_05a11_e.htm. Accessed 20 Feb 2013.

American Psychiatric Association. 2013. http://www.dsm5.org/Documents/changes%20from%20dsm-iv-tr%20to%20dsm-5.pdf. Accessed 9 Feb 2014.

B (Re). 2008. File No. TO-08-2193 (Consent and Capacity Board—Ontario).

Baker, N. J., V. L. Crowe, and A. Lewis. 2009. Making patient-centered care reliable. *Journal of Ambulatory Care Management* 32(1):8–15. http://www.ncbi.nlm.nih.gov/pubmed/19104289. Accessed 30 Dec 2012.

Ben-Moshe, L. 2005. "Lame idea": Disabling language in the classroom. In *Building pedagogical curb cuts: Incorporating disability in the university classroom and curriculum*, eds. L. Ben-Moshe, R. C. Cory, M. Feldbaum, and K. Sagendorf. New York: The Graduate School, Syracuse University.

Bill No. 625. 2013. An Act to Respect People with Intellectual Disabilities in Saskatchewan. http://www.davidforbesmla.ca/wp-content/uploads/2012/08/Bill-625-Intellectual-Disabilities.pdf. Accessed 7 Jan 2013.

Burkle, C. M., and J. J. Benson. 2012. End-of-life care decisions: importance of reviewing systems and limitations after 2 recent North American cases. *Mayo Clinic Proceedings* 87(11): 1098–1105. http://www.mayoclinicproceedings.org/article/S0025-6196(12)*00850-6/fulltext*. Accessed 30 Dec 2012.

C (Re). 2009. CanLII 64783 (Consent and Capacity Board—Ontario). *http://canlii.ca/t/26qht*. Accessed 3 April 2013.

C et al v Wren. 1986. 35 Dominion Law Reports (4th) 419–422 (Alberta Court of Appeal).

Charter of Rights and Freedoms, s 2, Part I of the Constitution Act. 1982. Being Schedule B to the Canada Act 1982 (UK), 1982, c 11. *http://laws-lois.justice.gc.ca/eng/Const/page-15.html*. Accessed 5 Jan 2013.

College of Physicians and Surgeons of Ontario. 2007. Determining capacity to consent: Guiding physicians through capacity and consent to treatment law. *Dialogue* July 2007:32–38.

Conrad, P. 1992. Medicalization and social control. *Annual Review Sociol* 18:209–232.

Conway, J., B. Johnson, and S. Edgman-Levitan, et al. 2006. Partnering with patients and families to design a patient- and family-centered health care system: A roadmap for the future, a work in progress. Institute for Patient- and Family-Centered Care. http://www.ipfcc.org/pdf/Roadmap.pdf. Accessed 31 Dec 2012.

Corker, M., and T. Shakespeare, eds. 2002. *Disability/Postmodernity: Embodying disability theory*. London: Continuum.

Dickens, B. M. 1994. Legal approaches to health care ethics and the four principles. In *Principles of health care ethics,* ed. R. Gillon, 305–317. University of Michigan: Wiley.

Dickens, B. 1997. Interfaces of assisted reproduction ethics and the law. In *Ethical dilemmas in assisted reproduction,* eds. F. Shenfield and C. Sureau, 77–81. Lancester: Parthenon.

Dickens, B. M., and R. J. Cook. 2005. Ethical and legal issues in reproductive health: Adolescents and consent to treatment. *International Journal of Gynecology and Obstetrics* 89:179–184.

E J G (Re). 2007. CanLII 44704 (Consent and Capacity Board—Ontario). http://canlii.ca/t/1tcqj. Accessed 22 May 2013.

Eldridge v British Columbia (Attorney General). 1997. CanLII 327 (Supreme Court of Canada), [1997] 3 SCR 624. http://canlii.ca/t/1fqx5. Accessed 5 Jan 2013.

Etchells, E., and P. Darzins. 1999. Assessment of patient capacity to consent to treatment. *Journal of General Internal Medicine* 14(1):27–34. http://www.ncbi.nlm.nih.gov/pmc/articles/PMC1496441/. Accessed 30 Dec 2012. The Aid to Capacity Evaluation tool and training materials are available online at http://www.jointcentreforbioethics.ca/tools/ace.shtml. Accessed 30 Dec 2012.

Etchells, E., G. Sharpe, M. J. Dykeman, and E. M. Meslin, et al. 1996a. Bioethics for clinicians: 4. Voluntariness. *Canadian Medical Association Journal* 155 (8):1083–1086. http://www.cmaj.ca/site/misc/bio_voluntariness.xhtml. Accessed 4 Jan 2013.

Etchells, E., G. Sharpe, and C. Elliot, et al. 1996b. Bioethics for clinicians: 3. Capacity. *Canadian Medical Association Journal* 155(6):657–661. http://www.cmaj.ca/site/misc/bio_capacity.xhtml. Accessed 4 Jan 2013.

Etchells, E., G. Sharpe, and P. Walsh, et al. 1996c. Bioethics for clinicians: 1. Consent. *Canadian Medical Association Journal* 155(2):177–180. http://www.cmaj.ca/site/misc/bio_consent.xhtml. Accessed 4 Jan 2013.

Epstein, R. M., and R. L. Street. 2011. The values and value of patient-centered care. *Annals of Family Medicine* 9(2):100–103. http://www.annfammed.org/content/9/2/100.full.pdf+html. Accessed 30 Dec 2012.

Fine, R. L. 2010. Keeping the patient at the center of patient- and family-centered care. *Journal of Pain and Symptom Management* 40(4):621–625. http://www.jpsmjournal.com/article/S0885-3924(10)00573-7/fulltext. Accessed 30 Dec 2012.

Finkelstein, V. 1 Dec 2001a. The social model of disability repossessed. Paper presented to the Manchester Coalition of Disabled People.

Finkelstein, V. 1 Feb 2001b. A personal journey into disability politics. Paper presented to the Centre for Disability Studies, Leeds University.

Fleming v Reid. 1991. 82 Dominion Law Reports (4th) 298 (Ontario Court of Appeal). http://canlii.ca/t/1p78q. Accessed 17 Feb 2013.

Gall, C., S. Kingsnorth, and H. Healy. 2006. Growing up ready: A shared management approach. *Physical & Occupational Therapy in Pediatrics* 26(4):47.

Garton G. (Updated April 2005). The Canadian Charter of Rights Decisions Digest. Justice Canada, Canadian Legal Information Institute. http://canlii.org/en/commentary/charterDigest/. Accessed 17 Feb 2013.

Gibson, B. E., B. Secker, and D. Rolfe, et al. 2012. Disability and dignity-enabling home environments. *Social Science & Medicine* 74:211–219.

G H (Re). 2012. CanLII 7682 (Consent and Capacity Board—Ontario). http://canlii.ca/t/fq6vq. Accessed 3 March 2013.

Gillick v West Norfolk & Wisbech Area Health Authority. 1986. 1 Appeal Cases 112, 3 All England Reports 402 (English House of Lords).

Gilmour, J. M. 2002. Children, adolescents, and health care. In *Canadian health law and policy,* eds. J. Downie, T. Caulfield, and C. Flood. 2nd ed., 205–249. Markham: Butterworths. http://drr.lib.athabascau.ca/files/hadm/400/caahc.pdf. Accessed 5 Jan 2013.

Goodley, D. 2011. *Disability studies: An interdisciplinary introduction*. London: Sage.

Greene, S. M., L. Tuzzio, and D. Cherkin. 2012. A framework for making patient-centered care front and center. *Permanente Journal* 16(3):49–53. http://www.ncbi.nlm.nih.gov/pmc/articles/PMC3442762/. Accessed 30 Dec 2012.

Groene, O. 2011. Patient centredness and quality improvement efforts in hospitals: Rationale, measurement, implementation. *International Journal for Quality in Health Care: Journal of the International Society for Quality in Health Care* 23(5):531–537.

Haller, B., B. Dorries, and J. Rahn. 2006. Media labeling versus the US disability community identity: A study of shifting cultural language. *Disability & Society* 21(1):61–75.

Handelman, M., and B. Parke. 2008. The beneficial role of a judicial process when "everything" is too much. *Healthcare Quarterly* 11(4):46–50.

Harrison, T. M. 2010. Family-centered pediatric nursing care: State of the science. *Journal of Pediatric Nursing* 25:335–343. http://www.ncbi.nlm.nih.gov/pmc/articles/PMC2965051/pdf/nihms240188.pdf. Accessed 6 Jan 2013.

Harrison, C., S. Albersheim, and L. Arbour, et al. 2004. Treatment decisions regarding infants, children and adolescents. *Paediatrics and Child Health* 9(2):99–103. Canadian Paediatric Society Reference Number B04-01. http://www.cps.ca/en/documents/position/treatment-decisions. Accessed 30 Dec 2012.

Harrison, C., N. P. Kenny, and M. Sidarous, et al. 1997. Bioethics for clinicians: 9. Involving children in medical decisions. *Canadian Medical Association Journal* 156:825–828. http://www.cmaj.ca/content/156/6/825.full.pdf+html. Accessed 22 Feb 2013.

Health Care Consent Act. 1996. SO 1996, c 2, Sch A. http://www.e-laws.gov.on.ca/html/statutes/english/elaws_statutes_96h02_e.htm. Accessed 4 Jan 2013.

Hébert, P. C., B. Hoffmaster, and K. C. Glass, et al. 1997. Bioethics for clinicians: 7. Truth telling. *Canadian Medical Association Journal* 156(2):225–228. http://www.cmaj.ca/content/156/2/225.full.pdf. Accessed 6 Jan 2013.

Hesson, K., D. Bakal, and K. S. Dobson. 1993. Legal and ethical issues concerning children's rights of consent. *Can Psychol* 34:317–328.

Hjorngaard, T. 2011. Family-centred care: A critical perspective. *Physical & Occupational Therapy in Paediatrics* 31(3):243–244.

Institute of Medicine. 2001. Crossing the quality chasm: A new health system for the 21st century. Washington, DC. http://www.nap.edu/openbook.php?record_id=10027&page=R1. Accessed 4 Jan 2013.

Kenny, N., J. Downie, and C. Harrison. 2008. Respectful involvement of children in medical decision making. In *Cambridge textbook of bioethics,* eds. P. A. Singer and A. M. Viens, 121–126. Cambridge: Cambridge University Press.

Kieckhefer, G. M., and C. M. Trahms. 2000. Supporting development of children with chronic conditions: From compliance toward shared management. *Pediatric Nursing* 26(4):354–363.

Koelch, M., and J. M. Fegert. 2010. Ethics in child and adolescent psychiatric care: An international perspective. *International Review of Psychiatry* 22(3):258–266.

Kuczewski, M. G. 2001. Disability: An agenda for bioethics. *American Journal of Bioethics* 1(3):36–44.

Langlois, S. 2012. Patient centred education: Lessons learned from a health mentor program. Presentation at the Centre for Faculty Development Best Practices in Education Rounds, University of Toronto. Toronto: 20 Nov 2012.

Larcher, V., and A. Hutchinson. 2010. How should paediatricians assess Gillick competence? *Archives of Disease in Childhood* 95:307–311.

Linton, S. 1998. *Claiming disability*. New York: New York University Press.

Little, M. O. 1998. Cosmetic surgery, suspect norms, and the ethics of complicity. In *Enhancing human traits: Ethical and social implications,* ed. E. Parens, 163–176. Washington, DC: Georgetown University Press.

MacKean, G., W. Spragins, and L. L'Heureux, et al. 2012. Advancing family-centred care in child and adolescent mental health: A critical review of the literature. *Healthcare Quarterly* 15(Special Issue):64–75. http://www.google.ca/url?sa=t&rct=j&q=&esrc=s&frm=1&source=web&

cd=1&ved=0CDkQFjAA&url=http%3A%2F%2Fwww.longwoods.com%2Fproduct%2Fdownload%2Fcode%2F22939&ei=ArXgUOfLCIiU2wXmqYG4CA&usg=AFQjCNG0J6GHf54geaBOLZ56KR3Bh86iQQ&sig2=tobshwue9NJKgc01VVOzNw&bvm=bv.1355534169,d.b2I. Accessed 30 Dec 2012.

Magasi, S. 2008a. Infusing disability studies into the rehabilitation sciences. *Topics in Stroke Rehabilitation* 15(3):283–287.

Magasi, S. 2008b. Disability studies in practice: A work in progress. *Topics in Stroke Rehabilitation* 15(6):611–617.

Malette v. Shulman. 1990. 67 Dominion Law Reports (4th) 321–339 (Ontario Court of Appeal).

McDonald, M. Z., L. Dade, D. Pape, and S. Bean. 2010. A framework for assessing capacity for treatment in children and adolescents (Adapted Aid to Capacity Evaluation). Document developed and used at Holland Bloorview Kids Rehabilitation Hospital (unpublished).

Myser, C. 2007. White normativity in U.S. bioethics: A call and method for more pluralist and democratic standards and policies. In *The ethics of bioethics—mapping the landscape,* eds. L. A. Eckenwiler and F. G. Cohn, 241–259. Maryland: The Johns Hopkins University Press.

Nyswander, D. 1956. ONCA 482 (Ontario Court of Appeal) Education for health: Some principles and their application. *Health Education Monographs* 14:65–70.

Oliver, M., B. Sapey, and P. Thomas, eds. 2012. *Social work and disability*. New York: Palgrave Macmillan.

Olkin, R. 2002. Could you hold the door for me? Including disability in diversity. *Cultural Diversity and Ethnic Minority Psychology* 8(2):130–137.

Personal Health Information Protection Act. 2004. SO 2004, c 3, Sch A. http://www.e-laws.gov.on.ca/html/statutes/english/elaws_statutes_04p03_e.htm. Accessed 4 Feb 2013.

Pope, T. M. 2007. Medical futility statutes: No safe harbor to unilaterally refuse life-sustaining treatment. *Tennessee Law Review* 75 (1):1–81. http://papers.ssrn.com/sol3/papers.cfm?abstract_id=989662. Accessed 30 Dec 2012.

Rasouli v. Sunnybrook Health Sciences Centre. 2011. ONCA 482. http://www.ontariocourts.ca/decisions/2011/2011ONCA0482.htm. Accessed 5 Jan 2013.

Registered Nurses Association of Ontario. 2002. Nursing best practice guideline: client centred care. http://rnao.ca/sites/rnao-ca/files/Client_Centred_Care_0.pdf. Accessed 30 Dec 2012.

Reibi v. Hughes. 1980. 2 Supreme Court Reports 880 (Supreme Court of Canada). http://canlii.ca/t/1mjvr. Accessed 4 Jan 2013.

Rosa's Law. 2010. Public Law 111–256 Oct 5, 2010 124 STAT. 2643. http://www.gpo.gov/fdsys/pkg/PLAW-111publ256/pdf/PLAW-111publ256.pdf. Accessed 20 Feb 2013.

Ross, L. F. 2003. Adolescent autonomy in health care? *APA Newsletter on Philosophy and Medicine* 2(2):193–200.

Scales, R., and J. H. Miller. 2003. Motivational techniques for improving compliance with an exercise program: Skills for primary care clinicians. *Current Sports Medicine Reports* 2:166–172. http://www.sneb.org/conference/documents/Mot_Interviewing_Scales.pdf. Accessed 4 Jan 2013.

Schalock, R. L., R. A. Luckasson, and K. A. Shogren. 2007. The renaming of mental retardation: Understanding the change to the term intellectual disability. *Intellectual and Developmental Disabilities* 45(2):116–124.

Selekman, J., and B. Snyder. 1996. Uses of and alternatives to restraints in pediatric settings. *AACN Clinical Issues* 7(4):603–610.

Sessums, L. L., H. Zembrzuska, and J. L. Jackson. 2011. Does this patient have medical decision-making capacity? *Journal of the American Medical Association* 306(4):420–427. http://www.gim.utoronto.ca/Assets/General+Internal+Medicine+Digital+Assets/residents_fellows/eresources/JAMA+-+Does+this+patient+have+medical+decision-making+capacity.pdf. Accessed 5 Jan 2013.

Sine, D. M., and V. A. Sharpe. 2011. Ethics, risk, and patient-centered care: How collaboration between clinical ethicists and risk management leads to respectful patient care. *Journal of Healthcare Risk Management* 31(1):32. http://onlinelibrary.wiley.com/doi/10.1002/jhrm.20077/pdf. Accessed 4 Jan 2013.

Smith, J. A. 1994. Ethical considerations of giving patient's choices. *Hospital Topics* 72(3):15–20. http://www.ncbi.nlm.nih.gov/pubmed/10137786. Accessed 30 Dec 2012.
Special Olympics. 2013. Fighting the R-word, Worldwide. http://www.specialolympics.org/03-31-09_Spread_the_Word.aspx. Also, R-Word: Spread the word to end the word. http://www.r-word.org/Default.aspx. Accessed 22 May 2013.
Substitute Decisions Act. 1992. S.O. 1992, c. 30. http://www.e-laws.gov.on.ca/html/statutes/english/elaws_statutes_92s30_e.htm. Accessed 6 Jan 2013.
Tremain, S. 2002. On the subject of impairment. In *Disability/postmodernity: Embodying disability theory*, eds. M. Corker and T. Shakespeare, 32–47. London: Continuum.
United Nations, General Assembly, Convention on the Rights of the Child. 20 Nov 1989. United Nations, Treaty Series, vol. 1577, p 3. http://www.unhcr.org/refworld/docid/3ae6b38f0.html. Accessed 21 Feb 2013.
United Nations General Assembly, Convention on the Rights of Persons with Disabilities: Resolution/adopted by the General Assembly. 24 Jan 2007. A/RES/61/106. http://www.unhcr.org/refworld/docid/45f973632.html. Accessed 21 Feb 2013.
United Nations General Assembly, Universal Declaration of Human Rights. 10 Dec 1948. 217 A (III). http://www.unhcr.org/refworld/docid/3ae6b3712c.html. Accessed 21 Feb 2013.
Van Mol v Ashmore. 1999. 168 Dominion Law Reports (4th) 637–641 (British Columbia Court of Appeal). http://www.canlii.org/en/bc/bcca/doc/1999/1999bcca6/1999bcca6.html. Accessed 8 Jan 2013.
Wade, L. 2011. Ideologies of intellect: A critical examination of the hype surrounding cognitive enhancement. MSc Thesis, McGill University (unpublished).
Wagner, E. H., S. M. Bennett, and B. T. Austin et al. 2005. Finding common ground: Patient-centeredness and evidence-based chronic illness care. *Journal of Alternative and Complementary Medicine* 11 (Suppl 1):S7–S15.
Wolpe, P. R. 2002. Treatment, enhancement, and the ethics of neurotherapeutics. *Brain and cognition* 50(3):387–395.
World Health Organization. 2002. Adolescent friendly health services: An agenda for change. WHO reference number WHO/FCH/CAH/02.14. http://whqlibdoc.who.int/hq/2003/WHO_FCH_CAH_02.14.pdf. Accessed 4 Jan 2013.

Chapter 18
Ethical Conflicts that may Arise When Caring for Dying Children

Christine Newman, Adam Rapoport and Gurjit Sangha

18.1 Paediatric Palliative Care

The Canadian Hospice and Palliative Care Association (CHPCA) defines paediatric palliative care as (CHPCA 2006, p. 8):

> an active, holistic approach to care which focuses on relieving the physical, social, psychological and spiritual suffering experienced by children and families who face a progressive, life-threatening condition, and helping them fulfill their physical, psychological, social and spiritual goals. Its philosophy is to provide optimal comfort and quality of life, and sustain hope and family connection despite the likelihood of death. Pediatric hospice palliative care aims to provide comprehensive care for children and their families through the living, dying and grieving processes. It affirms life and regards dying as a process that is a profoundly personal experience for the child and family. Pediatric hospice palliative care is planned and delivered collaboratively by an interdisciplinary team. It is a child and family centred approach to care that is based on shared decision-making and sensitivity to the family's cultural and spiritual values, beliefs and practices.

It has been suggested that there are four groups of diseases that affect children[1] where treatment following the principles of palliative care would be helpful (Steering Committee of the EPAC task force 2007). They are:

Group 1 – Diseases like cancer, which may be cured but have a possibility of death.

Group 2 – Diseases like cystic fibrosis, which are ultimately fatal but intensive medical therapy is required during the patient's life.

Group 3 – Neurodegenerative or metabolic diseases that are incurable but whose symptoms can be managed.

Group 4 – Severe neurological impairment from anoxic brain injury, where the condition itself is non-progressive but complications may lead to early death (Steering Committee of the EPAC task force 2007).

[1] Throughout this chapter, the term 'child' or 'children' will include children, infants and adolescents.

C. Newman (✉) · A. Rapoport · G. Sangha
Paediatric Advanced Care Team (PACT), Hospital for Sick Children, Toronto, Canada
e-mail: pact.info@sickkids.ca

R. Zlotnik Shaul (ed.), *Paediatric Patient and Family-Centred Care: Ethical and Legal Issues,* International Library of Ethics, Law, and the New Medicine 57,
DOI 10.1007/978-1-4939-0323-8_18, © Springer Science+Business Media Dordrecht 2014

Family-centred care has become a mainstay in the practice of paediatrics. Paediatric palliative care shares the feeling that the child is not just an individual patient, but is an integral part of a family unit. We aim to provide care to the entire unit. We also, however, must remember that the child is our patient and be mindful of the principles of patient-centred care. It is perhaps hardest to fully integrate these principles in cases where the family is facing the real possibility that their child will die. What the family most wants, for their child not to die, may not be an achievable goal.

The aim of paediatric palliative care is to maximize the quality of life for both the patient and their family, with an emphasis on control of the patient's physical and psychological symptoms. The goals of care must always focus on what is 'best' for the patient. Fortunately, there is usually synergy between family-centred care and patient-centred care in many aspects of end-of-life care when the family accepts that the child is dying. Both the family and the healthcare team want the child to be as comfortable as possible. We are also usually in agreement about the preferred location of death.

There are also times when providing paediatric palliative care will mean advocating to forego (i.e., either to withhold or withdraw) life-sustaining interventions or other aggressive medical therapies aimed at disease cure. Such decisions are in keeping with current ethical and legal principles. This is another potential point of tension between the principles of family-centred and patient-centred care. How can parents[2] and other family members accept that allowing death is the 'best' outcome for their child?

The role of the Paediatric Palliative Care (PPC) Team is to work with children and their families to establish goals of care that are in keeping with the reality of the child's current medical state. Ideally, we teach the child how to die and their family how to allow that to happen.

In this chapter we will use cases created by merging actual clinical experiences in order to explore three areas where families and healthcare providers may experience tension at the end of a child's life: foregoing life-sustaining treatment, truth-telling, and hearing the child's voice.

18.2 Case #1: Foregoing Life-Sustaining Treatment

At 4 months of age Patrick was diagnosed with Spinal Muscular Atrophy Type 1, a genetic condition that causes progressive weakness of all the muscles in the body and eventually leads to death. His parents first became concerned when they noted that he was "floppier" than his 3-year-old brother, Timothy. Upon learning of the diagnosis and its prognosis, Patrick's parents were devastated but vowed "God will help Patrick overcome this!"

[2] Throughout this chapter, the term 'parent' will be used to denote any legal guardian of a child for the purpose of medical decision-making.

In the months following his diagnosis, Patrick became weaker, as predicted. Though he remained interactive and happy most of the time, his physical development reached a plateau and then began to regress. At 6 months of age Patrick had a feeding tube inserted into his stomach because of increased choking and coughing with oral feeds; he has not had anything to eat or drink by mouth ever since. At 10 months of age, Patrick started using over-night BiPAP, an external machine that helps him breathe by compensating for his weakened respiratory muscles.

Now, at 11 months of age, Patrick has been in the Paediatric Intensive Care Unit for the last 2 weeks on mechanical ventilation. He presented to hospital in severe respiratory distress after vomiting and he is being treated for aspiration pneumonia—some of the vomit entered his lungs. Although the pneumonia appears to be improving on chest x-ray, efforts to take him off the ventilator have not been successful. It is believed that this acute illness has resulted in further deterioration of his respiratory muscles, making him 'ventilator dependent.'

There is no cure for Patrick's underlying illness. The options for treatment now are to continue mechanical ventilation (knowing that Patrick's illness will progress and he will ultimately die on the ventilator) or to remove the ventilator at this point and allow Patrick's illness to lead to his death now. Although a tracheostomy—a surgically placed breathing tube in his neck that will allow him to receive ongoing ventilation —is technically an option for Patrick, it is not one that the team recommends. While a tracheostomy would help keep Patrick alive and perhaps even allow him to be discharged home this option carries significant risks and complications. Patrick would constantly be attached to the machine and require an awake appropriately trained adult with him 24/7. The quality of both his life and that of his family would be negatively impacted and taking him out of the house would be much more difficult. The medical team feels that a tracheostomy would not be in Patrick's best interest.

Since his admission, Patrick's parents have been regularly praying at the bedside for him to recover. At times they have been accompanied by their church pastor and other members of the congregation. They have expressed to members of the healthcare team their deep belief that if given a chance, God will cure Patrick of his disease. Patrick's parents are adamant that a tracheostomy be inserted to allow more time for their prayers to be answered. They do not accept the healthcare team's assertion that Patrick is 'suffering.'

18.2.1 Discussion

Making the 'right' medical decision means different things to different people. Some consider the right decision to be the one with the best chance of restoring or promoting health; some regard the option that minimizes pain and suffering to be the best choice. Although a variety of interests might influence how one weighs the options, modern healthcare demands that each decision conform to ethical principles that guide decision-making in medicine.

Patient autonomy has emerged as the foremost ethical principle to be respected in medical decision-making for competent patients (Schattner et al. 2004). When a competent and informed patient is deemed to comprehend the consequences of their decision, the medical community is expected to uphold their request even if doing so might, in fact, compromise their health or hasten death.[3]

Young children like Patrick who have yet to develop the capacity to make autonomous decisions require substitute decision-makers, typically one or both parents, to make medical decisions on their behalf. The standard invoked to guide parental decision-making for children is referred to as *best interests*, and it compels parents to place the interests of their child above all else when choosing among available medical options (American Academy of Pediatrics (AAP) Committee on Bioethics 1995; Canadian Pediatric Society (CPS) Bioethics Committee 2004).

Despite substantial support for the best interests standard when decision-making for the paediatric patient, much debate persists about how one might operationalize the standard to choose the 'best' option among a variety of choices (Cantor 2005; Wilkinson 2006). According to this standard, substitute decision-makers must consider the potential harms and benefits of all options when determining an incapable patient's best interests. However, medical uncertainty around prognosis can make this imperative extremely challenging. Furthermore, the social context and cultural beliefs of the family, particularly around end-of-life issues, often demand basing decisions on more than just net benefits and harms to the individual patient.

Widespread agreement exists that parents ought to have authority over medical decisions (AAP Committee on Bioethics 1994; CPS Bioethics Committee 2004; UN Convention on the Rights of the Child), even when decisions are made for critically ill children. In order to facilitate their ability to make these decisions, the health-care team must fully inform the family about the medical situation, including the child's prognosis, the potential options under the circumstances, and their expected outcomes. What is less clear, however, is whether decision-makers should arrive at the child's best interests by taking a patient-centred or a family-centred approach.

18.2.1.1 Paediatric Patient-Centred Decision-Making

When the term 'patient-centred' is invoked to describe the decision-making paradigm, it suggests that external influences are either disregarded or relegated to secondary considerations only. As such, factors including the family's cultural approach to decision-making, religious or spiritual guidance, and even parental preferences are taken into account only if they align with what is deemed to be in the child's best interests.

A patient-centred approach to medical decision-making is in keeping with the guidance provided by western paediatric organizations. The Canadian Paediatric Society (CPS) believes that "although family issues are important and must be

[3] This may not apply to all jurisdictions. In some jurisdictions the age of the patient is also taken into consideration when determining decision-making capacity.

considered, the primary concern for health professionals who care for children and adolescents must be the best interests of children and adolescents" (CPS Bioethics Committee 2004, p. 99). The CPS believes that in discerning a child's best interests, a parent or other proxy decision-maker should consider such factors as: chances of survival; the harms and benefits of treatment; evidence regarding long- and short-term medical outcomes; and long-term implications for suffering and quality of life. Although the American Academy of Pediatrics seemingly acknowledges that proxy decisions account for more than simply medical harms and benefits, they reaffirm that when parental wishes appear to conflict with patient medical needs, "the pediatrician's responsibilities to his or her patient exist independent of parental desires or proxy consent" (AAP Committee on Bioethics 1994, p. 315).

Healthcare teams might advise a family to forgo life-sustaining treatment for a child when they believe that prolonging his or her existence would result in an unacceptable quality of life, or when employing such treatments would be futile. Most parents facing such painstaking decisions have little or no experience concerning the realities of caring for a child dependent on life-sustaining technology and thus depend on the medical team to share accurate information. Indeed, many parents value recommendations to guide them in their decisions (Carroll et al. 2012; Madrigal et al. 2012; Meert et al. 2000). The clinical experience of the healthcare team coupled with their knowledge of the medical literature make them uniquely well suited to helping parents appreciate the probable outcomes of the options and why some treatments may be futile in achieving specific goals of care.

While there is little doubt that the healthcare team plays a crucial role in paediatric end-of-life decision-making, the question as to whether or not death could actually represent a child's best interest is perhaps less evident. The answer largely depends on whether one believes that there are fates worse than death. Many adults view living in certain health states, such as severe pain or in a permanent coma, to be worse than dying (Ditto et al. 1996; Patrick et al. 1997). However, these same studies also demonstrate that no health state is unanimously rejected by all. Moreover, when evaluating the lives of individuals with chronic conditions, family members and healthcare professionals frequently underestimate a patient's quality of life (Farsides and Dunlop 2001). Severely disabled children and adults dependent on medical technology, such as tracheostomies, generally report being satisfied with their quality of life (Carnevale et al. 2006; Noyes 2006). Healthy individuals, in contrast, tend to be much more concerned about the suffering presumed to be associated with these conditions (Klein 2011).

The issue of medical futility as a guide to the appropriateness of treatment remains controversial. Practically, a futile treatment is one that offers "…no therapeutic benefit to a patient" (Schneiderman 2011, p. 128). However, the 'benefit' of a given treatment must be assessed in light of the treatment goals (Kasman 2004). In the case of Patrick, a tracheostomy will not prevent his eventual death from the SMA. However, a tracheostomy would likely prolong Patrick's life, facilitate a return to his home, and possibly improve his quality of life.

Although the allocation of potentially scarce and costly resources may be a real concern, such issues are best addressed at the level of healthcare policy and should be avoided when deliberating the right course of action for an individual patient (AAP Committee on Bioethics 1996). Rather, members of the healthcare team should advocate for necessary resources and supports in order to mitigate the social determinants of health and to ensure that decisions are based on the merit of the proposed treatment for the child, not the family's social situation.

18.2.1.2 Family-Centred Decision-Making

As a result of competing obligations, both in and outside of the family, as well as complicating social and cultural values, parents may find it difficult to make medical decisions solely in consideration of the child's best interests, as traditionally defined. A common criticism of the best interests standard is that it is too individualistic (Kopelman 1997); in guiding decision-makers to consider only the patient-centred interests of the child, the best interests standard may obligate parents to override their own self-interests or those of the family unit. Family-centred decision-making has been defined as, "a way of caring for children and their families within health services which ensures that care is planned around the whole family, not just the individual child/person, and in which all the family members are recognized as care recipients" (Shields et al. 2006, p. 1318).

Central to the idea of family-centred decision-making are: the recognition of the family as a constant in the child's life; acceptance of the racial, cultural, religious and socioeconomic diversity of families and their influences on child-rearing practices; and designing healthcare that is flexible and responsive to unique family needs. Although tolerance towards the wide array of parenting styles and family dynamics is chief among the tenets of the family-centred model, limits do exist. Parents must continue to make decisions for their children congruent with the traditional expectation of what an 'average person' in the same situation might choose for him or herself (or child) (Kopelman 1997); family-centred decision-making does not condone decisions which clearly violate the child's well-being. Decisions and actions consistent with abuse or neglect, sometimes committed under the guise of established cultural or religious norms, are not tolerated by the family-centred decision-making model, and healthcare providers have a fiduciary duty to their patients to invoke established mechanisms to protect the child.

Questions regarding the limits of family-centred decision-making emerge when family values and medical opinions conflict with one another. Unfortunately, such conflicts are not uncommon near the end-of-life, when subjective impressions about another individual's quality of life become the focus (Bowman 2000). In the absence of overt signs of distress or discomfort, who decides if another individual's life is overburdened by suffering? How much responsibility do healthcare providers have to help families achieve their spiritual goals, including miracles? Traditionally, potential treatments have been assessed based on their therapeutic value—how much consideration should be given to the reason behind a treatment preference? Satisfactory responses to these questions will likely depend on the context and details of any particular case, however some general points are worthy of consideration.

When members of the healthcare team believe that prolonging a child's life will only lead to greater suffering, they should ask themselves whether the suffering they are concerned about will be experienced by the patient or themselves. In an effort to empathize with patients, it is not uncommon for healthcare providers to project a state of suffering onto those with a degree of impairment that we, ourselves, are uncomfortable with. Suffering is often invoked as a reason to withdraw life-sustaining therapy in patients who lack the consciousness required to perceive this experience, such as in a persistent vegetative state (Schnakers and Zasler 2007; The Multi-Society Task Force on PVS 1994). Claims of suffering in these situations most likely reflect the healthcare team's own torment resulting from any one of a number of factors: promoting what is believed to be a poor quality of life; maintaining a therapy that is felt to be futile; or engaging in patient care that is deemed inappropriate.

In Patrick's case his underlying disease leaves no doubt in his ability to perceive pain, however there is no mention that he is feeling any discomfort. Experience tells us that Patrick will continue to deteriorate and he will likely require additional hospital admissions in the future, yet his overall quality of life and degree of suffering during the time that remains is uncertain. A child's suffering is a worthy concern, but in the absence of pain, discomfort, or other clear signs of suffering, potential sources of a child's satisfaction or pleasure are germane considerations in assessing a child's quality of life. For many people, real 'benefits' may come from simply knowing that in some way we enrich the lives of those around us (Breitbart 2003). As such, provided that a child's suffering is not objectively evident, it may be reasonable to believe that when a family values the life of their severely disabled child, there is a child who is living a meaningful existence.

Spirituality and religion remain important sources of hope to families facing the mortality of a child (McSherry et al. 2007). When one's faith in medicine does not provide adequate answers, it is natural to look elsewhere. Continued hope for a miracle in the face of certain death does not necessarily indicate that parents do not understand their child's poor prognosis (Reder and Serwint 2009); it may be their way of coping with the situation or delaying the inevitable. Some parents need to experience the 'journey' of caring for their severely disabled child for varying durations before reaching a conclusion that such an existence is inconsistent with their view of an acceptable quality of life.

The suffering of healthcare providers can have long-lasting effects on members of the healthcare team; however, a family must live with the death of a child forever. So long as continued life-saving treatments do not clearly breach child-protection obligations, attention should be aimed at relieving objective suffering through aggressive symptom management.

18.2.2 Resolution of Case #1

Patrick's parents remained steadfast in their belief that their son would be healed. The PICU staff arranged a consult with the hospital's bioethics team to help them work through the distress they were feeling over this case. Six weeks after his admission to the PICU Patrick underwent a tracheostomy without complication. His care was easier after the surgery (without the concern of dislodging the endotracheal tube when his head was moved) and his family delighted in being able to see his beautiful face without the tube and tape. Patrick's parents, an aunt and both of his grandmothers were trained in care of his tracheostomy tube and ventilator. Three weeks after his surgery he was transferred to the children's rehabilitation hospital as a transition to home while the family learned to care for the new tracheostomy.

Over the next 18 months Patrick had three further admissions to the PICU for treatment of respiratory deterioration secondary to acute viral illnesses. The first two stays were for 8 and 5 days respectively. Patrick seemed much sicker on the third admission and was still in the PICU 3 weeks later. His parents shared with the PICU staff that he seemed much weaker overall in the past 2 months and they also felt he was not as happy as he once was. While their faith remained strong, they now allowed that God's plan for Patrick might not include either a cure or a long life. Both the hospital chaplain and the family's church pastor supported them in this view.

The hospital's PPC team, who had met Patrick and his family at the time of his initial PICU admission, helped the family establish new goals of care for Patrick based on how they were thinking about things now. The overriding goals became comfort, enjoyment of life and a desire for Patrick to die at home without having to endure further deterioration.

Patrick was transported home 4 days later while still being ventilated through his tracheostomy. He was greeted by a large number of family and friends and had a surprise early third birthday party, complete with chocolate cake for him to feel with his fingers and even taste (only a small amount was given to prevent aspiration). He then had a bath and was dressed in his favourite Diego pyjamas. He said goodnight to his grandparents and went to his bedroom with mom, dad, a nurse and a paediatric palliative care physician. He was given sedation through the intravenous still in his left arm and a morphine infusion was started to ensure his continued comfort. Once sedated he lay on his 'big boy bed,' on his mom's lap, while the tracheostomy was disconnected from the ventilator and an oxygen 'mask' was fitted over the opening. The nurse and doctor titrated the morphine infusion based on Patrick's work of breathing and he died in his father's arms 40 minutes later.

18.3 Case #2: Truth-Telling and Hearing the Child's Voice in Decision-Making

Lucy is a 14-year-old girl with a diagnosis of stage IV osteosarcoma, a metastatic bone cancer which originated in her pelvis 2 years ago. After the diagnosis, Lucy underwent chemotherapy and radiation treatments which led to remission of the cancer. However, her cancer returned after a few months and Lucy underwent treatment once again. This cycle of treatment and remission has continued for a period of 2 years. Throughout this time Lucy has continued to attend school and participate in various extra-curricular activities. She and her parents have gone on vacations and spent most of their time together.

Now, 2 years since her initial diagnosis, Lucy's cancer has recurred for the fourth time. This time it is in her bones and lungs and the healthcare team has no further curative treatments to offer. Lucy's parents have been told that since there is no curative treatment available, she will die from her disease. They do not want the healthcare team speaking to Lucy about this fact. The healthcare team respects their wishes and allows Lucy's parents to tell her what the recent set of investigations has shown. Lucy's parents tell her that her cancer is back and that they will continue to fight.

Lucy continues to come to the hospital for follow-up appointments on a biweekly basis for an oral chemotherapy that the team has offered in the hope that it may slow the spread of her cancer. Lucy is now requiring pain medication to treat the pain she is experiencing in her pelvis and her legs. She is no longer attending school and is unable to walk. She now uses a wheelchair when leaving the house, which she does chiefly for her hospital appointments.

Lucy and her parents have met the hospital's PPC Team, a team Lucy was told "provides help and support to kids and teens like you who have had cancer which keeps coming back." During one of her hospital clinic visits, Lucy asks the PPC team, "How long do I have to live?" As the team starts to answer, her mother cuts in and says, "I don't want her talking about or thinking about dying or death." The team offers to do a home visit during which they could help answer any questions on Lucy's mind. Her mom agrees to that but when the team calls to arrange the home visit Lucy's mom declines the offer as "things are fine." When asked why she doesn't want the team to speak with Lucy, especially since she seems aware of what is happening to her, Lucy's mother explains that she doesn't want her daughter to be sad all of the time. Attempts to help Lucy's mother understand the potential benefits of allowing Lucy to speak about her fears and worries and to have any questions answered were unsuccessful.

A few weeks later Lucy is admitted to hospital for pain management. A few days into the admission the team appreciates that Lucy's death is likely just days to weeks away. The team shares this information with Lucy's parents and also informs them that Lucy has told one of her nurses that she wants to remain in hospital and not go home. Lucy stated that she is comfortable knowing that the nurses and doctors are available in hospital to help her when she needs it. On hearing this Lucy's parents indicate that they want to take Lucy home to die. They believe that if Lucy dies at home, after death her spirit will return there, but if she dies elsewhere her spirit will not return to their home. Lucy's mother indicates that she could not live without having her daughter's spirit with her.

18.3.1 Discussion

Decisions regarding sharing of health information and decision-making can be complex, especially in the case of adolescents and their parents. Perhaps ethical principles in healthcare guide us to share information about diagnosis, prognosis and treatment options with our patients regardless of the patient's age. However, when the patients are children, this information is also shared with the child's family.

The ethical principal of truth-telling is important in Western society, yet there are many cultures and families in which withholding the truth from the ill person is the preferred method of communication (Kagawa-Singer and Blackhall 2001). Ethics guide us to share diagnosis, prognosis and treatment information with patients, but we must do this in a manner which is beneficial to the child and does not cause harm. Families are instrumental in helping to ascertain the best way to share information. Each family will have its own decision making process and views on how best to share information within the family.

In cases of young children the information may be shared solely with the parents and family. However, as children get older it becomes less clear what information, if any, should be withheld from the children themselves. As healthcare providers we feel an obligation to share information with our paediatric patients and to provide them with the opportunity to ask questions. Sometimes, however, parents try to prevent this from happening. This may be because the parents want to protect their child by preventing them from being given information that may scare or upset them or cause them to lose hope. Trying to protect a child in this way is often unsuccessful and comes at a cost to both the child and the family. Alternatively, parents themselves may be struggling with their own grief and inability to accept their child's impending death, and are therefore unable even to think about this information being shared with the child. Regardless of the reasons, keeping information from the child about what is happening to them can negatively impact the child's experience, the relationship of the child with his/her parents, the relationship of the child with the healthcare team, and the relationship of the family with the healthcare team (Beale et al. 2005).

Research has shown that families who chose not to speak to their child about dying, but in retrospect believe that the child was aware of what was happening, regretted not talking with them (Kreicbergs et al. 2004). However, families who did speak to their child about dying had no regrets about having done so. The benefit of allowing children to hear the information and to ask questions is that it enables us to hear their wishes and to know what is important to them. In situations such as Lucy's it enables the child and family to make plans for the future. By not engaging in the conversation, the opportunity to plan together is lost. Sharing literature which is based on the experience of other families with respect to truth-telling gives healthcare providers a way to provide families with all of the information they need to make the best decision.

18.3.1.1 Paediatric Patient-Centred Decision-Making

Adolescence is a particularly challenging time for both the adolescent and their parents. Taking a patient-centred care approach with adolescents and their parents can also be challenging. While adolescents are developing their autonomy and should be involved in decision-making, parents are often learning what the increasing autonomy means for the adolescent and also for them as parents. Parents may believe that the adolescent is not yet able to fully comprehend the severity of a situation and thus their involvement should be limited, which can lead to conflict between adolescents and parents. As healthcare providers, we are often in the position of helping to ease any conflict by respecting the autonomy of the adolescent, but also by helping parents to gain confidence in their child's ability to participate in decision-making.

The need to respect the developing adolescent's autonomy is critically important to being able to provide effective end-of-life care to an adolescent. By not respecting an adolescent's autonomy we risk them becoming withdrawn, isolated or angry (Lyon et al. 2004). Most adolescents indicate that they would like to be involved in decision-making about their end-of-life care, even if a small percentage of adolescents would prefer that their family make decisions on their behalf (Lyon et al. 2004). Hence it is important to determine how much information the adolescent wants to receive, how they want to receive it, from whom they wish to receive it, and what level of involvement they want in the decision-making.

Children who want to be involved in decision-making have the ethical right to information about their health so that they can make the most informed decision. In particular, they have a right to have their questions answered, as this can alleviate fears that they may have and clear up misconceptions. Most importantly, it can give them a sense of having some control over the situation. Adolescents who learn that they are dying soon can find the time they have remaining to be very meaningful: they may use that time to say goodbyes or create legacies for their family and friends.

When healthcare providers do not share information with their patient it creates mistrust, in particular when the child knows that information is being withheld. Furthermore, the child may then look to other sources for the information they seek and the answers to their questions. Unfortunately, this can lead to them getting incorrect information or living with the uncertainty that comes with unanswered questions. This, in turn, can lead to suffering which may not be seen physically but which the adolescent experiences in silence and alone.

When there is a lack of information-sharing within the family, the relationship between the child and his/her family members can become strained. Adolescents can sense when parents are not talking about something. Adolescents may then want to protect their parents by not bringing up a conversation that they think will upset their parents or that their parents do not want to have. This can lead to the adolescent choosing to spend their time with friends or others, whom they feel are not hiding information or to whom they feel they can speak openly. Parents can become distressed with their adolescent's lack of interest or involvement in their usual activities. They may not understand that this behavior is related to the lack of sharing of information with the adolescent.

18.3.1.2 Family-Centred Decision-Making

Family-centred decision-making encompasses the principles that the child exists within the context of his/her family and that members of the family should be involved in decision-making. As the child is part of his/her family, we rely on the expertise that parents have about their child and believe that parents know their child best. Parents raise their children and care for them when they are ill. They know how they respond to new people, new information and new situations. We as healthcare providers rely on parents to help us help their child. While we encourage families to tell children about changes in their health and when death is a possibility or likely outcome, we recognize that some families will not do so.

We try to help families by suggesting words that they can use to talk to their child; we provide resources such as books to help start the conversation. But when families decide not to engage in the conversation with their child, we feel challenged, disappointed, and sometimes even like we have failed. However, if we are practicing family-centred care and truly believe that families know their children best, then we must trust that their decision to not engage in this conversation is best for their child. Thus while it may be difficult not to have open communication with the child, it is important to ensure that open communication with the family continues. We must remember that there may be opportunities in the future to revisit this discussion with the family.

Decisions about end-of-life cannot be made without the inclusion of parents. Parents are the ones who will often provide a great deal of the care required by the child. But when parents are the ones making the decisions without including the child, it is questioned whether or not the child would be in support of the decisions. The true challenge comes when the child's wishes become known and they are in conflict with the parents' wishes.

In the case of Lucy, her parents' beliefs about the importance of a death at home, as opposed to the hospital, were not explored with their daughter. This led to conflict between the staff, who wanted to carry out Lucy's last wishes by having her remain in hospital for her last hours, and her parents, who wished for her to die at home. The cultural and religious factors that influence a family's decisions about end of life care are important. Had this information been shared with Lucy as well, the team could have explored Lucy's concerns about going home. The team could have shared with Lucy the supports that would be available at home, including nurses providing her care at home, a visiting palliative care physician, medications and equipment (including oxygen, intravenous pumps, etc.) to keep her comfortable, and emergency contact numbers should they need more assistance. In addition, seeing the benefits of this conversation for Lucy, her parents may have been more open to allowing Lucy to have an opportunity to discuss any other concerns she had.

It is the families who are left with the memories of their child's final days and hours. For this reason it is so important to explore, early on, what their hopes are for their child during those final moments. Those discussions can help to alleviate families' concerns about the child dying at home rather than in the hospital, and can help to ensure that the correct supports are put into place so that the experience can be the best it can, for both the child and family.

18.3.2 Resolution of Case #2

Lucy remained in hospital for the next week. Her pain management was optimized and oxygen was started once she showed signs of dyspnea due to the progression of the disease in her lungs. Her parents stayed at her bedside and they watched movies together and spoke about past family vacations. Lucy's teacher and a few close friends also visited. Lucy seemed content and asked no further questions of her nurses. Each night she would tell her parents she loved them before falling asleep. Her appetite dwindled to the point where she took only a few bites of her favourite foods each day. Her parents raised the question of a feeding tube with the healthcare team and accepted the recommendation that it not be used, as Lucy's death was now very close.

The parents and healthcare team both met with the hospital's bioethics service to help resolve the tension that existed around the question of where Lucy should die. Her parents acknowledged to the team that they did not want to have a conversation with Lucy about her concerns with respect either to going home or why they wanted to take her home. Rather, they decided that the decision to take Lucy home would be made at the last minute, likely when Lucy was no longer alert. The team understood the family's perspective and felt reassured that Lucy could be kept comfortable at the end of her life at home. Thus preparations for appropriate home care support were begun so that Lucy could be brought home quickly when the family decided it was the right time. A few days after, Lucy became unresponsive to voice and touch. She was then transported home, where she died 12 hours later, with her parents at her bedside.

18.4 Summary

Generally speaking, parents make decisions for their children based on love. They have an intense desire to protect their child from dying and death. The job of paediatric palliative care providers is to help parents understand that they cannot protect their child from death. What they can do, however, is to prepare their child for what is to happen.

References

American Academy of Pediatrics Committee on Bioethics. 1994. Guidelines on foregoing life-sustaining medical treatment. *Pediatrics* 93 (3): 532–536.

American Academy of Pediatrics Committee on Bioethics. 1996. Ethics and the care of critically ill infants and children. *Pediatrics* 98 (1): 149–152.

Bowman, K. W. 2000. Communication, negotiation, and mediation: Dealing with conflict in end-of-life decisions. *Journal of Palliative Care* 16 Suppl: S17–23.

Beale, E. A., W. F. Baile, and J. Aaron. 2005. Silence is not golden: Communicating with children dying from cancer. *Journal of Clinical Oncology* 23 (15): 3629–3631.
Breitbart, W. 2003. Reframing hope: Meaning-centered care for patients near the end of life. Interview by Karen S. Heller. *Journal of Palliative Medicine* 6 (6): 979–988.
Canadian Hospice Palliative Care Association. 2006. *Pediatric hospice palliative care guiding principles and norms of practice*. Ottawa.
Canadian Pediatric Society Bioethics Committee. 2004. Treatment decisions for infants, children and adolescents. *Paediatrics & Child Health* 9 (2): 99–103.
Cantor, N. L. 2005. The bane of surrogate decision-making: Defining the best interests of never competent persons. *Journal of Legal Medicine* 26 (2): 155–205.
Carnevale, F. A., E. Alexander, M. Davis, et al. 2006. Daily living with distress and enrichment: The moral experience of families with ventilator-assisted children at home. *Pediatrics* 117 (1): e48–60.
Carroll, K. W., C. J. Mollen, and S. Aldridge, et al. 2012. Influences on decision making identified by parents of children receiving pediatric palliative care. *AJOB Primary Research* 3 (1): 1–7.
Committee on Bioethics, American Academy of Pediatrics. 1995. Informed consent, parental permission, and assent in pediatric practice. *Pediatrics* 95 (2): 314–317.
Ditto, P. H., J. A. Druley, and K. A. Moore, et al. 1996. Fates worse than death: The role of valued life activities in health-state evaluations. *Health Psychol* 15 (5): 332–343.
Farsides, B., and R. J. Dunlop. 2001. Is there such a thing as a life not worth living? *British Medical Journal* 322 (7300): 1481–1483.
Kagawa-Singer, M., and L. J. Blackhall. 2001. Negotiating cross-cultural issues at the end of life: "You got to go where he lives". *The Journal of the American Medical Association* 286 (23): 2993–3001.
Kasman, D. L. 2004. When is medical treatment futile? A guide for students, residents, and physicians. *Journal of General Internal Medicine* 19 (10): 1053–1056.
Klein, D. A. 2011. Medical disparagement of the disability experience: Empirical evidence for the "Expressivist Objection". *AJOB Primary Research* 2 (2): 8–20.
Kopelman, L. M. 1997.The best-interests standard as threshold, ideal, and standard of reasonableness. *The Journal of Medicine And Philosophy* 22 (3): 271–289.
Kreicbergs, U., U. Valdimarsdóttir, and E. Onelöv, et al. 2004. Talking about death with children who have severe malignant disease. *New England Journal of Medicine* 351:1175–1186.
Lyon, M. E., M. A. McCabe, and K. M. Patel, et al. 2004. What do adolescents want? An exploratory study regarding end-of-life decision-making. *Journal of Adolescent Health* 35:529.e1–529.e6.
Madrigal, V. N., K. W. Carroll, and K. R. Hexem, et al. 2012. Parental decision-making preferences in the pediatric intensive care unit. *Critical Care Medicine* 40 (10): 2876–2882.
McSherry, M., K. Kehoe, and J. M. Carroll, et al. 2007. Psychosocial and spiritual needs of children living with a life-limiting illness. *Pediatric Clinics of North America* 54 (5): 609–629, ix–x.
Meert, K. L., C. S. Thurston, and A. P. Sarnaik. 2000. End-of-life decision-making and satisfaction with care: Parental perspectives. *Pediatric Critical Care Medicine* 1 (2): 179–185.
Noyes J. 2006. Health and quality of life of ventilator-dependent children. *Journal of Advanced Nursing* 56 (4): 392–403.
Patrick, D. L., R. A. Pearlman, and H. E. Starks, et al. 1997. Validation of preferences for life-sustaining treatment: Implications for advance care planning. *Annals of Internal Medicine* 127 (7): 509–517.
Reder, E. A., and J. R. Serwint. 2009. Until the last breath: Exploring the concept of hope for parents and health care professionals during a child's serious illness. *Archives of Pediatrics & Adolescent Medicine* 163 (7): 653–657.
Schattner, A., D. Rudin, and N. Jellin. 2004. Good physicians from the perspective of their patients. *BMC Health Services Research* 4 (1): 26.
Schnakers, C., and N. D. Zasler. 2007. Pain assessment and management in disorders of consciousness. *Current opinion in neurology* 20 (6): 620–626.

Schneiderman, L. J. 2011. Defining medical futility and improving medical care. *Journal of Bioethical Inquiry* 8 (2): 123–131.
Shields, L., J. Pratt, and J. Hunter. 2006. Family centred care: A review of qualitative studies. *Journal of Clinical Nursing* 15 (10): 1317–1323.
Steering Committee of the EPAC task force. 2007. IMPaCCT: Standards for paediatric palliative care in Europe. *European Journal of Palliative Care* 14 (3): 109–114.
The Multi-Society Task Force on PVS. 1994. Medical aspects of the persistent vegetative state (1). *The New England Journal of Medicine* 330 (21): 1499–1508.
UN Convention on the Rights of the Child.
Wilkinson, D. 2006. Is it in the best interests of an intellectually disabled infant to die? *Journal of Medical Ethics* 32 (8): 454–459.

Afterword

I invited this expert group of authors to write this collection of essays with the goal of highlighting ethical and legal issues associated with the models of patient-centred care and family-centred care in paediatrics. They shared analyses and reflections from their vantage points as clinical ethicists, physicians, nurses, genetic counselors, lawyers, philosophers, administrators, educators, researchers, trainees and parents as well as provided a window into paediatric patient and family-centred care in both urban and rural settings. The roles of and attention to both patients and families is of vital importance to optimizing paediatric health care. That is why the ethical and legal issues associated with the models of paediatric care warrant such fulsome attention.

I approached this project with some hesitation, not wanting the challenges to be interpreted as a critique of the important consideration that must be given to family relationships and context when caring for children. I am thrilled to see that this unique collection highlights the significance to paediatric patients of having their own as well as their families' values and circumstances respected and the challenge for paediatric patients when patient best interests may not be consistent with families' values.

Notwithstanding 18 chapters of breadth and depth, as a bioethicist in a paediatric hospital, I am still being exposed to new contexts where the tensions and synergies between these models play out in unique and challenging ways. My hope is that with the strategies and analyses provided, readers will be equipped to engage with the challenges of new contexts and further contribute to the discussions and literature in this area.

My purpose was not to settle this discussion but rather to draw attention to the complexities and vulnerabilities of the model options. My objectives will have been met if readers come away sensitive to the fragility of paediatric patient rights and appreciative of the implications of context, the extent to which the well being of children can be dependent on family well being and the impact of language on the way roles and rights of all parties are operationalized.

Notwithstanding the challenges of integrating the two models, I am not worried that concern for the significance of attending to the values of patients or families in healthcare is in jeopardy. While research derived evidence would undoubtedly be

R. Zlotnik Shaul (ed.), *Paediatric Patient and Family-Centred Care: Ethical and Legal Issues,* International Library of Ethics, Law, and the New Medicine 57,
DOI 10.1007/978-1-4939-0323-8, © Springer Science+Business Media Dordrecht 2014

helpful in guiding processes and infrastructure, the preceding chapters reflect how the value of each, is entrenched in organizational priorities and professional practice norms. To appreciate the scope of the benefits and challenges associated with these models, it was vital that we recognize the extent to which they are products of historical, social and political contexts. Such appreciation generated through the preceding chapters increases our ability to develop strategies that are ethically defensible and legally informed.

This book may have been the reader's introduction to the models of patient-centred care and family-centred care or it may have enriched the experience of working with the models clinically and/or academically. Whatever the case, after completing this book, readers will have a deep appreciation of the values at stake, the potential challenges associated with their integration as well as realistic suggestions for practice.

In terms of next steps, readers from organizations who have embraced models of patient and family-centred care are now better equipped to understand challenges that may have already arisen in trying to operationalize the models. Readers from organizations considering the implementation of these models are now better equipped to appreciate the benefits associated with these models as well as to preempt challenges through practice informed policies and infrastructure.

While attending to the values and interests of paediatric patients and their families may to many, seem like an uncontroversial objective, how one should go about doing this remains less of a fixed formula than a goal in need of ongoing reflective practice. Many goals grounded in ethics find their richness in their application and do not easily lend themselves to mathematical resolution. Because ethical and legal professional accountability in paediatric healthcare is essential to quality care, careful ongoing attention must be paid to the implications of the models we put forward to inform practice. Given the profound significance of models of care to all parties, we must remain attuned to the dynamic contexts in which models are applied, cognizant of the perspectives of those who want or need a voice and attentive to unintended ethical and legal complexities. It is only through a fulsome appreciation of the care models being promoted and the potential challenges to their application, that strategies can evolve that are supportive to families and in the best interest of padiatric patients.

Index

R. Zlotnik Shaul (ed.), *Paediatric Patient and Family-Centred Care: Ethical and Legal Issues,* International Library of Ethics, Law, and the New Medicine 57,
DOI 10.1007/978-1-4939-0323-8,

MIX
Papier aus verantwortungsvollen Quellen
Paper from responsible sources
FSC® C105338

If you have any concerns about our products,
you can contact us on
ProductSafety@springernature.com

In case Publisher is established outside the EU,
the EU authorized representative is:
Springer Nature Customer Service Center GmbH
Europaplatz 3, 69115 Heidelberg, Germany

Printed by Libri Plureos GmbH
in Hamburg, Germany